THE FLAWED GENIUS OF WILLIAM PLAYFAIR

The Story of the Father of Statistical Graphics

The Flawed Genius of William Playfair

The Story of the Father of Statistical Graphics

DAVID R. BELLHOUSE

UNIVERSITY OF TORONTO PRESS
Toronto Buffalo London

Toronto Buffalo London
utorontopress.com

ISBN 978-1-4875-4503-1 (cloth)
ISBN 978-1-4875-4504-8 (EPUB)
ISBN 978-1-4875-4505-5 (PDF)

Library and Archives Canada Cataloguing in Publication

Title: The flawed genius of William Playfair : the story of the father of statistical graphics / David R. Bellhouse.
Names: Bellhouse, D. R., author.
Description: Includes bibliographical references and index.
Identifiers: Canadiana (print) 20230182402 | Canadiana (ebook) 20230182437 | ISBN 9781487545031 (cloth) | ISBN 9781487545055 (PDF) | ISBN 9781487545048 (EPUB)
Subjects: LCSH: Playfair, William, 1759–1823. | LCSH: Economists – Great Britain – Biography. | LCSH: Statisticians – Great Britain – Biography. | LCGFT: Biographies.
Classification: LCC HB103.P55 B45 2023 | DDC 330.092–dc23

Cover design: John Beadle
Cover image: *Revenues of England and France*, courtesy of Stephen M. Stigler, University of Chicago

We wish to acknowledge the land on which the University of Toronto Press operates. This land is the traditional territory of the Wendat, the Anishnaabeg, the Haudenosaunee, the Métis, and the Mississaugas of the Credit First Nation.

This book has been published with the help of a grant from the Federation for the Humanities and Social Sciences, through the Awards to Scholarly Publications Program, using funds provided by the Social Sciences and Humanities Research Council of Canada.

University of Toronto Press acknowledges the financial support of the Government of Canada, the Canada Council for the Arts, and the Ontario Arts Council, an agency of the Government of Ontario, for its publishing activities.

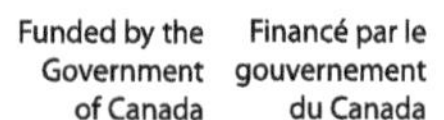

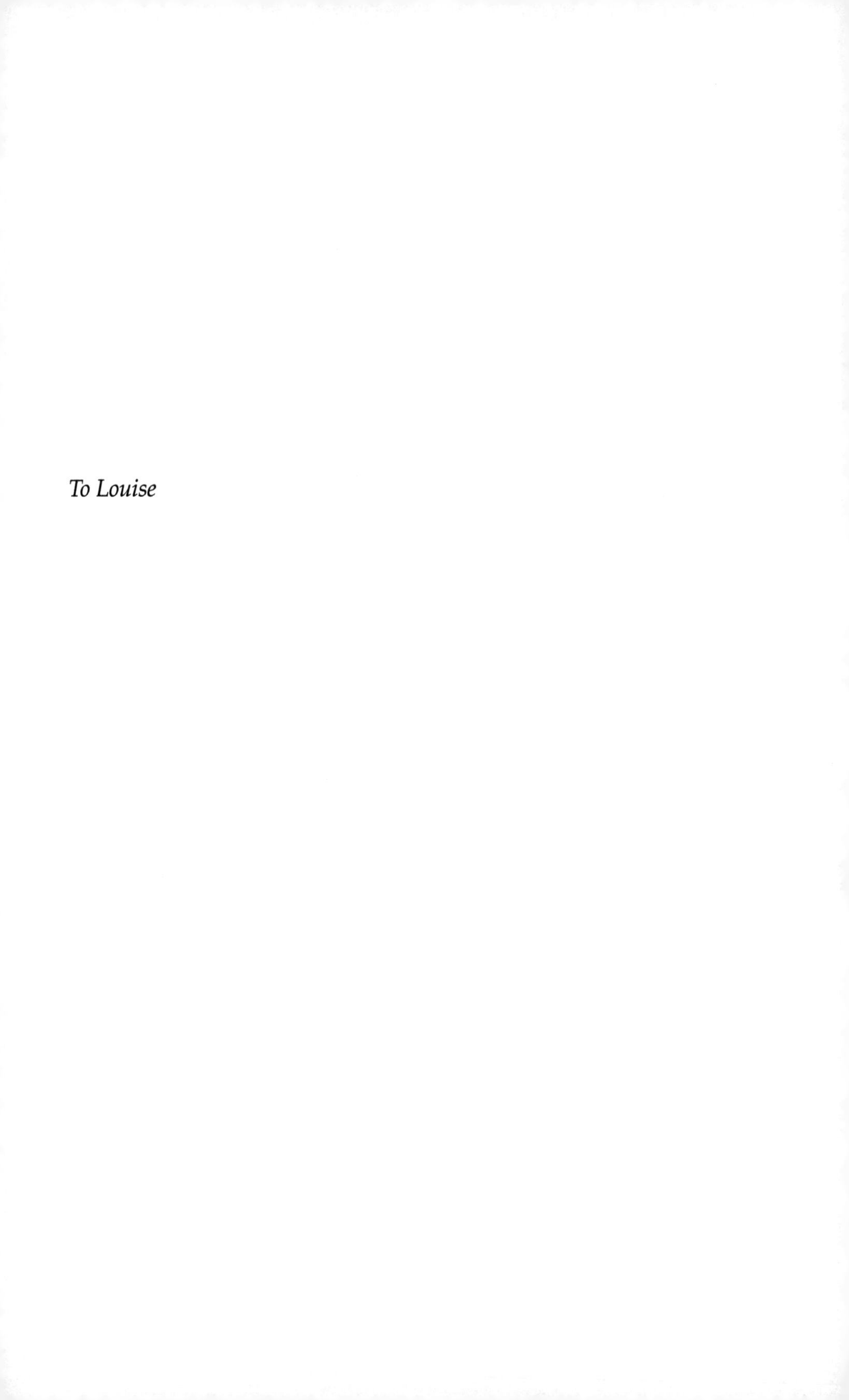

To Louise

Contents

Illustrations

Preface
Playfair Is Introduced

Today, William Playfair (1759–1823) is known as the father of statistical graphics. That is a twentieth-century attribution from statisticians and psychologists to someone who died nearly two centuries ago. In London, where Playfair lived and worked for many years, the Royal Statistical Society came into being only a little more than a decade after Playfair's death. Statistics as a discipline was barely a toddler when Playfair was active. What motivated Playfair's graphical work was economic data, and what piqued his interest in economics was Adam Smith's *Wealth of Nations*. It is not surprising then that the earliest biographies of Playfair, and assessments of his work, appear in the economics literature (Edgeworth 1899; Funkhouser and Walker 1935). The latter article shows a budding interest by the statistical community. It was written not by an economist, but by two historians of mathematics. One of them, H. Gray Funkhouser, soon followed the 1935 article with a major study on the use of graphical techniques up to the 1930s (Funkhouser 1937). Playfair is now nearly forgotten among the economists, but significant interest in his life and graphical work has picked up and accelerated in the statistical literature over the last thirty years or so.

Throughout this book I will try to put Playfair into his historical context. He was a product of the Scottish Enlightenment and it shows in much of his writing. He worked during a period of major political and economic upheaval. He began his career as the Industrial Revolution was taking hold and as the American Revolution was concluding. He went to France as the French Revolution was unfolding. As a writer on economic and political issues, he was most active in England during the Napoleonic Wars. I have tried to weave all this into the narrative of his life.

Playfair's life was more than his work in statistics and economics. He also worked as an engineer and inventor, banker, scam artist, land speculator, historian, and political propagandist. He began his career as a young engineer with much promise, working for Matthew Boulton and James Watt in Birmingham. He was involved in the early production of Boulton and Watt's steam engines. After launching out on his own, Playfair saw many ups and downs throughout his career that ranged from being moderately well off to bankruptcy and imprisonment. With about forty publications of books and pamphlets during his lifetime, he ended his career as a writer.

In some of Playfair's activities, history, along with some of his contemporaries, has not been kind to him. Some of it was self-inflicted. When desperate for money, he resorted to extortionary scams among the highly placed. Many were unsuccessful and earned Playfair an unsavoury reputation in some quarters. Other more legitimate undertakings were a mixture of bad planning and bad luck. Early in the twentieth century Playfair was vilified by American historians for an ill-fated scheme to settle French émigrés in Ohio following events in France in 1789 (Belote 1907a, 1907b, for example). More recent scholarship by a French historian has shifted the blame for the immigration fiasco more to the ineptitude of the American entrepreneurs involved and has given Playfair a notch below a full exoneration for his activities in the scheme (Moreau-Zanelli 2000).

In the first full-length biography of Playfair, it has been posed that he was a spy for the British government (Berkowitz 2018). The claim has been promoted to the extent that he is described in Wikipedia as "served as a secret agent on behalf of Great Britain during its war with France." Certainly, Playfair claimed to have connections or acquaintances in high places such the Prince of Wales, later King George IV. If that were the case, why didn't his handlers, or perhaps former handlers, help him in some way when he ran into trouble, especially financial trouble? The simple answer is that Playfair was not a spy – perhaps a wannabe spy, and very briefly a courier, but never a spy. Several of the historical sources that I have examined, and will present here, support this simple answer.

Two centuries after his death, Playfair's lasting contributions are in statistics, where he is known as the father of statistical graphics, and in economics, where he is best known for his contributions to the first posthumous edition of Adam Smith's *Wealth of Nations*, in 1805. Putting Playfair's contributions succinctly, the Anglo-Irish statistician and economist Francis Ysidro Edgeworth wrote at the end of the nineteenth century, "[He] evinces some acumen as an economist as well as some originality as a statistician" (Edgeworth 1899, 117). Playfair's other major contribution to economics appears in his book *Decline and Fall of*

Powerful and Wealthy Nations, also published in 1805. There, according to the Polish economist Henryk Grossman, he made a notable contribution to economic theory. Grossman argues convincingly that Playfair was the first theorist of capitalist development (Grossman 1948). Turning to his graphical contributions, Playfair invented the now ubiquitous bar chart and the pie chart. He is best known for his time series charts, which first appeared in 1786 in his *Commercial and Political Atlas*; and he is also known for a way to display multivariate data, which appeared in his *Statistical Breviary* in 1801. Playfair also pioneered the use of colour in his graphs, when colour was rarely used in the printing process.

Playfair's graphs have kept his legacy alive. He was on the leading edge of statistical graphs that took flight at the beginning of the nineteenth century (Friendly 2008). The first half of the nineteenth century has been called the "age of enthusiasm" for statistical graphs, which turned into a "golden age" in the last half of the century.[1] In Britain, Playfair was relatively forgotten until the English economist William Stanley Jevons discovered his work in 1861 near the beginning of the golden age when Jevons was promoting graphical methods (Jevons 1886, 158; see also Spence and Wainer 2005). After Jevons, with a handful of exceptions up to the 1930s, Playfair's work continued to be forgotten by both economists and statisticians (Mitchell 1927, 191, 209; Symanzik, Fischetti, and Spence 2009). Once the golden age lost its lustre, interest in developing new graphical methods declined. As a capstone to this era, Willard Brinton produced a lengthy and insightful compendium of graphical methods up to the beginning of the Great War (Brinton 1914). Beginning in the 1950s, interest in graphs took off again with the computer revolution (Friendly 2008). It was the American statistician, John Tukey who popularized the use of graphical methods in data analysis in the late 1960s (Tufte 1983, 53). The American political scientist Edward Tufte rediscovered Playfair's work in the 1980s (Tufte 1983). This was followed by examinations of Playfair's life and work, mainly by Ian Spence and Howard Wainer, as well as others, in the closing decade of the old millennium and beyond (Costigan-Eaves and Macdonald-Ross 1990; Palsky 1996; Wainer 1996; Spence and Wainer 1997a, 1997b, 2000; Spence 2004; Friendly and Wainer 2021). Graphs are now ubiquitous, and Playfair is no longer forgotten in some quarters. In addition to Playfair's graphs, others have taken an interest in other aspects of Playfair's career. Jean-François Dunyach, for example, has examined Playfair's work in the context of the Scottish Enlightenment (Dunyach 2010, 2014, 2016).

Playfair has no longer been forgotten; but some aspects of his career have been overlooked in the enthusiasm for his graphs. I hope to change that over the pages of this book. Playfair crossed paths with many of the leading lights of his day. In a sense, his was a life of adventure as he often went into, and then tried to stay out of, prison. He hobnobbed with the low-lifes and the upper crust in both English and French society. From the time of his birth in 1759 in the village of Liff, Scotland, where his father was the local Presbyterian minister, until his death in 1823 in poverty in London, England, he was a slightly damaged survivor following a very shaky and sometimes questionable but interesting lifestyle. His career illustrates the precariousness of making a living in late eighteenth- and early nineteenth-century Britain when one was ambitious and talented but lacking in wealth or patronage connections. It also illustrates the lengths one had to go to in making ends meet when one was lacking an important skill – the ability to keep on top of one's finances no matter how large or small. This is the story of a Scot working in the middle ranks of literary and industrial society in England. I will describe both the peaks and troughs of his career, how he ascended to the peaks, and why he fell to the troughs, rising again until his final crash to poverty at his death.

Acknowledgments

About six years ago, after I finished writing *Leases for Lives*, I was discussing with a colleague and friend, Duncan Murdoch, about what to do next. He is keenly interested in statistical graphics and suggested that I look into William Playfair, someone he heard was involved in graphical methods early on. Neither of us knew much about the man, so I did a little preliminary research. He turned out to be a fascinating character who had many twists and turns to his life. I developed a mild obsession for him and his work.

As I carried out my research and writing, I was given encouragement and help by several people. Stephen Stigler and Howard Wainer commented, and put me in the right direction, on some chapters. Professor Stigler has also provided me with digital copies of several of Playfair's graphs. Ian Spence commented on some of the statistical and biographical parts of the manuscript and made a copy of Playfair's "Memoir" available to me. David Laidler read my manuscript, guiding me through some of the economic issues that Playfair wrote about. Christian Genest reviewed the manuscript and helped me with the intricacies of the French language that allowed me to completely unmask one of Playfair's scams. Vivian McAlister gave me his expert medical opinion on Playfair's illness and death. Jérôme Malhache obtained information and photographs for me from the Archives nationales de France when I was unable to visit the archives as the result of COVID. Barbara Budnick proofread my first draft, a thankless task for someone not remotely connected to the subject material, and Ian MacKenzie carefully copy edited the final draft. I am thankful for everyone's input and help. Finally, I am grateful to my editor, Len Husband of University of Toronto Press, for believing in this project and seeing it through to publication.

THE FLAWED GENIUS OF WILLIAM PLAYFAIR

Chapter One

Playfair Is Sent to Newgate Prison

On 27 April 1809, as the prisoner stood in the dock at the Court of King's Bench sitting in London's Westminster Hall on the parliamentary estate, Justice Nash Grose, known for his clear and concise judgments, delivered his verdict from the bench:

> William Playfair you have been convicted of an offence very alarming to Society and for which some of your co-defendants have already received the Judgment of the Court. Upon that occasion I very fully expressed what it was fit for me to express as to the great criminality of their conduct and in consequence the Judgment of the Court will be very different from what we are inclined to inflict upon you. We have heard the circumstances which are stated in your favour as tending to mitigate the sentence and taking all those circumstances into our consideration we do order and adjudge that for the offence you have committed you be imprisoned in his Majesty's Gaol at Newgate for three months and that you then be discharged.[1]

Upon receiving this sentence, Playfair was handcuffed (in early nineteenth-century terms, he was placed in irons) and taken the two miles from Westminster Hall to his place of confinement near where the Old Bailey stands today. When Playfair entered Newgate, the keeper wrote in his register that Playfair was there for "certain conspiracies."[2] He did not elaborate on what these conspiracies were.

Newgate had been a prison since the twelfth century. The latest version of the prison, an imposing solid stone building, was erected in 1782 and still relatively new in 1809. There was a section for female prisoners and two sections for male prisoners with several prisoners per room or ward. Male prisoners were divided into debtors and felons. Since Playfair had been convicted of a conspiracy, he was incarcerated as a felon.

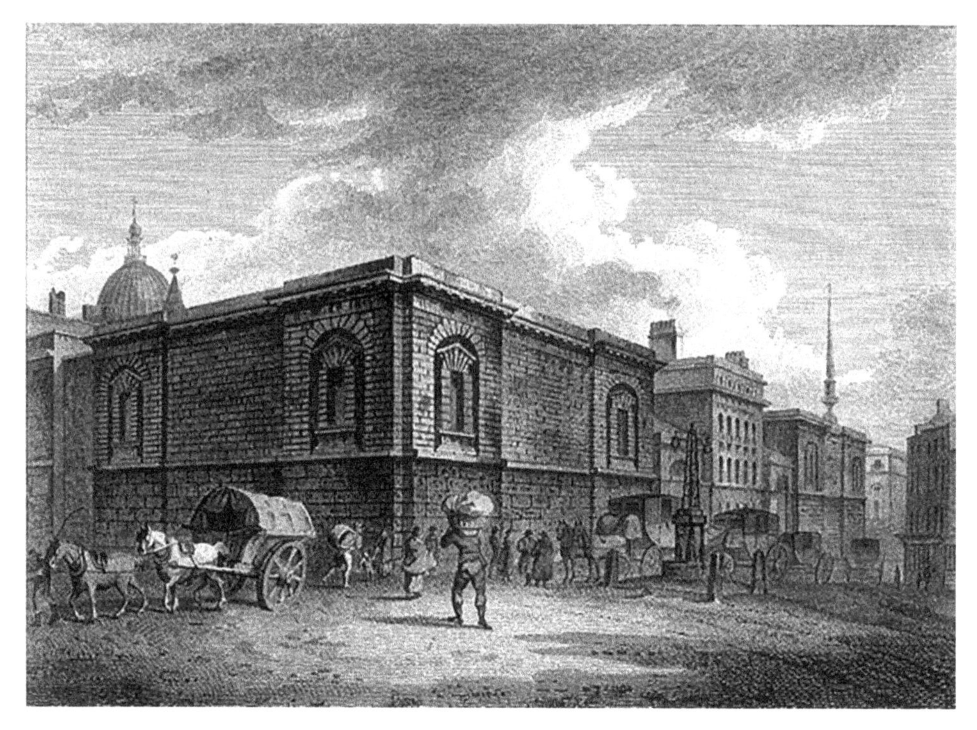

Figure 1.1. Newgate Prison. Alamy Stock Photos.

Despite the relative newness of the building, the prisoners were held in appalling conditions. It was overcrowded, holding two and often three or more times the number of prisoners for which it was originally built. The only food provided by the prison was ten ounces of bread a day and water (Great Britain 1814b). Illness in the prison was rampant. In 1818, about one third of the 1,500 prisoners, who were there over a seven-month period, were admitted to the prison's infirmary with serious illnesses. It was also reported that the women's side of the prison was especially bad. Many were in rags and slept on the floor with no covering. The roof in that section of the prison leaked. Groups of women cared for their children, cooked, ate, washed, and slept all in the same room or ward. The place was filthy and stank. A grand jury, held in 1813 to discuss the prison conditions at Newgate, noted the crowding and the problems in the women's section. Despite these problems, they concluded at the end of their less-than-one-page report, "In other respects the Grand Jury have the pleasure to state, that the whole conduct of the Prison gave them great Satisfaction" (Shelton 1814, 88).

There were three types of confinement in the felons' wards on the men's side of Newgate: Common, Master, and Superior Master (Great Britain 1813). In the Common wards, the state provided each prisoner with a coarse woollen cloth at no cost. The prisoners, with several to a room, slept on a sloped floor. The sleeping arrangement was called a barrack-bedstead. There were fees with accompanying benefits for the other two types. To be lodged as a Master felon, it cost the prisoner 13s. 6d. initially and then 2s. 6d. weekly after that. At that price, the prison keeper supplied the prisoner with a bed, although if the room was crowded there might be two prisoners to a bed with no reduction in the fee. To be lodged as a Superior Master felon, it cost the prisoner 2 guineas or £2 2s. initially and 7s. per week thereafter. Superior Master felons' rooms were less crowded (three or four to a room), but the prisoner might again have to share a bed because of the crowding problem, with once again no reduction in the fee. All prisoners had access to outside yards, which also might be crowded. Those lodged as Superior Master felons were allowed into the Press Yard, to which a maximum of about thirty prisoners had access. They could also meet guests in the keeper's parlour. Both Master and Superior Master felons could bring their own beds and sheets, but there was no reduction in the fees. Anyone could supplement his ten ounces of bread per day by buying his own food.

On entering Newgate, Playfair was taken to a Common ward and then offered the choice of paying the two-guinea fee to be in a Superior Master ward, followed by the option to pay the lesser amount for the Master ward if he did not have sufficient funds. Playfair did not get accommodation in the Superior Master ward. In a letter of 13 May 1809 written from prison, he claimed that only those awaiting execution or transportation got the better accommodation.[3] This was clearly a lie; money spoke to the accommodation, not the severity of the sentence. Those under sentence of execution, execution commuted to transportation, or just transportation were almost always poor and would normally be housed in the Common felons' ward. Playfair was trying to cover up his money woes to the letter's recipient, Spencer Perceval, who was chancellor of the exchequer at the time, soon to become prime minister.

This was not Playfair's first trip to Newgate. He had been held there briefly in 1798.[4] As a result, it is no wonder that he tried to pull strings in high places to ameliorate his situation. Playfair thought he had an ace up his sleeve. Prior to his sentencing day in court, he desperately tried to avoid going to prison; after sentencing he tried to play the ace again to get out of prison. The ace was in the form of help he said that he had given earlier to George, Prince of Wales. Playfair claimed that he had saved the prince from some embarrassment. When all other avenues failed, Playfair said that he was going to write to the prince for help.[5] Either his card was never played or it was ineffectual. Playfair remained in prison for his three-month sentence.

Playfair had tried to plug into his network in a vain attempt to avoid prison, and once in, to get out. His network, which included political, social, business, and literary connections, was extensive. How he was successful in assembling this network, which he said was done through chance meetings, is best described in his memoirs (surviving only in fragmentary form) written late in his life.

> When I went to France [he arrived in 1787] I had very few introductions [letters of introduction] and these I had were of no use. On this occasion as nearly all through life I have found chance provided me the most useful connections and in conversing with other people who have been abroad or gone into places where they were strangers in their own country I found the same held true with them.
>
> Finding that the case was general I concluded there must be a general cause for a general effect and I think it is this.
>
> When a friend gives an introduction it is not so much because the person to whom one is introduced is likely to be serviceable but because one man can only introduce another to those with whom he is acquainted. It is done from friendship or politeness and an invitation to dinner and some civilities are all the fruits of such acquaintances. It is not one time in ten that two persons so introduced find any inclination to keep up the connection. When people meet promiscuously or by chance if they form any connection it is because they find that they suit each other in some way either by similarity of taste or views or opinions and such connections when formed are always of more or less duration. (Playfair n.d.)

Of course, the chances of a productive connection being made are improved if the individual is actively engaged with a large group of people who generally have similar occupations or interests. A beggar is unlikely to have a chance meeting with a monarch. Playfair seems to have been proactive in seeking out connections, and in using those made to extend his network.

How did Playfair's misfortune come to pass? He was a relatively talented and prolific author whose publications cover a wide spectrum of topics: biography, economics, history, political science, and statistics. He was an inventive manufacturer, with five patents to his credit. So, what was his problem? Was it bad luck? Bad decisions? Or bad company that he kept? Was it some flaw in his character that made him a cheat or a scoundrel as some modern biographers have claimed? Or was he a victim of forces beyond his control? The answer is "yes" to all these questions. Playfair was a flawed genius. I use "genius" in a common eighteenth-century sense – natural ability. He seems to have had lots of it, and it was the Scottish Enlightenment in which he grew up that shaped how he applied his genius. The flaw that constrained his genius was his continued inability to deal with his personal and professional finances. He was financially inept. Almost always short of money, he spent his life clawing away, trying to keep the wolves from the door. In his desperation to make ends meet, he often resorted to highly questionable actions, mainly extortion. I leave it to you to decide whether Playfair was a genius in the common modern sense of the word.

Chapter Two

Playfair Goes to Birmingham to Work for Boulton and Watt

It had all started so well. Apprenticed to the prominent Scottish engineer and millwright Andrew Meikle, William Playfair was an ambitious young man. Playfair's family and family friends were also ambitious for him to succeed. With the Industrial Revolution in its infancy, there could be many opportunities for a man with good connections and talent, mixed with some luck, to make his fortune.

Having recently served his apprenticeship with Meikle, Playfair arrived in Birmingham in the fall of 1777 to take up a position with Boulton and Watt of steam engine fame. He was still a teenager, having just turned eighteen. He arrived in Birmingham when it was a town of little more than 42,000 people. It was also just two decades into the Industrial Revolution (Hopkins 1989, 31). The town was to experience near exponential growth during the nineteenth century as the revolution changed the face of Birmingham, as well as the rest of Britain (see, for example, Ashton 1969; Flinn 1966).

Central to the development of William Playfair's career are two people and one location – Boulton, Watt, and Birmingham. Matthew Boulton and James Watt were leading figures in the Industrial Revolution, and Birmingham was a major industrial centre in the revolution from its very beginning. Although Boulton had been courting Watt for nearly a decade, the two partners had been in business together in Birmingham for only about two years prior to Playfair's arrival. Playfair came in on the ground floor of what was to be a highly successful enterprise.

Boulton was Birmingham born and Watt arrived to live in Birmingham in May 1774 at Boulton's instigation. The formal business partnership of Boulton and Watt dates from 1 June 1775. Boulton brought his business acumen, innovativeness, and salesmanship to the partnership, while Watt brought his inventiveness in the form of a twenty-five-year extension to his patent on improvements to the steam engine and the promise of future improvements and inventions. The extension to the patent was granted about one month prior to the beginning of the partnership. Playfair arriving as early on the scene as he did, likely stirred his own ambitions to make his way as a manufacturer.

During the 1770s and early 1780s, the steam engines produced by Boulton and Watt were used mainly to pump water out of copper and tin mines in Cornwall; the steam-powered railway and other applications of the steam engine were yet to come. In Boulton and Watt's shop, the manufacture of these engines was similar, in a sense, to the manufacture of automotive engines today. Different suppliers manufactured various parts, which were then assembled in one location. Precision parts were made by Boulton and Watt in Birmingham; all other parts were manufactured elsewhere. With respect to precision parts, there was one important exception. The steam cylinders were manufactured by the ironmaster John Wilkinson, who had a factory about eighty miles away in Bersham, Wales. Wilkinson had developed a very precise method of boring gun barrels, which could be applied to the boring of cylinders.

Once the parts were all manufactured, the steam engine was assembled at the site where it would be used. This was done under the supervision of engineers employed by Boulton and Watt. The partners supplied their customers with plans of their engines, as well as instructions for their use and maintenance. This was where Playfair was to fit into the process. In 1777, Boulton was looking "for some ingenious young man that might go to distant places & superintend the erection of them [steam engines]."[1] Playfair was recommended and subsequently employed, but not in that capacity. The person whom Boulton was looking for turned out to be William Murdoch, who supervised the erection of steam engines in Cornwall for about ten years. Unlike Playfair, who failed as a manufacturer as well as in other businesses, Murdoch went on to become a successful engineer and inventor, making improvements to steam engine design, self-propelled vehicles, machine tools, and gas lighting. Such was his success that a statue of him, together with Boulton and Watt, stands in central Birmingham. An inscription at the base of the statue says that it "commemorates the immense contribution made by Boulton, Watt and Murdoch to the industry of Birmingham and of the world." Although William Playfair did make significant contributions in other areas, there are no statues of him or even likenesses, painted or drawn. All that we know of what he looked like is taken from a brief physical description of him when he was taken to Newgate Prison in 1798. He had a fair complexion, brown hair, and blue eyes.[2] It is not much to go by.

Playfair's first position at Boulton and Watt was as a draughtsman, making drawings of engine works and engine parts. Initially, Watt had made his own drawings for his engines, while working out of his own house about two miles from the factory where the engine works were situated. When work pressures increased on Watt, a position was created for an assistant who would make the drawings. It was to this position that Playfair was hired, becoming Watt's first assistant in this capacity. Like Watt, Playfair first worked out of Watt's house. About a year later, he had relocated to the factory to carry out his work in the counting house.[3] Although Watt initially thought that Playfair's talents in drawing were overrated, to my eye they appear quite reasonable. To make my point, an example of Playfair's work on a barley mill is shown in figure 2.1.[4] The barley mill drawing also shows elements that Playfair used in his later graphs: hachure (in the cross section ends of horizontal pieces in the bottom illustration to show solidity of the pieces), shading (in both the top and bottom illustrations to give a three-dimensional look), and stippling (in the top illustration within the wheel to show depth). (See, for example, Spence 2004; and Playfair 2005, 14–17, for descriptions of the graphical elements that Playfair used.)

The steam engine was only one of several business ventures taken on by Boulton. At the time of his partnership with Watt, Matthew Boulton had been in business for more than twenty-five years. Many of his other business interests were carried out both before and during the work on the steam engine. After he left Boulton and Watt, Playfair's first business ventures were closely related to Boulton's manufacturing activities that had begun in the late 1740s and were continuing through the time that Playfair was in Birmingham.

Originally, Matthew Boulton was a toymaker, working with his father and then inheriting the business from him. Toymakers of the eighteenth century are different from toymakers today.[5] Eighteenth-century toymakers are defined as makers of "small steel articles, as hammers, pincers, buckles, button-hooks, nails, etc." (*Oxford English Dictionary*) – hardware manufacturers in modern parlance. The Boultons manufactured steel buckles for shoes and knee breeches; later, Boulton also manufactured buttons. Typically, Birmingham toy manufacturers were small operations, sending their wares to middlemen in London for sale there or for the export market on the Continent. It was a highly competitive market. Boulton was enough of an entrepreneur and innovator that he could break into this market and dominate it. Working with John Fothergill in 1761 and partnering with him the next year, Boulton mechanized and expanded the manufacturing operation in order to produce his goods more cheaply and with greater precision. Initially, Boulton and Fothergill leased a cottage and a water-driven metal-rolling mill for

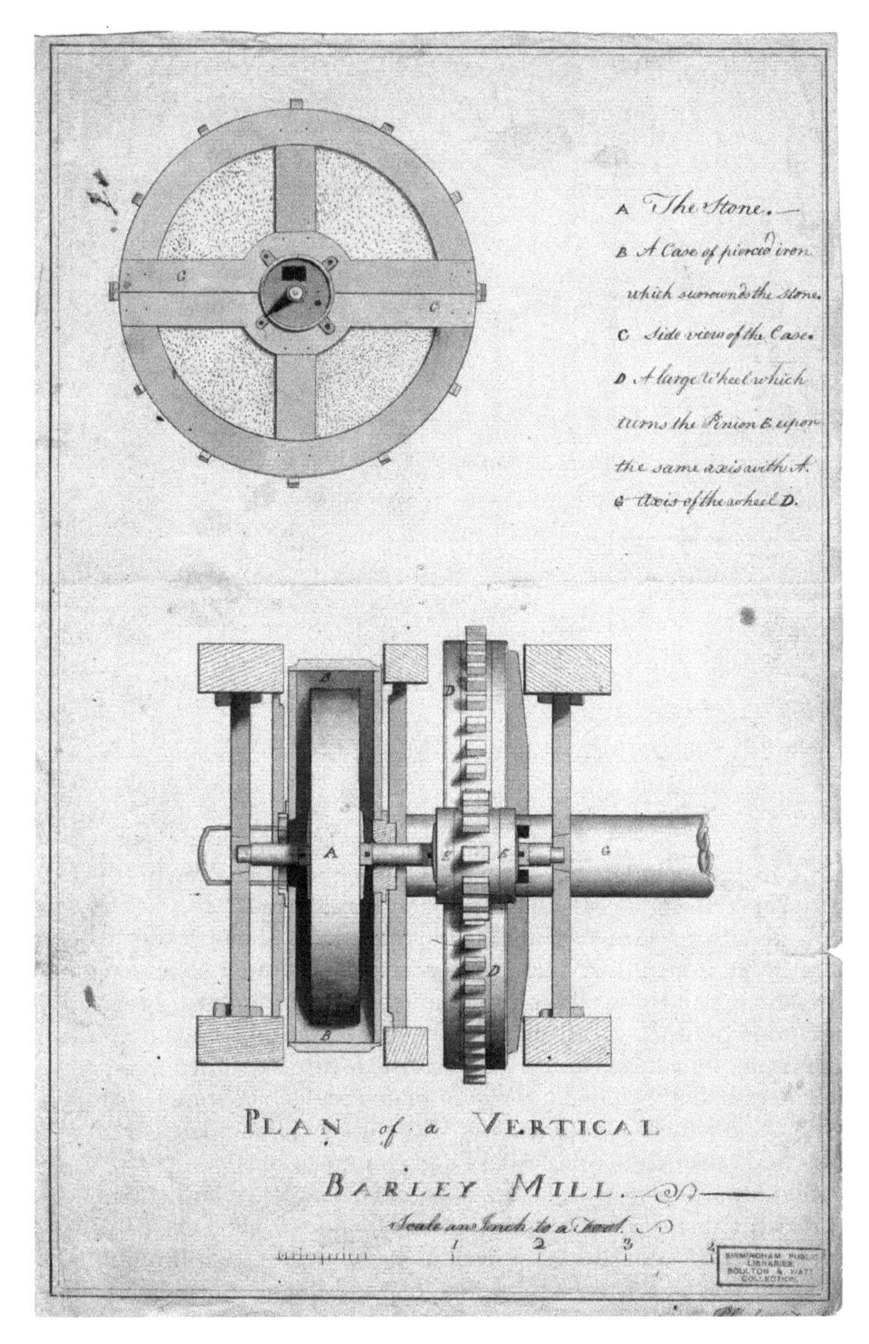

Figure 2.1. Plan of a vertical barley mill. Courtesy of Library of Birmingham, Birmingham, United Kingdom.

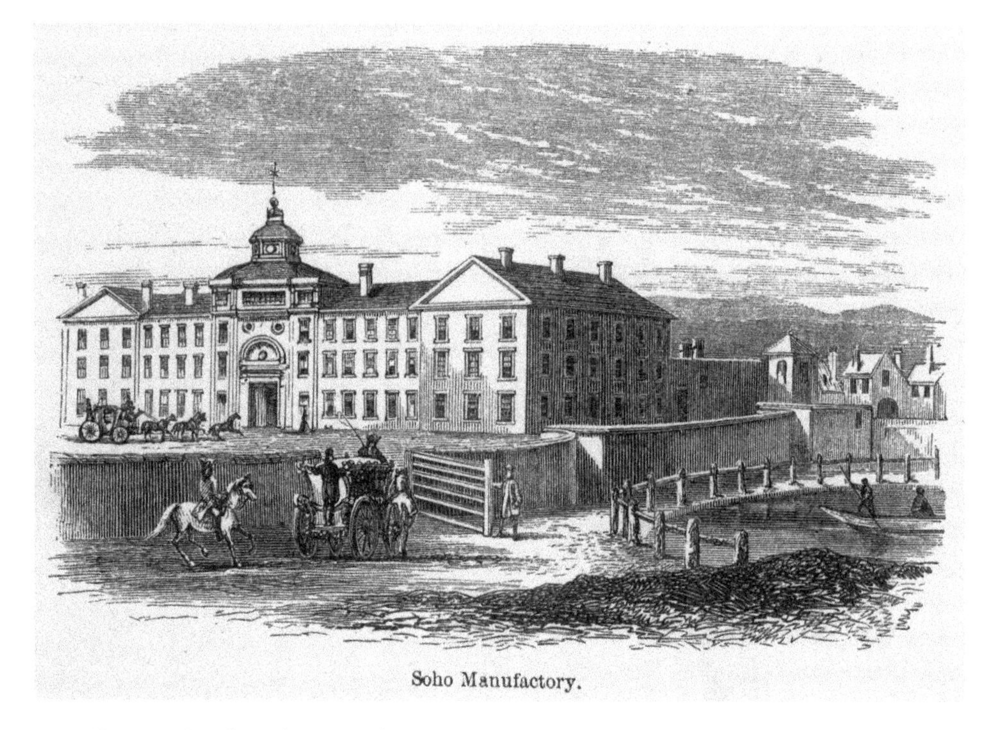

Figure 2.2. The Soho Manufactory. Alamy Stock Photos.

their operation. In 1766, they replaced the cottage with a newly built factory, called the Soho Manufactory. Shown in figure 2.2, the factory was powered by a waterwheel and employed 1,000 workers. It was in Soho that Boulton and Watt made precision parts for their steam engines. Boulton expanded his toy product line by going into luxury items. He was soon producing gilt and silver personal items that included watch chains, toothpick boxes, and snuffboxes. He also expanded into household items that included a range of plated and silver ware to compete with the Sheffield manufacturers. This product line ranged from teapots and tankards to candlesticks and chimneypieces.

Boulton was also a marketing wizard. He employed an agent in London, kept abreast of new fashion trends, and travelled extensively to meet old and new clients, as well as having them visit Birmingham. He courted the aristocracy as potential and ongoing customers; they also served Boulton as a network of wealthy individuals who could casually promote his products and provide information on the latest fashions. One such aristocrat was William Petty, 2nd Earl of Shelburne, and later 1st Marquess of Lansdowne, who was to cross paths with Playfair in the 1780s.

It is uncertain whether Playfair met or corresponded with Shelburne while he was in Birmingham or if the connection was made on Playfair's initiative after he moved to London. What is certain is that Playfair made some connections in Birmingham during his time there, and they all relate to his employment with Boulton and Watt. In addition to Boulton and Watt, of course, Playfair's network that turned into business connections included James Keir, Joseph Gerentet, and William Wilson. Keir was a chemist and an industrialist. He operated for a time during the 1770s as the factory manager for Boulton and Watt during Boulton's frequent absences from Birmingham. In the 1780s, Keir built an alkali factory in Tipton, about ten miles from Birmingham. Later he added soap making to the factory's activities. What little is known of Gerentet in Birmingham is that he is listed as a "foreign foil maker" in a 1781 business directory for Birmingham (Bailey 1781, 149). The next year he was buying large quantities of perna, commonly called "pearl shell," which was used to obtain carbonate of lime.[6] All that is known of Wilson is that he, like Playfair, worked as a draughtsman for Boulton and Watt.[7]

Keir, Boulton, and Watt were all members of the Lunar Society – an informal scientific society that grew up around Matthew Boulton, his friends, and scientific associates (Schofield 1963; Uglow 2002). Meeting monthly (the Monday nearest to the full moon and hence the name), usually at Boulton's house, they discussed a range of scientific issues and how they could be applied to various aspects of society. Other members of the society whose paths crossed, or may have crossed, Playfair's include Joseph Priestley (theologian, dissenting minister, and chemist), William Small (a physician with widespread interests in science), and Josiah Wedgwood (a master potter who brought many innovations to the industry). There is some speculation that Playfair got some of his ideas for statistical graphics from Priestley's work. Priestley joined the society after he moved to Birmingham in 1780 following a stint as Lord Shelburne's librarian, with Priestley describing himself as "a guest in the family" during his time of employment with Shelburne (Rutt 1831, 205). Playfair's window of opportunity to meet Priestley in Birmingham was about one year. Rather than in Birmingham, Playfair probably first met Wedgwood in London when attending a meeting of the Royal Society with his brother, John Playfair (see, for example, Smiles 1895, 266–7). William Small died before Playfair arrived in Birmingham so that any connection was post mortem.

William Playfair does not appear ever to have been a member of the Lunar Society, nor is there any evidence that he attended any Lunar Society meetings after he arrived in Birmingham. During his sojourn in Birmingham, he was an employee, not an employer, and he was not an eminent scientist, inventor, or industrialist – terms that would describe the members. It was, however, Lunar Society connections that got him his position with Boulton and Watt. Robert Small, brother of William Small and the man who arranged for Playfair to apprentice with Andrew Meikle, visited Birmingham at some point after the death of Small's brother in 1775. While in Birmingham, it was Robert Small to whom Matthew Boulton expressed the need to have a young engineer to supervise on location the erection of Boulton and Watt's steam engines.

During August and September 1777, Robert Small wrote letters to James Keir highly recommending Playfair for a position at Boulton and Watt.[8] Keir passed the letters on to Matthew Boulton, who subsequently hired Playfair as a draughtsman. He was given a three-year contract and paid sixteen shillings a week. Playfair probably arrived in Birmingham in late September or early October. As I have already said, Watt thought that Playfair's talents in drawing were overrated. On the other hand, Keir was more favourably impressed, although he thought that Playfair's wage was a little high. There was another issue beyond wages. When Small wrote to James Keir, what he had in mind for Playfair was not only immediate employment but also a path for the development of Playfair's career. In a letter dated 2 September 1777, Small writes:

> I beg you also to believe that the other question [the first was wages] proposed about Mr Playfairs prospects at the expiration of the three years was not at all meant to bring Mr Boulton or Mr Watt under the slightest appearance of an engagement. But that his friends considering that his attention was to be turned to one particular branch of machinery wished to have some information how far the knowledge he might acquire would be useful to him in future life. But as this in the end must depend chiefly on his own good conduct and abilities they do not think it needful that anything more should be said about this matter than you have already told them.[9]

Small's wish was not immediately fulfilled. Until about 1780 Playfair worked mainly as a draughtsman. He was trusted, at least by Boulton. When Boulton was away, he would report to Boulton by letter about what was happening at Soho with regard to the business. As an underling, he was also required to do some work marginally related to his job. Once, during a school break, Playfair and at least one other draughtsman were told to teach arithmetic and drawing to Boulton junior and Watt junior for a couple of hours a day.[10] There were a few occasions when Playfair did some hands-on engineering work. While Murdoch handled the faraway engine erection work in Cornwall, Playfair was sent to some jobs nearby. In a letter of 15 August 1778, it is stated only that "Joseph Harrison and Mr Playfair are not yet return'd Home." Another letter of 17 November 1778 is much more explicit. Playfair had been sent to Bedworth, about twenty-five miles from Birmingham, to see what tests had been made on the engine erected there. Tests were made to see whether the machinery was working properly and to estimate how much coal was used to run the engine. The latter issue was important. The price charged for a Boulton and Watt steam engine was based on the savings made on the reduction of coal needed to run the engine compared to other steam engines.[11]

Business was brisk during 1778 and so there were some production problems. Either Watt was not giving clear instructions or one of his engineers, a man named Hall, was negligent in getting some parts made for some steam engine orders. It is unclear where the blame lay. It was later found that Hall had been stealing iron for his own use. On the other hand, Boulton had the eye for detail in production. Playfair also figures into the problem, perhaps as a bottleneck, in producing his drawings used in the production of the parts. Some of Watt's accusations and expressions of frustration are given in a letter from Watt to Boulton dated 19 June 1778:

> I have written very fully this post to Mr Walker on these heads so you need not write but do advertise or take some other method to find a Clerk that will give himself the trouble of considering what is to be done. I am very sensible that in ye mult^{r}. of things I have to think of & the vexations I some times meet with I am not so accurate in giving out orders as I might be but in ye present case I did my part, & the Blunder lies among Mr H^{n}: [Henderson] Playfair & Hall. And at any rate I see a determined resolution in him [Hall] to understand nothing without as much as more drawing or explanations as would serve a workingman who had never seen one of our engines. In so much that I should save myself trouble by giving the orders to the Engineers upon ye spot instead of him and it is really impossible for me to comply with all that accuracy in giving orders to our own workman who ought to understand these minutiae better than me otherwise are not fitt for their places. (Dickinson and Jenkins 1981, 285)

A week or so after Hall's thieving had been discovered, Watt wrote to Boulton saying that Hall should be terminated and replaced by Playfair.

It would finally give Playfair the opportunity to use the engineering skills that Robert Small had wanted honed. Of course, just as Watt had trashed Playfair's drawing skills, he was also highly critical of what Playfair could do as an engineer, writing to Boulton: "I would recall Playfair who can do part of the business & I think now that you are at home you can contrive to gett him proper assistance. I must warn you that Playfair is a blunderer but I dare say he will be assiduous and obedient and plain directions must be given him."[12] At least Playfair was only a "blunderer." Watt referred to Hall as a "very great blunderer."[13] Despite Watt's misgivings, that is how Playfair got to go to Bedworth later that year in November and perhaps on some other engineering outings.

It must be said that, despite Watt, Playfair deserves some credit for his engineering abilities. And Watt may also have changed his mind about Playfair. In a letter of 6 October 1778, James Keir reported to Boulton that Playfair had found an error in one of Watt's drawings for a new steam engine for the Poldice mine in Cornwall:

> I have received Mr Watt's letter ordering a gigantic engine for Poldice mine. Playfair has copied the drawing & sent the original to Bersham [where John Wilkinson manufactured the cylinder for the engine] with orders to give preference to this engine over all others. Playfair observes that in Mr Watt's drawing of the bottom of the air-pump he has not represented that part by which it is joined to the Eduction-pipe. Although there is little probability of Mr Watts making such omission, yet as it is possible for any man to make omissions, I thought it not amiss to send the copy of that drawing in which the black lines are Mr Watt's & the black lead lines are what Playfair supposes to be omitted.[14]

Keir, along with Playfair, seem to have handled the situation discreetly (Hills 1993, 157, interprets the situation as Playfair's error, not Watt's).

It has been reported that it was at around this time, perhaps during 1779, that Playfair married Mary Morris and that they had their first son, John, in 1780 or thereabouts (see, for example, Playfair 2005, 6). The problem for me is that I cannot find their marriage record or the birth records of any of their children. I can find a few of them in censuses and in burial records, but nothing that would tie down exact dates of births or christenings, or of their marriage.[15] And there is an interesting footnote in Douglas Hamilton's *Jacobitism, Enlightenment and Empire, 1680–1820*: "In 1780 he [Playfair] was even suspected by Boulton's wife Ann of 'scandalous conduct' with women, though he had married Mary Morris in 1779" (Dunyach 2014, 257). The solution to this conundrum is that perhaps he did not marry Mary Morris at that time and that all, or most, of the children were born out of wedlock. This would explain Ann Boulton's perception of Playfair's scandalous behaviour. Entry no. 876 in the *Register of Banns of Marriage Published in the Parish Church of St. Mary le Strand* in London shows that the banns of marriage for William Playfair and Mary Morris, bachelor and spinster respectively of that parish, were read in the church on the three consecutive Sundays of 29 November, 6 December, and 13 December 1795.[16] If this is indeed the correct William and Mary, this was probably just after their daughter Zenobia was born and when the Playfairs were living in London. One further wrinkle is that although there is a record of the banns being read, there is no record of the actual marriage either in the St. Mary le Strand register or any other church. The marriage would have taken place on 14 December or shortly thereafter. It is possible that the Playfairs never married; it is much more likely that they did marry and that the marriage was not recorded properly.[17]

In around October 1780 Playfair's three-year contract would have expired. And yet he continued to work for Boulton and Watt. What probably kept him on was either the attraction of a new venture for him, or a shortage of skilled workmen. The new venture came with a new invention. Initially inspired by suggestions from Erasmus Darwin at a Lunar Society meeting, Watt invented a machine that would make copies of his correspondence and other handwritten documents. He made application for the patent on 14 February 1780; Playfair was one of the witnesses to Watt's signature on the application.[18] Watt was the inventor, Boulton the organizer and promoter. About six weeks after submitting the patent application, Watt, Boulton, and Keir formed a partnership under the name of James Watt & Co. to manufacture and market the new copier. Boulton, the marketer, took the copier to London to show potential customers who were likely to have extensive correspondence, including parliamentarians, bankers, and merchants. Playfair was given a new job with a new contract, in addition to his continued work as a draughtsman. He became the manager for the production and sale of the new copier.[19]

The invention was simple in theory, but more difficult in implementation. The basic idea was to have the original document written in slow-drying ink. The paper on which the copy was to be made would be moistened, placed on top of the original and passed through a pair of rollers to squeeze the papers together in order to have the copy absorb some of the ink of the original. Part of the trick was in the ink to be used; it had to be semi-soluble. Another part was in the rollers that were used. Three types were tried – wood (lignum vitae), iron, and brass – before brass was finally settled upon much farther down the road. Most of the original machines used wood, which tended to crack. The main trick was in the copy paper. The moistening agent had to be such that the ink would be absorbed leaving the copy and original undamaged. The paper also had to be thin so that the imprint showed on both sides, one side a negative image where the original and copy came together, and the other side a positive image that would be readable.

Although about 150 copiers were made and sold in the first year of operation, Playfair thought that business was slow in these early days. In September 1780, he wrote to Boulton: "The copying business does not go on so well, there are not many more than 200 sent off from hence, not half that Number delivered and still fewer payed for so that tho' that business is not quite dead it has not much life and is carried on with no spirit."[20] There were two probable reasons for the slowness of business. One was in the rollers made of wood – lignum vitae in particular. The other was that Watt's first machines were not portable, but had to be fixed to a table, so that there was some loss in convenience of use (Muirhead (1854, 30–1). Watt's portable copier would come later.

Three early customers are of interest here. The first is John Smeaton, a civil engineer whose business connections to Boulton went back to the 1760s (Uglow 2002, 65 and 78). Boulton had demonstrated the machine to Smeaton, who became interested enough to obtain a copier in August 1780. Boulton and Watt offered the copier to Smeaton for free, but because of the cost of the material in the making of it, Smeaton refused the offer and asked for a bill. Smeaton put the copier through a number of experiments and wrote to Playfair about his results in November 1780.[21] Smeaton's main insight was that the most important factor in obtaining a good copy was in the moisture of the copy paper. The second customer is Playfair's brother Robert, an Edinburgh lawyer.[22] Robert Playfair bought four copiers, probably for resale in Edinburgh; about three or four years later, Robert was William's agent in Edinburgh after William had established a business in London. The third customer, Joseph Priestley, was not a paying customer; Watt had ordered a copier to give to Priestley. This may have been one occasion when William Playfair came into contact with Priestley in Birmingham.

About a year after being placed in charge of the copying business, Playfair was making plans to leave the employ of Boulton and Watt. He gave three months' notice in a letter to Boulton dated 15 October 1781.[23] He gave two reasons for leaving. On the positive side he was going into business with the assistance of his brothers. On the negative side he was concerned about the copying business. About a month before,

Keir had told Playfair that he thought the production of the copying machines should be stopped until they were able to sell some of the stock in hand. Playfair came to believe that, at least in the short term, the copying business would not "pay for me attending to it."

Playfair continued working at Boulton and Watt until nearly the end of 1781. He was still making drawings and supervising some of the steam engine work within the factory.[24] Since his major work was the copying machine, he was also putting this business in order for James Keir either to take it over or for whoever would replace him. Keir was also getting Playfair to finish as many machines as possible before he left and noted to Watt, "The manufactory of the machines is so well regulated by Playfair that it will be very easy to continue in the same order." Keir was sorry to see Playfair go, but recognized that Playfair was keen to go out on his own. He also expressed the opinion to Watt that "there will be little difficulty in continuing the business" and suggested some changes that should be made to improve the business model for the copier.[25] The manufacture of copying machines by James Watt & Co. continued until 1794 (Dickinson 1937, 209).

Chapter Three

Playfair Goes to London to Set Up His Own Business

By late December 1781 Playfair had established himself in London. He was living at No. 2 Portland Road, Great Portland Street, near Cavendish Square.[1] Within the next year, or perhaps even weeks or months, he was in business nearby, off Tottenham Court Road. The location of the business is circled in figure 3.1. Some street names have changed since the late eighteenth century. For example, Quickset Row is now part of Euston Road. Officially, William Playfair and his brother James took possession of this property on 6 June 1785.[2] William was definitely on the property prior to this date. A letter written by Playfair on 12 September 1782 puts him at 9 Howland Street.[3] There is some circumstantial evidence that William was on this leased property shortly after his arrival in London. The term of the lease on the brothers' property was 95½ years less twenty-one days. Typically, a lease for that length of time is 99 years. If we assume that the original lease was for 99 years, then the original purchase would have been on approximately 27 December 1781, shortly after William Playfair's arrival in London. Either William obtained the original lease and renegotiated it with his brother when the brother arrived 3½ years later, or else the two brothers purchased the lease from someone else 3½ years into the 99-year lease. In either case, William Playfair was on the property, perhaps as a tenant, prior to the brothers' taking the lease.

For some time prior to leaving Birmingham, William Playfair must have quietly prepared for his new business venture. On 19 December 1781, just prior to his leaving Birmingham, he obtained a patent for a new way of manufacturing tongs, spoons, knives, forks, and other objects made from silver or other metals (Great Britain Patent 1781/1309). He was given the sole right for fourteen years to produce and sell the articles made by his new process. These were exactly the sorts of articles that Boulton produced in his toy business at Soho. Playfair claimed

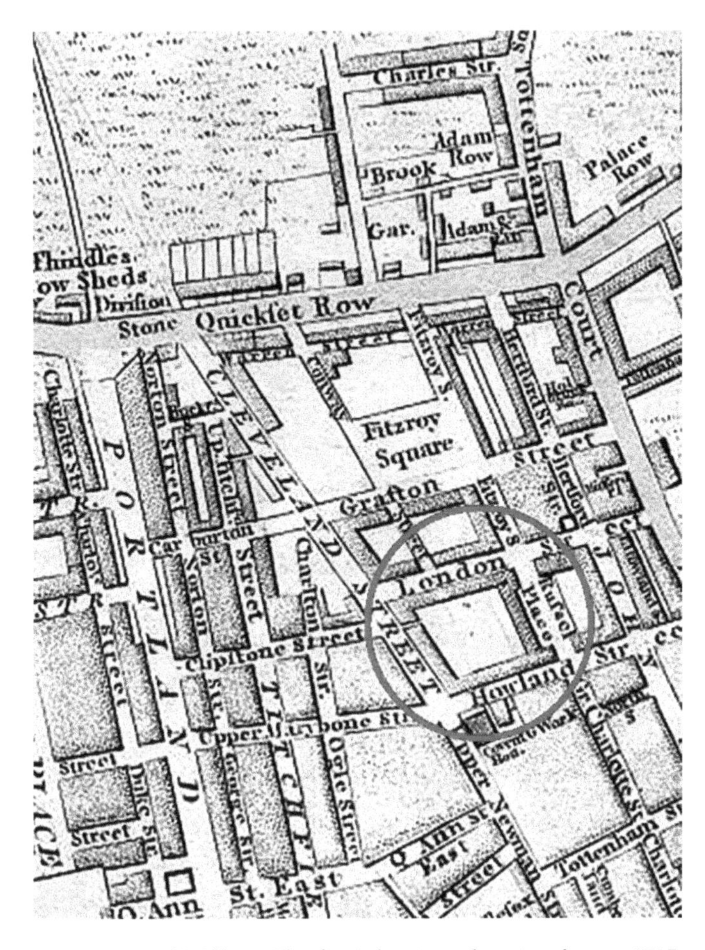

Figure 3.1. James and William Playfair's business location from a 1795 map of London. Courtesy of David Hale and the MAPCO: Map and Plan Collection, http://mapco.net.

that his invention would reduce the cost of production of these articles. He must have carried out some experiments at Soho before leaving, in order to develop his new method. Metal, rather than paper as in the copying machine, was passed under pressure between two steel or iron cylindrical rollers. These rollers had impressions in them that would produce the raised surface on the finished product. The novelty, it appears, was in how the impressions were made in the rollers to take into account the stretching or lengthening of the metal that occurred

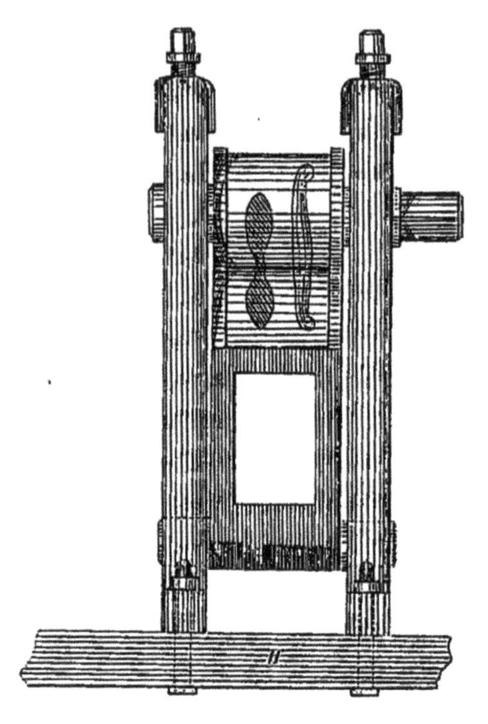

Figure 3.2. Playfair's rolling mill for the manufacture of tongs, spoons, knives, and forks. British patent specifications, public domain.

when passing through the rollers. In addition, the material was not completely sent through the rollers, but oscillated back and forth before being extracted. An example of this machine, or rolling mill as it would have been called in the eighteenth century, taken from Playfair's patent specification and showing the impression in the rollers, is given in figure 3.2.

Very quickly, Playfair's business in London was up and running. And he was poaching one of Boulton and Watt's employees to help him run it. Playfair wrote to Boulton in Birmingham on 9 March 1782 informing him of William Wilson's intention to leave Boulton and Watt to work with Playfair in London.[4] He had tried to see Boulton about it, expecting to have Boulton visit him on a trip to London. The visit did not occur and by the time Playfair went to Boulton's hotel, he had gone. In the letter, Playfair tried to placate Boulton: "I do not wish to do any thing in an underhand way to a person who has behaved in a manner

Figure 3.3. Maker's mark for Playfair and Wilson. Courtesy of Phil Osbourn, owner of the website https://www.silvermakermarks.co.uk.

that will always preserve my respect & esteem." By May, Wilson was in London, and on 16 May the two had registered their maker's mark, shown in figure 3.3, for their silverware (Chaffers 1883, 191).

After leaving Birmingham, Playfair continued to maintain contact with Boulton. At Playfair's request, Boulton had sent him a pattern book for silverware from which Playfair could make his own goods. Playfair had been out and about with the book trying to drum up business. In January 1783, he contacted the Websters of Leadenhall Street in London, later to become Wedderburn, Webster, and Co., traders in the West Indies who were closely involved in the slave trade.[5] The Websters were interested in Playfair's silverware, but he had come to them just after their fleet had sailed for the West Indies. They were interested in keeping the pattern book and wanted another to send to Jamaica. As a result of this interest, Playfair offered to be a retailer for the silverware that Boulton produced in Soho. Boulton responded that he was interested and wanted to know if the Websters were also interested in buttons that he produced. He added that he could only pay a 2 per cent commission. As a sign of difficult times on the horizon for Playfair, Boulton told Playfair that the trade in general was slow and had very narrow profit margins, hence the low commission that he offered.[6]

Playfair enlisted some of his family in marketing his silverware. Advertisements for his goods, done "in the neatest taste and in superior workmanship to any hitherto made," appear in the *Caledonian Mercury* on 19 and 24 January 1784. Brother Robert Playfair, the Edinburgh lawyer, is given in the advertisement as the agent who would take orders for the silverware.

In London, a new contact was made, or an old one renewed. Josiah Wedgwood was a member of the Lunar Society that met in Birmingham, but it is uncertain whether his and Playfair's paths ever crossed there. They definitely crossed in London. And it led to Wedgwood buying a substantial amount of silverware from Playfair during 1784 and 1785, over 100 pieces in all. The purchases include one punch bowl and many ink stands, cream jugs, sugar basins, teapots, candlesticks, egg-cups and dishes. The total bill seems insubstantial today, about £47.[7]

Wedgwood and Playfair met in London on 9 May 1782 at one of the weekly meetings of the Royal Society. Visiting from Scotland, Playfair's eldest brother John was in town and attended several meetings between February and June. William attended the 9 May meeting with his brother. This was possibly William's only encounter with the Royal Society.[8] At the meeting, Wedgwood presented a paper on the design of a thermometer that would be useful for measuring high temperatures. Previously, high heat was only descriptive, with terms such as "red hot," "bright red hot," and "white hot" for the object being heated. Anything above 674°F, the boiling point of mercury on the Fahrenheit scale, could not be measured with a mercury thermometer, the standard thermometer of the day. Wedgwood (1782) based his high temperature scale on the shrinkage of clay at high temperatures. Playfair was intrigued by the presentation that he saw and wrote to Wedgwood about it on 12 September 1782.[9] In addition to wanting to know where he could purchase such a thermometer, Playfair suggested that Wedgwood make a connection between the commonly used Fahrenheit scale and the scale that Wedgwood was using. He enclosed a two-page paper suggesting a method to find the relationship. Two years later Wedgwood obtained and published the simple relationship (at least in terms of example numbers instead of a formula) of $°F = 1077 + 130 \times °W$, where °W are the degrees on Wedgwood's scale (Wedgwood 1784). Playfair is not mentioned in the paper and his suggested method appears incorrect.

Playfair continued to build on his industrial experience at Soho. On 24 May 1783, Playfair submitted a patent application for the manufacture of bars for sash windows, curtain rods, and mouldings to make borders for trays or tea caddies, all in metal (Great Britain Patent 1783/1373). The idea was similar to his first invention. Lengths of metal to be ornamented would pass through a rolling mill as Playfair had done before, but this time constructed specifically to take the metal rods. The major difference was that lengths of metal for a certain shape, such as flat on one side and convex on the other, would be drawn through a metal plate with a hole in it of the appropriate shape. Later that year, Playfair followed up with another patent application that would use a rolling mill to manufacture lengths of metal that could be any uniform shape such as a cylinder or that could be tapered. In a wild flight of fancy, he envisioned a wide range of application from mouldings for doors, to looking glass frames, to bannisters for stairs, to ornaments for coaches and carriages. Playfair's was one of the early inventions for making window and door sashes and frames that were not made of cast iron (Louw 1987).

By June 1783 there was trouble. Playfair wrote to Boulton about his money problems; he needed extra time to pay one of Boulton's bills. He accused James Keir of spreading rumours about his inability to pay his creditors. The result was, Playfair claimed, that he was obliged to set days when payments would be made to his creditors. Claiming that there was imminent cash flow for his business, he asked Boulton for a week's grace to pay his bill.[10]

Despite the bad blood, it was probably within a year of this episode that Playfair went into partnership with Keir and Alexander Blair. Earlier, Blair had partnered with Keir to build the alkali and soap factory in Tipton (Schranz 2014). The new partnership with Keir and Blair was all about the metal ingredients Playfair wanted to use in his windows and mouldings. Advertisements in newspapers between June and August 1784 show Playfair making windows, sky lights and fan lights, as well as sashes for greenhouses, from iron and a new product called "El Dorado" metal, the latter for use in elegant and fashionable houses (*Morning Herald and Daily Advertiser*, 30 June 1784, 5 July 1784, and 10 August 1784). The problem Playfair faced was that the metal used, an alloy of copper, zinc, and iron, was the same, or very similar, in composition to an alloy that Keir had patented in 1779 (Great Britain Patent 1779/1240). Playfair needed either permission or partnership to manufacture his windows with this metal at his London works. Keir was also making window sashes with a rolling mill in Tipton using his patented metal (Prosser [1881] 1970, 143). Playfair's brother John, who had had a good professional relationship with Keir from at least the early 1780s, may have been the one to broker the partnership (Schranz 2014, 251). Through all of this, Robert Playfair served as William's agent in Edinburgh for orders for windows (*Caledonian Mercury*, 30 August 1784).

The partnership quickly fell apart with Keir holding the upper hand. Formal notice of the termination of the partnership appeared in the *London Gazette*, 26 April 1785. All accounts receivable and all accounts payable were to be handled by Keir at the business premises in London, given as London Street and shown on the map in figure 2.3. Animosity and bitterness continued, at least on Playfair's part, for a few months. In a letter to Boulton in September, Playfair referred to Keir as ill-natured.[11] A week later, Playfair wrote to Watt, mentioning his quarrel with Keir. According to Playfair, the nature of the quarrel appears to have been over the cost of alkali and how it affected their business. What continued to rankle Playfair was Keir's claim that Playfair was in debt to Blair and that he had made free with Keir's money – claims that Playfair denied vehemently. There may be some truth to Keir's claims. Playfair was never on top of the money, writing to Watt, "Had our books been more exactly kept it w^{d} have been better for me but I never kept them myself."[12] On 20 September 1780 Watt had written to Boulton about Playfair's abilities as an accountant: "Playfair does not understand bookkeeping nor method" (Lord 1966, 201).[13] This lack of attention to financial detail would return to haunt Playfair in future businesses and put him on the road to debtors' prison. It was a road without turnoffs that he followed for the rest of his life. Keir seems not to have harboured bad feelings towards the rest of the family. Despite the failure of the partnership, Keir remained on good terms with John Playfair. In 1794, Keir entered into a partnership with John in a colliery business (Schranz 2014, 251).

During his brief partnership with Keir and Blair, Playfair had at least two major commissions. One was for door sashes for the British Museum and the other was for several windows in Carlton House, the London residence of George IV when he was prince regent ([Byerley] 1823, 171). No information that I can find exists for the British Museum work. Carlton House is similar with one minor difference. George carried out extensive renovations on Carlton House beginning in 1783. Playfair shows up on a list of about fifty people for expenses paid for work done on Carlton House between 1783 and 1786. He was paid £331 8s. 11d.[14] His was not a particularly large bill; the largest was over £4,000; George was an extravagant man.

Very soon after his partnership with Keir and Blair was dissolved, William Playfair's partnership with William Wilson was also dissolved. It seems that Keir also handled the winding up of this partnership, probably wanting to make sure he recovered whatever of his money that he could.[15] In 1786, Josiah Wedgwood paid one of his bills to "James Keir & Co. Trustees for Wm. Playfair & Co." The bill was probably for silverware, not window sashes.[16] In the notice of the Playfair-Wilson dissolution, unlike the other partnership, Playfair carried on the business "as usual," which was not a good sign. He tried to capitalize on a new patent that he had obtained in February 1785 for silver plating or gold plating small articles such as shoe and knee buckles (Great Britain Patent 1785/1466). Advertisements for these articles appeared in newspapers until mid-July 1785, while advertisements for the articles produced under the two failed partnerships ceased in April 1785.[17]

It was at about this time that the Playfair brothers James and William signed the lease for 95½ years less twenty-one days on the property on which William was already running his business. Records of the Sun Insurance Office provide information on the rise of James's architectural practice and the decline of William's manufacturing businesses.[18] The first insurance policy, taken out in 1784, shows William as a spoon maker and manufacturer, with no professional information on James. William had some workshops and warehouses in London Street that were made of brick. The house at 9 Howland Street was new and also made of brick. Both William and James appear to have been living comfortably. Their household goods were valued at £350, their books at £50, and their clothing at £50 each. The working utensils and the goods to be sold that were spread over the house, the workshops, and the warehouses were valued at £1,000. A year later, in 1785, there is no mention of William Playfair's stock in trade on the policy, but his workshops and warehouses were valued at £200. The buildings were also mortgaged. With a career on the rise, James was building two houses in Russell Place, described as "not quite finished." The houses were just around the corner from Howland Street. By 1787, James had finished the two houses and was living in one of them. William still had his workshops and warehouse; no house was mentioned in the policy. His occupation as silversmith from two years before had now changed to button maker. A letter to Playfair, dated 16 October, puts his residence in Russell Place so that he and his family were probably living with his brother James.[19]

One other notable event in 1787 was the visit of economist Adam Smith to London and his meeting William; Smith was a leading light of the Scottish Enlightenment. Eighteen years later, when Playfair produced a posthumous edition of Smith's *Wealth of Nations* in 1805, he briefly recounted his meeting with Smith (Smith and Playfair 1805, xxxiv). Playfair's connection to Smith was probably through his brother John. In Edinburgh, John Playfair was a member of the Oyster Club, a dining club of intellectuals organized by Smith and two others (Rae 1895, 334). When William Playfair met Smith in London in the summer of 1787, Playfair commented that Smith was not well (Smith and Playfair 1805, xxxiv). Although Smith was in London between late March and early August 1787, Playfair's recollection of their interaction implies that Smith and Playfair met only once. Smith had come to London to consult an eminent physician about an inflammation of the bladder and a very bad case of haemorrhoids. Although he seems to have spent most of his time in London convalescing, Smith was able to meet with several of the leading politicians, including William Pitt and Henry Dundas (Campbell and Skinner 1982, 202).

William Playfair's businesses went into decline and soon after his meeting with Adam Smith, Playfair increasingly turned his attention to writing and to setting up a business in France. There are several indications over the next few years that he preferred a return to business over life as a writer. When he did return to business, his businesses all failed, sometimes through bad luck, but more often through poor business decisions.

Chapter Four

Playfair Evolves into a Writer by Profession

Six years after William Playfair composed his first publication in 1785, an anonymous writer defined the term "author by profession" as a person who makes a living from publishing literary works. The same anonymous writer considered "author by profession" a derogatory term. It meant that the author had to "minister to the pleasure of the public" (A.D. 1791). There were others, not authors by profession, who published their literary efforts, but had other means of employment. In the eighteenth century, these ranged from self-financing authors among aristocrats and landed gentry to those who depended on this group's patronage, or government patronage, in order to publish, as well as many others in between.[1]

Initially, Playfair was not, in terms of this definition, an author by profession. When he first started writing, he saw himself as an engineer by profession. He continued to carry on that description of himself when he arrived in France in 1787, and perhaps as late as 1792 when he returned from France. It was after his bankruptcy in 1797, after the collapse of the Original Security Bank that he had established in London, when he truly became a writer by profession. Bankruptcy closed a number of doors to him or made them very hard to open. There was a very strong stigma attached to bankruptcy at this time. As Adam Smith said, "Bankruptcy is perhaps the greatest and most humiliating calamity which can befall an innocent man. The greater part of men, therefore, are sufficiently careful to avoid it. Some, indeed, do not avoid it; as some do not avoid the gallows" (Smith 1776, 2:106).[2] The stigma ran deep. The bankrupt's reputation was usually destroyed and the merchant community typically shunned him (Servian 1985). One door that was open to Playfair was writing. He crossed that threshold and spent the rest of his life trying to make a living writing articles, books, and pamphlets, but it was not an easy way to make a living. He was in and out of debtors' prison several times. When desperate (and sometime when not), he resorted to schemes involving extortion and other scams to put food on the table. As with much of his writing, these schemes and scams had a creative side to them.

The difference between Playfair as an author and Playfair as an author by profession can be seen in his first and probably what became his last published work. His first publication, which came out in July 1785, carries the weighty title *The Increase of Manufactures, Commerce and Finance with the Extension of Civil Liberty, Proposed in Regulations for the Interest of Money* (*Morning Post and Daily Advertiser*, 28 July 1785). The last publication attributed to him is a graph in the book *Chronology of Public Events and Remarkable Occurrences within the Last Fifty Years; or, from 1774 to 1824*. This was a posthumous publication, as Playfair had died in 1823, the year before this version of the *Chronology of Public Events* came to press.

Regulations for the Interest of Money contains Playfair's suggestions for changing the British usury laws and for opening a bank that would make loans at various rates of interest. This would, Playfair claimed, stimulate the manufacturing sector of the economy, resulting in increased revenues for the state. At the time, Playfair was working as an engineer; his workshops producing plated silverware were still in operation. While Playfair had, or claimed to have had, the interests of the public in mind in *Regulations for the Interest of Money*, his proposal might also have helped him to continue in his business. He divided the business world into merchants and manufacturers – those who sold goods and those who produced them. He argued that merchants had no need for this kind of bank, as it was easier for them to obtain credit, while manufacturers found it harder to obtain credit. Manufacturers would benefit the most and, in a reciprocal way, benefit the country. According to Playfair, since the larger established manufacturers had easy access to capital, it would be the smaller manufacturers who would really benefit from the proposal. Playfair was one of those smaller manufacturers.

Playfair opens the *Regulations for the Interest of Money* with an abstract discussion: "The dismemberment of America from the British empire has roused the public attention to those resources which yet remain to England. As a person who has embarrassed his circumstances by exceeding his income, or by engaging in projects beyond his abilities to accomplish, endeavours to retrieve his affairs by lessening his expences, or striking into new paths of industry and less precarious adventure, so Great Britain naturally seeks, in her present situation, the means of retrenching the public expenditure and of increasing the public revenue" (Playfair 1785, 1).

This paragraph may have been written in the abstract, but the situation concerning the "person who has embarrassed his circumstances" applies directly to Playfair. By the middle of 1785 his partnerships had been dissolved. He was also described as a man "unfortunate in business" (Dunyach 2014, 167). Unfortunate though he was, as late as September 1785, he was still known as a "manufacturer of plate ware" (Dempster and Ferguson 1934, 145).

Regulations for the Interest of Money was written anonymously, but soon after its publication Playfair was identified as the author in advertisements for his next book, *The Commercial and Political Atlas*, which was published by Debrett and others (*Public Advertiser*, 14 September 1785).

The publication of *Regulations for the Interest of Money* follows the general path taken by some of Playfair's associates, as well as his family, and perhaps several others: write a book, and probably finance it mostly yourself, to self-promote your work and abilities. Playfair's brother James, the architect, did it. Two of his mentors in manufacturing, James Watt and James Keir, also did it. James the architect wrote *A Method of Constructing Vapor Baths* in 1783 that describes a method of providing his clients with a convenient-to-use hot indoor bath at a cost of about five guineas (about £750 today). Around 1770, fairly early in his career, James of steam engine fame wrote a short tract titled *A Scheme for Making a Navigable Canal from the City of Glasgow to the Monkland Coalierys*, which displayed his engineering talents. Demonstrating his abilities in chemistry, the last James wrote *A Treatise on the Various Kinds of Permanently Elastic Fluids, or Gases* in 1777. This was again fairly early in his career.

Flushed with the success of his *Commercial and Political Atlas* published later in 1786, Playfair showed his ambitions for a literary career beyond promoting his career as an engineer. Presumably, the two would run in tandem, with his engineering side providing the funds and the literary side providing the fame. On 12 July 1786, Playfair wrote to Thomas Townshend, Viscount Sydney, about his latest literary project, a history of the American Revolution, which had officially concluded three years before.[3] The letter appears to be an early example of Playfair making cold calls, either to promote his career or to carry out a scam, whichever was appropriate to the situation. This one was literary, and the cold call was the first of many in the years to come. At the time, Townshend held a senior cabinet position, secretary of state for the Home Office, or home secretary. Townshend had been close to William Petty, 2nd Earl of Shelburne, who figures prominently in the promotion of Playfair's *Commercial and Political Atlas*. Playfair thought his project on the

American Revolution would take several months to complete. Asking for no monetary support, he apparently wanted Townshend's blessing for his project. But no blessing seems to have been forthcoming and it was Playfair's second time of asking. Nothing came of the project.

Late in his life, Playfair wrote for the *Monthly Magazine*, which was owned and edited by Sir Richard Phillips. It was an unlikely match between editor and author. After the early stages of the French Revolution, Phillips had been seen as pro-Jacobin and Playfair was definitely anti-Jacobin (Seccombe and Loughlin-Chow 2008).[4] On the other hand, Phillips and Playfair shared the experience of bankruptcy. When he held some power as high sheriff of London, Phillips took an interest in improving prison conditions for debtors. Phillips also had very unconventional views about science. What Phillips brought to Playfair was work that Playfair very desperately needed. Phillips probably commissioned Playfair to draw a graph for each of his annual books that carry the title *Chronology of Public Events and Remarkable Occurrences within the Last Fifty Years*. Various editions were published over 1820 to 1824. For each, Playfair's graph appears at the beginning of the book and is acknowledged in the foreword: "The scale of the progressive variation of certain prices and amounts, connected with public finance, for which the Editor is indebted to the ingenious Mr. WILLIAM PLAYFAIR, will serve to check any errors of the press in regard to those subjects; while, as a picture, it cannot fail to interest the politician and philosopher." For his part, Playfair was ministering to the pleasure of the public using one of the best tools in his literary toolbox – his graphs. And then getting paid for it.

Playfair's graph from the 1821 version of the *Chronology* is shown in figure 4.1. This graph builds on Playfair's long track record in statistical graphics that show trends in economic data. The graph shows the annual values of six economic series: national revenue in millions of pounds, national expenditure in millions of pounds, exports in millions of pounds, the British national debt in tens of millions of pounds, the price of the 3½-pound loaf of bread in farthings, and the price in pounds of 3 per cent consols (perpetuities used to finance the British national debt and redeemable at the option of the government). Shown on the time scale are the American War of Independence, the French Revolutionary War, and the Napoleonic War so that the reader can see their effect on each of the six series. This was a graph done for hire by a man who died in poverty soon after he or someone else completed the graph for a later edition.

Playfair's incarceration in 1809 in Newgate Prison had its origins in 1805 when his trial for conspiracy began. He skipped his court hearing in 1805 until the law caught up with him four years later. In the meantime,

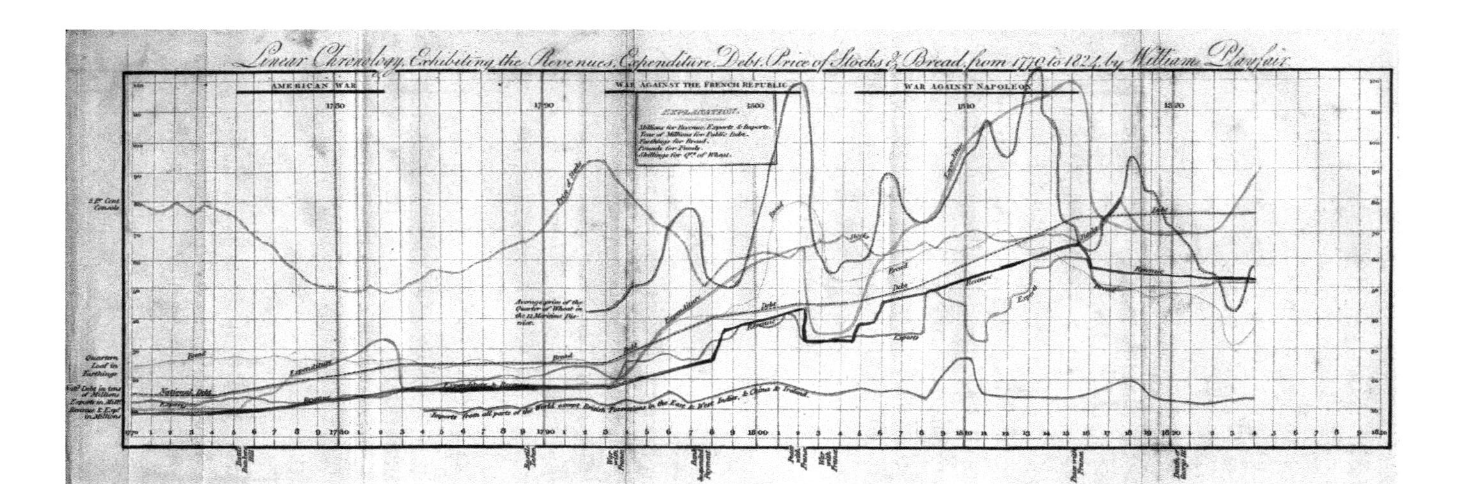

Figure 4.1. Playfair's graph showing six economic time series. Courtesy of the British Library, digitized by the Google Books project.

Playfair had financial problems. David Williams had founded the Literary Fund (now the Royal Literary Fund) in 1790 to help writers in financial need.[5] Playfair applied for financial help in 1807. He had made an earlier application to the fund in 1802 from which he received £10 (or about £900 in today's money). As a result of his 1807 application, Williams required Playfair to list his publications in order to receive financial support.

Playfair's publication list is revealing. For all his publications up to 1787, the year he left for France, Playfair gives John Debrett as his publisher (although this was not totally accurate). After his return from France in 1792 and prior to 1800, his publisher for the most part was John Stockdale. The change is a subtle indication of Playfair's political leanings before and after the French Revolution. Two quotations, one from a description of booksellers in 1802 and the other from Debrett's obituary in 1822, illustrate the probable change.

The first is taken from *The Picture of London, for 1802*: "At the west end of the town are booksellers shops (particularly *Debrett's, Stockdale's, Ginger's* (late *Wright's*) in Piccadilly; *Ridgway's*, York-street, St. James's square; the *Hookhams*, in Old and New Bond-street; *Earle's* Albemarle-street; and *Lloyd's*, Harley-street) furnished with all the daily newspapers, which are much frequented about the middle of the day by fashionable people, and are used as lounging-places for political and literary conversation" (Feltham 1802, 27).

The second is from Debrett's obituary in *Gentleman's Magazine*: "He was formerly a very eminent Bookseller in Piccadilly, where he succeeded the well-known John Almon; and his shop was the general resort of gentlemen of the first consequence in the Whig interest, and in opposition to the measures of Mr. Pitt; whose friendly admirers at the same period were to be found in the shop of his neighbour Stockdale" (*Gentleman's Magazine* 1822b, 474). In his *Memoirs*, written about 1822, Playfair wrote of his experience in making connections when he arrived in France: "On this occasion as nearly all through life I have found chance provided me the most useful connections" (Playfair n.d.). Likely many of Playfair's early political connections were made initially at Debrett's and later at Stockdale's bookshops.

The reference to Whigs and Pitt may require some further explanation. It was not until the nineteenth century that parliamentary factions coalesced into formal political parties in Britain. Initially, the factions were called Whigs and Tories. From the time of the accession of George I in 1714, the Whigs typically held power for the remainder of the century and the Tories were always in the minority. There were splits and cliques within the Whig faction that changed as the century progressed.

Whig ministries were formed as the balance of power between cliques in the Whig faction changed. The Tories as a faction disappeared mid-century. In 1783, in the aftermath of the American Revolution, William Pitt and his faction took power and the faction headed by Charles James Fox was in opposition. Pitt's ministry continued uninterrupted until the turn of the century. The Whigs in Debrett's obituary refer generally to Fox and his supporters. Pitt's faction eventually became the political party known as Tories in the nineteenth century.

Playfair was raised and initially worked in the milieu of the Scottish Enlightenment. Elder brother John was a leading figure of this phenomenon. The main themes and ideals of the Scottish Enlightenment are commercial modernity and its progress, coupled with a strong and stable central state. Enlightenment discussions surrounding economic and political questions were typically framed by empirical evidence taken from historical examples, and much of Playfair's written work comes under this influence. The development of his proposal on interest rates in *Regulations for the Interest of Money* is a typical example. Playfair begins his argument by citing several historical examples dating from antiquity up to the Renaissance to demonstrate that manufacturing, trade, and commerce are all beneficial to a country. After this lengthy preamble, he launches into his proposal.

Throughout his career, Playfair received mixed reviews of his books. Some were praised; some were panned. For example, most reviewers were very positive about *Regulations for the Interest of Money*. The *Monthly Review* was generally positive but thought the book too wordy; it could have been reduced to a much smaller book and sold at a third of the price (two shillings rather than six, the publisher's advertised price). Perhaps the reviewer did not buy into the Scottish Enlightenment technique of citing several historical examples.

Beyond criticisms of the contents of his work, the most common criticism by reviewers was Playfair's writing style. Some early reviewers, who were probably London based, tended to sneer at Playfair's obvious provincial origins. This comes out in the reviews of Playfair's *Joseph and Benjamin* (1787d) written anonymously just prior to his embarkation for France. The book is a report of a fictitious meeting between Emperor Joseph II of Austria and American founding father Benjamin Franklin. At heart, it is an exposition of Playfair's political views on government. The reviewer for the *English Review* commented, "From the provincial phrases which frequently occur, it appears to be the composition of a North Briton" (*English Review* 1787, 27). The reviewer for the *Monthly Review* was also patronizing about the writer's origins, guessing that the writer came from "beyond the Tweed" while saying that he wanted "to put young writers on their guard with respect to purity of language" (*Monthly Review* 1788a, 257). One would think that Playfair's writing style would improve, but he continued to write hurriedly and took little care over looking at the page proofs. Twenty years after *Joseph and Benjamin*, a reviewer in *Monthly Review* wrote of Playfair's 1805 work, *An Inquiry into the Permanent Causes of the Decline and Fall of Powerful and Wealthy Nations*, "His composition also admits of much polish and correction" (*Monthly Review* 1807, 238). The reviewer for the *Critical Review* went much further: "Upon the whole, we have derived very considerable profit and instruction from the perusal of Mr. Playfair's work. It is evidently the production of one who thinks, though not of one who thinks with clearness or precision, or has the habit of expressing himself with ease and perspicuity. The arrangement is highly defective, and the whole view of the subject much less comprehensive than we expected to have found in so bulky a volume" (*Critical Review* 1806, 170). Playfair was never a writer in the first rank of literary society: his greatest strength was in his graphs.

In the late eighteenth and early nineteenth centuries there were several ways for an author to finance the publication of a book:

1 Self-finance the whole project or have a patron pay for it
2 Sell the book by subscription in advance of publication
3 Contract with the bookseller/printer to write a book of a given length and get paid by the bookseller for the number of pages, or amount, written (this can also apply to writing an article for a newspaper or magazine)
4 Sell the copyright, or a part interest in it, to the bookseller/printer

Options 1 and 2 were the typical ways to have something published prior to the Copyright Act of 1710 (Statutes of the Realm. 8 Ann c. 21; see Great Britain 1822b, 256–8), and both continued well into the eighteenth century and beyond. They became rarer over time and option 4 was typical for authors during the eighteenth century (Griffin 2014, 173–8). Prior to 1710, once the author submitted or sold a manuscript to the printer or bookseller, that printer or bookseller typically owned the manuscript from which the printed publication was made. After 1710 the new law stipulated that only the author and the printers to whom they chose to license their works could publish an author's work. The copyright term was set at fourteen years, which could be renewed for another term. In view of the printing tradition, it was natural for the printer to buy an author's copyright. At one time or another, Playfair probably exercised all four of options 1 through 4.

One technique that Playfair used throughout his career was to publish a relatively short newspaper article on a particular subject and then expand it into a book. This would have given him two sources of revenue for the same subject: one for the article and one for the book. This method might fall into option 3 under ways to make a literary living. Near the beginning of his career in letters, Playfair used this technique in the publication of his 1786 *Letters of Albanicus* published by John Debrett. The book was anonymously written in defence of Warren Hastings who had served in the East India Company and then as the British government's first governor-general of Bengal. At the time that Playfair wrote his book, Hastings was being impeached by Parliament. Some of the material in *Letters of Albanicus* first appear in one of the London newspapers. There are six letters in the book, neatly numbered I through VI. The first half of Letter I was published in the *Morning Post and Daily Advertiser* on 24 July and the second half on 28 July. This was a Whig newspaper, probably in line with Debrett's political leanings. The *Morning Post* was also an eighteenth-century form of scandal sheet. In *Letters of Albanicus*, Playfair split the difference and dated Letter I to 26 July 1786. Letters II through VI never seem to have hit the newspapers. Late in his career in 1819, Playfair wrote *France as It Is, Not Lady Morgan's France*, in response to a popular book *France* by Sydney, Lady Morgan. The book was her impression of France after a six-month stay there. It is apparent that, in his approximately 730-page response to Lady Morgan, Playfair hated her book (Playfair 1819). Two years earlier, when he was still in Paris, Playfair wrote a short review of the book as editor of *Galignani's Messenger* ([Playfair] 1817). He trashed the book, concluding his review with: "Indecent and insulting to the feelings of all good Frenchman, the political tenets of Lady Morgan will only form an object of their contempt, while her mistaken and distorted ideas respecting much of their internal economy, can only reflect upon herself the censures of those whom time or chance will find ultimately deceived by her erroneous representations."

At least two of Playfair's books were sold by subscription: *History of Jacobinism* published in 1795 by John Stockdale, with an expanded second edition in 1798 published by John Wright, and *British Family Antiquity* published in nine volumes between 1809 and 1811 by Thomas Reynolds and Harvey Grace. Both books have interesting publication histories. Though the subscription list is fairly lengthy, *British Family Antiquity* was such an ambitious and expensive undertaking that it drove Reynolds and Grace into bankruptcy. The subscription list for *History of Jacobinism* is a much more modest list of sixty-five names. Of those, nine were London booksellers who purchased eighty copies.

There was also a bookseller from Cambridge who bought six copies, another from Bristol who bought two, and one from Newcastle who bought a single copy. When the single subscription books are factored in, the total number of books sold by subscription amounted to 142. The subscriber price was 8s 6d per copy. For non-subscribers, the price was 10s 6d (*Courier and Evening Gazette*, 8 May 1795). On the political side, there were eight MPs on the subscription list. They could all be described as minor politicians who were usually inconspicuous in the House of Commons, but generally supportive of Pitt's government (political biographies of the MPs are in Thorne 1986). It was not a ringing endorsement by those in power. The second edition of *History of Jacobinism* was also to be sold by subscription by Stockdale with profits to be donated to the Bank of England "in aid of government against the threatened invasion" (*Sun*, 9 February 1798). During the previous year there had been a failed French invasion. When it was finally available, the second edition of *History of Jacobinism* was printed for the London bookseller John Wright. The subscription list that appears in the 1798 edition of *History of Jacobinism* is exactly the same as the original one in the 1795 edition. For some reason, Stockdale abandoned the project. Presumably there were no new subscribers, and no contributions were made to the Bank of England.

At least part of Wright's operation is an example of patronage at play. During 1797 and 1798, Wright was publishing *The Anti-Jacobin, or, Weekly Examiner*, a satirical anti-Whig literary journal. It was supported, definitely in a literary way and probably also financially, by Pitt and Pitt's protégé at the time, George Canning. Canning later held several senior cabinet positions and was briefly prime minister in 1827. Playfair's second edition of *History of Jacobinism* may be an example of second-hand government patronage. Wright could have paid Playfair for his copyright after Wright received the government subsidy to print the book. Technically, Playfair would not have received any funding from the government.

The only hard evidence of how Playfair was paid for his literary work is related to his imprisonment in the Fleet for debt. When applying for release in 1801, he listed one of his two assets as half the profits to be earned from the edition of the *Commercial and Political Atlas* published in 1801.[6] He was not to get the entire half; there was also an unspecified lien on his profits, which had been lodged with his lawyer. This is a variation on option 4 related to how authors were paid for their work.

As we have already seen from Playfair's marital situation, or lack thereof, in the 1780s, from his poaching of an employee from Boulton and Watt, and from his business dealings with James Kier, Playfair was already of bit of a rogue. The world of publishing helped to develop his roguishness further in what he hoped would provide him with a source of income. It was probably from newspaper proprietors of the mid-1780s that Playfair learned the art of extortionary scams, which he carried on periodically almost up until his death. Beyond advertising and sales, there were three income streams open to newspaper publishers in the 1780s (Werkmeister 1963, 90). A paper could create a puff piece for a price. The same paper could, again for a price, contradict a story that had previously been published in its own or a rival publication. The last was through extortion. A paper could demand payment to suppress an embarrassing story detrimental to some prominent person. John Benjafield, proprietor of the *Morning Post*, the London daily that published Playfair's first instalment of *Letters of Albanicus*, made these income streams into a high art form. These streams became major, but undoubtedly unwritten, line items in the *Morning Post*'s balance sheet. Benjafield's victims included leading figures of the day, such as Josiah Wedgwood and George, Prince of Wales. Many of Playfair's scams follow a similar extortionary path taken by the London press: find or concoct information that is prejudicial to the victim (purportedly found by someone else), and offer to help suppress the information for monetary or other reward. Unfortunately, but only for Playfair's finances, very few of his scams worked.

Chapter Five

Playfair Expresses His Early Political Views

After George III called on William Pitt to head his government late in 1783, Pitt began pursuing a commercial policy that was influenced by Adam Smith's *Wealth of Nations*. As part of that policy, Pitt tried to increase British production of consumable goods through a trade treaty with France, remarkably so after Britain had been in an off-and-on war with France since 1688. This reciprocal agreement was known as the Eden Treaty of 1786, named after the British negotiator William Eden, later elevated to the title of Baron Auckland. The treaty was ratified in Parliament, both the Lords and Commons, on 8 March 1787. By the treaty, there were reduced tariffs on a number of goods: French wine, vinegar, brandy, and olive oil into Britain, and British manufactured goods into France, including hardware and cutlery, glassware, earthenware, and pottery (Gifford 1809, 1:507–31). Other trade restrictions remained in place; for example, the export of certain types of machinery and blueprints was forbidden (Henderson 1957, 108). This trade agreement was concluded after Pitt's failed attempt to change the trade policy and excise duties related to trade with Ireland. The main opposition to Pitt's Irish policy had been through the General Chamber of Manufacturers of Great Britain. The Chamber, consisting primarily of manufacturers in cotton, iron, hardware products, and pottery was formed in about 1785 in reaction to the proposed Irish trade policy (Bowden 1924). Playfair's silverware customer, Josiah Wedgwood, and his former employers, Matthew Boulton and James Watt, were all prominent members of the Chamber (Bannerman and Schonhardt-Bailey 2016, 2).

Despite support by the likes of Josiah Wedgwood, Matthew Boulton, and James Watt, there was early opposition to the Eden Treaty (Henderson 1957, 109). And Playfair took a leading part among smaller manufacturers in this opposition. Initially, Playfair publicly supported the treaty and then changed his mind. His reversal occurred at a meeting of the Chamber in London on 10 February 1787.[1] Earlier on 9 December 1786, the Chamber had endorsed the treaty during a meeting in which the president, Josiah Wedgwood, was in the chair (*Public Advertiser*, 12 December 1786). At the February meeting, Wedgwood was not present and the person who chaired the meeting, the Manchester cotton merchant and founder of the Chamber, Thomas Walker, was strongly opposed to the treaty (for a biography of Walker, see Davis 2009). The February meeting was called in response to recent developments in Parliament. On 5 February 1787, Parliament decided to have a full debate on the treaty a week later. Walker claimed that the December meeting of the Chamber had been improperly called, as not all members had received notice of it and members who were not present had written in with objections to the treaty. At the February meeting, Walker put forward a motion contrary to the Chamber's earlier decision under Wedgwood. In 1809, the political writer and biographer of Pitt, John Gifford, succinctly put the motion and its effect on Parliament: "As soon as the House was opened on that day, a petition was presented from certain manufacturers, requesting that the House would come to no final decision on the treaty until *they* should have the leisure to *understand it*. – It met, of course, with the fate which a petition so framed, deserved; and the House proceeded to the discussion" (Gifford 1809, 1:336–7). One news report has Playfair seconding Walker's motion; other reports have a different seconder.[2]

Playfair spoke to the motion for thirty to forty-five minutes. His arguments in support are interesting. He did not think that any European country could compete in manufacturing with Britain and disparaged the inhabitants of these countries. The common man, for example, in Russia, Sweden, and Denmark was rude in manners and no better than a slave to the nobility. France, on the other hand, was different. In versatility, activity, and industry, the French had all the necessary attributes to be commercially successful. What they lacked was the proper machinery to manufacture goods. The problem could be traced back a century to when the Huguenots fled France, hollowing out an entire class of people highly successful in business. But the religious persecution begun under King Louis XIV had ended under Louis XVI, and now the latest Louis was promoting manufacturing to improve the economy of France. In view of the rivalry between France and Britain over several centuries, and especially considering what France had done against Britain during the American Revolution, France could not be trusted, Playfair said. He concluded the treaty "to be fraught with every thing pernicious to English greatness" and "ruinous to English liberty" (*World and Fashionable Advertiser*, 12 February 1787). Within a year, Playfair was to change his tune and go to France to set up his own business in the new climate that Louis XVI's government was trying to bring to France.

At about this time, there is some evidence that Playfair was writing a political pamphlet or a book about providing the middle class, or at least the manufacturers and perhaps merchants in the middle class, with more political power. It was apparently never published; no reference to it is made on Playfair's 1807 list of publications drawn up for the Literary Fund. The suggestions for this political reform were made in a letter to Patrick Colquhoun, a prominent Glaswegian merchant. Only Colquhoun's response is extant. In a letter to Playfair dated 16 October 1786, Colquhoun writes, "I perfectly understand your Idea with regard to the proposed alteration in the Constitution and certainly there is not a Clearer Proposition in Euclid than that which you state with regard to the proper share of political weight which the Spirit of the Constitution gives to the Commercial Part of the body politic but which from the present state of things this order in Society are in a manner totally deprived of."[3] A few years later, the French Revolution sharpened Playfair's thinking in a different direction about how the British should be governed.

Some inkling of Playfair's suggested political reforms can be gleaned from Colquhoun's letter. He appears to have had in mind a politically influential bloc put together by a union of the trading towns of Britain. Colquhoun commented that Playfair would be up against the jealousy of the landed interest, and went on to say that the commercial interests need "a proper head, a Proper System and a Collected Strength to Combat so formidable a Phalanx." He illustrated the daunting task in this political struggle by pointing to the difficulty faced in the establishment of the General Chamber of Manufacturers in London. He wrote, "Many were intimidated, some were persuaded and others were induced from political Influence to withdraw their support till the few that remained vexed by the opposition they experienced in so truly disinterested and patriotic pursuit became disgusted and gave the matter up."[4] Playfair stuck to his beliefs and remained a member of the Chamber; he was active at the meeting held on 10 February 1787. Whatever argument Playfair was making, it was in the style of the Scottish Enlightenment. Colquhoun advised him not to use historical examples in his argument since he felt that political solutions are fluid and thus depend on current circumstances, not historical precedents.

Playfair's first actual publication concerning political issues was *Letters of Albanicus*, published by John Debrett in 1786, with a second edition in the same year. Authorship has usually been attributed to David

Steuart Erskine, 11th Earl of Buchan.[5] At least the incorrect attribution was made to the correct country of origin: "Albanicus" means one who was born in Scotland. The correct Scot is, of course, William Playfair. The 1807 list of publications drawn up for the Literary Fund includes "The Letters of Albanicus in Defence of Governor Hastings, Debrett, 1787," which had been written anonymously. In view of the fact that it was twenty years later and that he was writing from memory, Playfair could be excused for missing the publication date by one year.

The Governor Hastings mentioned in Playfair's Literary Fund submission was Warren Hastings who had been with the East India Company from 1750 to 1786. In 1773, the British government passed the East India Regulating Act that changed how the East India Company operated in India, putting the administration of India under a supreme council appointed by the British government. Hastings was made governor-general of Bengal. In his new position, he revised the taxation system to try to increase revenue; he overhauled the judicial system, which became a mixture of British and Indian law; and he set up a diplomatic structure to deal with Indian states that had emerged as the Mughal Empire in India had gone into a tailspin. During the 1770s until his resignation in 1785, Hastings had two major problems. The first was a series of continuing wars with Indian states carried out to maintain British influence in the area. The second was that Hastings often clashed with his ruling council. In all this activity, he made several enemies.

Hastings's major enemy in India was Philip Francis, one of the original council members appointed by the British government. They had such a severe falling out that in 1780 they engaged in a duel, which resulted in Francis being slightly wounded. Francis returned to England near the close of 1780 where he began to carry out a campaign against Hastings. He soon became allied with the politician Edmund Burke, who had become interested in Indian affairs dating from the 1770s. In 1784, Burke was made a member of a parliamentary select committee on British governance in India. He soon became a leading member of the committee and, on the basis of his investigations, was convinced that Warren Hastings should be impeached. At the same time, Francis was often one of Burke's sources of information on the situation in India. Burke worked hard to make the impeachment come before Parliament and set the wheels in motion in February 1786. Hastings was formally impeached on 10 May 1787. The case dragged on into the mid-1790s, at which point Hastings was found not guilty. After Burke put the process of impeachment into play in 1786, Playfair as Albanicus came to Hastings's defence. His publication had no effect, as the impeachment process continued for several years; the whole trial process seems to have been shaped by politics (Rudd 2011, 26). Rather more probably, the main effect of the publication was to ingratiate Playfair to a patron who supported Hastings.

Playfair took a variety of tacks to defend Hastings. They ranged from the unfairness of the proceedings (several accusations against Hastings were misrepresentations of the facts) to national security (other countries helped the colonies in America to break away from Britain; India is now vulnerable to dismemberment and loss to other European powers unless there is a united front) to the inability to hire good people in the future (an honourable person likely would not take over the position of another honourable person who had been sacked for questionable reasons). Another line in his defence of Hastings was to consider the role of the governor-general. Playfair likened it to that of a judge. If a judge makes a wrong decision, is he impeached? No, the judgment is overturned in a higher court. He then likened the position to that of an agent of a company working at a distance from the head office. Decisions have to be made with good sense in light of current developments, not strictly by rules set by a distant head office with which direct communication is difficult.

What Playfair wrote in *Letters of Albanicus* seems completely at odds with what appears in his *Commercial and Political Atlas*, published in the same year. The *Commercial and Political Atlas* was with the booksellers by mid-July 1786, only a week or two before Playfair's first letter in defence of Hastings appeared in the *Morning Post and Daily Advertiser*.[6] With perhaps a hint of the Hastings affair, Playfair opens his discussion of India in the *Commercial and Political Atlas* with "The Chart of our trade with India is given at this time, because the affairs of that country are likely soon to become objects demanding the utmost attention" (Playfair 1786, 31). After that opening, Playfair claims that trade as given in the chart (see figure 5.1) must be understated. It shows that trade with India was not substantial, although many had returned from India with large fortunes. No mention is made of Hastings; but Playfair goes on to say that the servants of the East India Company "plunder" the local population. He uses words like "rapacious," "oppression," and "unjust" to describe the whole system under which the employees of the East India Company, along with the British government, operated. Playfair attributes the problems in India to their unfettered pursuit of vast wealth, knowing that the potential to obtain such wealth was great. In view of the language that Playfair used, he does not seem sympathetic to the unnamed Warren Hastings. And yet in the same year, perhaps within months, he was. In the later pages of *Letters of Albanicus*,

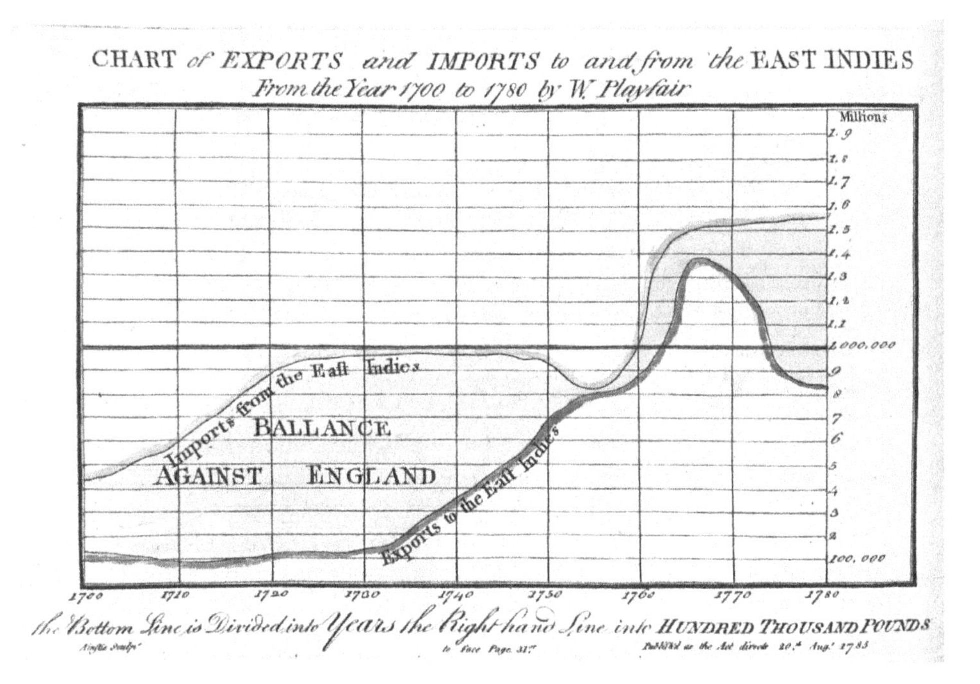

Figure 5.1. English trade with India. Courtesy of Stephen M. Stigler, University of Chicago.

Playfair seems to square this circle. The problem is the East India Company itself, Playfair wrote. And the faults of an imperfect government in India have been heaped onto one man. Playfair expresses this point rhetorically: "If a pilot is ordered to sea in a ship which is rotten and full of holes, is the sinking of the ship to be laid to his charge?"

With Playfair's apparent chameleon-like appearance between *Letters of Albanicus* and *The Commercial and Political Arithmetic*, it is no wonder that *Letters of Albanicus* was written anonymously. The perceived inconsistency in what he wrote in the two publications might be attributed purely to patronage, or as the saying goes, "He who pays the piper calls the tune." Playfair claimed in *Letters of Albanicus* to "never have had even the most distant connection with Mr. Hastings" ([Playfair] 1786, iii). I would accept that, taken literally, Playfair's claim is true.

Playfair's most likely patron was the MP George Dempster, one of Playfair's earliest connections in his growing network of friends and acquaintances. Dempster had been a shareholder in the East India

Company since 1763 and served as a director of the company in 1769 and 1772. Throughout his career he continued to be interested in the affairs of the company. During 1786 and 1787 he was a "vehement defender" of Warren Hastings (Namier and Brooke 1964, 317). As mentioned in chapter 4, there is also a connection to John Benjafield, proprietor of the *Morning Post and Daily Advertiser.* The first letter in *Letters of Albanicus* was published earlier as two letters in the *Morning Post.* Likely Dempster paid Benjafield to print Albanicus's letters in the *Morning Post* and paid Playfair to write them. Perhaps with some help from Dempster, Playfair may have then sold the idea of a lengthier publication to John Debrett. Debrett had numerous titles on the Hastings affair in his bookshop and had already published some of Playfair's work (see, for example, *Morning Post and Daily Advertiser,* 20 December 1786, for a list of Debrett's titles). The affair was topical enough that Debrett considered Playfair's book a good investment. And he was correct; *Letters of Albanicus* went through two editions in 1786.

Playfair had another good reason to publish *Letters of Albanicus* anonymously. In 1786 he was still living, and perhaps working, with his brother James in Howland Street and then Russell Place, in Bloomsbury. At the time, Henry Dundas, later elevated to the peerage as Viscount Melville, employed James Playfair to design Melville Castle near Dalkeith in the Midlothian district of Scotland. Dundas was an MP and part of William Pitt's government when Pitt formed his ministry in 1783. Prior to coming to office, Dundas had been an expert on Indian affairs and had been against Hastings's actions in Indian affairs since at least 1781 (Fry 2009). Claiming authorship of *Letters of Albanicus* might be bad for his brother's business.

While the reviews of *Letters of Albanicus* were either positive or lukewarm, those of Playfair's next publication in politics, *Joseph and Benjamin,* were much more negative. The *Critical Review* devoted one line to the book: "The effusion of some inexperienced writer, who retails his own conceptions in the character of personages which he does not support with consistency" (*Critical Review* 1788, 395). The reviewer for the *Monthly Review* took more space, but was just as scathing: "Some young adventurer, we suppose, here steps forth to try his strength, by endeavouring to bend the bow of Ulysses; but his arms are yet feeble, and the sturdy yew refuses to yield to his most strenuous efforts. Without a figure, this writer surely mistook his powers, when he judged himself able to support, with spirit, a dialogue between the Emperor Joseph and Benjamin Franklin, for not less conspicuous are the personages here introduced. Their characters however are not well supported" (*Monthly Review* 1788a, 257).

The writer for the *English Review* spilt much more ink on his review than the other two combined. He was kinder, but still leaned towards the negative side: "Our author seems as zealous to inculcate a jealousy of French politics, as good protestants have been, for a hundred years, to propagate the terror of popery. There are many just reflections and ingenious hints in this performance; but, upon the whole, it holds more of the imagination than the judgement, and contains more theory than investigation (*English Review* 1787, 27).

Putting aside these negative reviews, the full flowering of Playfair's political beliefs prior to the French Revolution may be found in *Joseph and Benjamin*.

Joseph and Benjamin is written in the form of a dialogue between Benjamin Franklin, as American ambassador to France at the time the dialogue was set, and Joseph II of Austria, Holy Roman Emperor (Playfair 1787d). The setting of the dialogue is Joseph's arrival in Paris incognito to meet privately with Franklin. A younger philosopher, Joseph in this case, asks questions of an older and more experienced philosopher, Franklin. In the process, the older philosopher imparts his wisdom to the younger one.

The title page of *Joseph and Benjamin* claims that the publication was a French manuscript translated into English. The fiction continues in the first pages of the book, where it is claimed that the origin of the manuscript must be kept from the public until "the heads of some persons concerned shall be laid low" (Playfair 1787d, ii). That the work is a piece of fiction was hinted at by the first reviewers and can be easily confirmed by a close reading of the text.

Sprinkled throughout *Joseph and Benjamin* are Playfair's political ideas, usually expressed through Franklin, on a range of topics. For example, Playfair throws out a few ideas on the ebb and flow of the prosperity of countries. These were later fleshed out and developed in his 1805 book, *An Inquiry into the Permanent Causes of the Decline and Fall of Powerful and Wealthy Nations*. He also outlines a simplistic theory of government, which was to change radically after he had first-hand experience of the French Revolution. All the ideas he presents are brought to bear on the political and economic situations in Britain, France, and America.

Playfair's first statement on the ebb and flow of the prosperity of countries is a truism: when prosperity comes to a nation, it thinks it will endure forever. According to Playfair, what brings down prosperity is "indolence, selfishness and indifference to the public good" (Playfair 1787a, xii). Playfair sees a natural evolution in prosperity. Industrial activity is the first step. It promotes and maintains prosperity. However, prosperity is self-defeating; it tends to destroy virtue. Prosperity results in the accumulation of capital; what follows is indolence and luxury, and a decline in prosperity. Playfair's position is a natural progression from what he wrote in *Regulations for the Interest of Money* where he was concerned about the promotion of industry through the establishment of a bank that would offer loans at interest rates above the legislated maximum of 5 per cent. In 1787, in *An Essay on the National Debt*, Playfair expanded on his theme of indolence and luxury (Playfair 1787c). He was concerned that, with an increasing national debt, more people would invest in 3 per cent consols and live off the interest earned. Increased investment in the consols, Playfair feared, would result in an increasing number of people, who were previously active, becoming idle and unproductive.

Playfair presents only the bare bones of his theory of government. A country is governed through a king and a parliament. The king heads the government. Parliament exists "to support the rights of the people against the encroachments of the king" (Playfair 1787d, 48–9). Reading between the lines, my interpretation of Playfair's pronouncements on the subject is that parliamentary representation should be broadly based, depending upon what he means by "the people." Taking into account Playfair's correspondence with Patrick Colquhoun, by "people" he probably meant merchants and manufacturers, in addition to the landed families who held power at the time. Playfair's one comment on parliament is that if it were properly run in support of the people, it should have the power to declare war. He naively concludes that, unlike the experience of the American Revolution, the people would support a war declared by such a parliament, since that parliament represents the people.

Flowing from his general ideas about prosperity and governance, Playfair has specific comments about the state of affairs in Britain. On the positive side, Britain has taken a leading role in manufacturing and commerce, which is well oiled by its system of credit. The rest is negative, with predictions of possible doom and gloom. The king's courtiers are concerned mainly with making the internal affairs of the country run smoothly and managing Parliament. They tend to ignore external threats unless those threats are extreme. The main external threat comes in French designs on India, along with those of other countries. The difficulty in countering the threat is the distance of India from Britain and the rebelliousness of the Indian princes. The French also pose a second threat with their recent commercial treaty with Britain, which, according to Playfair, would encourage manufacturing in France to the detriment of British manufacturers. The internal threat is the national debt, which could worsen because of the external threats. If Britain went to war to protect her interests in India, the national debt would increase substantially. The situation would be exacerbated if the French developed their manufacturing capabilities, thus leading to a decline in British manufacturing. Then the British would be less able to service their own debt, a situation that may lead to revolution. The weakness of the whole political system, Playfair claims, is in the cabals of party politics and the inability to make bold moves, such as responding to possible French aggression and dealing properly with the national debt. This weakness is directly related to the patronage system, which results in unmerited promotions that "fill the minds of all men that the more they can get by the state the better it is, and that it is all fair game" (Playfair 1787d, 44).

Playfair took from history the fact that every prosperous nation eventually goes into decline. A corollary was that Britain, being prosperous, would follow suit and eventually go into decline. In a summary of all his arguments, Playfair predicted how Britain would fall: "Perhaps the territories in India may be taken from Britain, with a great increase in the national debt; other nations, and in particular France, may be able in time to undersell British merchants in foreign markets, and the balance of trade turn against Britain. There would then be an impossibility of paying the immense taxes, a revolution might ensue, after which things would begin on a new footing; but in my opinion never could they return to their present state of prosperity, for other nations would be rising in their fall" (Playfair 1787d, 185–6). Playfair feared that the decline would come sooner than later. He was off in his prediction by well over a century.

France, Playfair observes, is a very different country:

> In France, the aggrandisement of the Royal Family is the first object. The happiness of the people, their riches, and every thing else, is but a secondary object of attention. It may be compared to the haughty proprietor of an estate, who improves his lands, or depopulates them; or does whatever seems to be most for his own interest, without considering the happiness or the misery of his people, and more than he does the preservation of the weeds that spring out of the soil. Such is the reigning view of the French Court; to be the first family in Europe is the point of ambition; and whether this is to be accomplished by lifting up the power of France or depressing other nations, or by both; it has been, and it will continue to be, the constant view, and uninterrupted aim of the House of Bourbon. (82)

Ringing alarm bells even louder, Playfair gives definition to the Bourbons' desire to be "the first family of Europe." The ambition of France is actually to establish a universal monarchy, with the French king sitting on the throne. However, this cannot be done in Europe, Playfair claims. Generally speaking, if territory is acquired by war, it will eventually decay and be lost to the conqueror. With particular reference to Europe, there is a supplementary reason. Europe is "civilized" and "civilization cause[s] a strong attachment to ancient boundaries" so that "where men are most civilized, they have been most tenacious of their boundaries" (114 and 116). Rather than pursue direct all-out war, the French will bring down the British through subterfuge. As they had done in America since 1768 (so Playfair claims), the French will foment discord in India between the native princes and the East India Company. They will also bring in the Dutch and the Spanish to harass the British. The ensuing war, fought mostly by proxy on the French side, will be costly to the British. Since the French national debt was smaller than the British debt and since France had more resources, the French could easily finance their side of the war while the British would be driven deeper into debt and possible collapse. With the French Revolution two years away and the near collapse of the French economy, Playfair could not have been more wrong in his assessment.

Despite the fact that France had assisted the Americans in their war with Britain, Playfair was certain that France and America would never be friends. For unspecified reasons, he claimed that the Americans were jealous of the French and disliked them. Further, because of their help in the past, the French would expect the Americans to grant them exclusive trading rights. This could not happen. Playfair attributes this to the Earl of Shelburne. It was Shelburne who was the British negotiator for the 1783 Treaty of Paris that recognized, among other things, America's independence. Playfair noted that the treaty "was concluded very favourable for Britain, and …, for a while at least, will stop the career of French ambition" (Playfair 1787d, 93). One result of the treaty was that the Americans and the British continued to be major trading partners.

Although highly supportive of Shelburne without ever identifying him by name in *Joseph and Benjamin*, Playfair did have one criticism of him. When Shelburne's ministry fell in 1783 and after a brief interlude was replaced by Pitt's, rather than a cabinet position under Pitt, Shelburne was rewarded with the title Marquess of Lansdowne. Playfair thought that Shelburne "should not have accepted of a title that rather diminished than added to his career" (Playfair 1787d, 96).

According to Playfair, speaking through Franklin, America's prospects for the near future were dim. His assessment of the situation was far off the mark. Playfair stated that the main economic activity in America was farming and the manufacturing of "coarse materials which are necessary to the lower class" (Playfair 1787d, 37). He then recommended that the manufacture of luxury items be left to the British. This theme recurs in Playfair's later publications. The people of America, Playfair claimed, were too poor and so thinly dispersed over the countryside that they could not be properly governed. He likened the governance of America to remote areas of Russia where local lords and chiefs directly controlled their tenants with little input from the central state. Without proper governance, he predicted that America would be divided into different parts that would each have its own government. He feared that internal dissensions in the former colonies might lead to some European powers dividing up parts of America among themselves. What America needed was a steady form of government and Playfair (or Franklin incredibly, as expressed in *Joseph and Benjamin*) had the solution. In line with his simplistic theory of government, he proposed that a son of King George III be made king of America, on the condition that he would never hold the crowns of England and America.

Playfair ends *Joseph and Benjamin* with a bit of self-promotion of his *Commercial and Political Atlas*. In a letter from Franklin to Joseph placed at the end of the dialogue, Franklin mentions a new invention he received from a friend in London "in which the trade and finances of England were delineated in charts upon paper, like a map of a country" (231).

Chapter Six

Playfair Makes His Mark on Statistical Graphics

Regulations for the Interest of Money was Playfair's entrée into the world of patronage connections that helped with the promotion of Playfair's first and major work in statistical graphics, *The Commercial and Political Atlas*. As mentioned in chapter 4, *Regulations for the Interest of Money* contains Playfair's take on the need for changing the British usury laws and ways to carry out his suggested reforms by opening a bank that would make loans at varied rates of interest. The bank would tie the interest rate offered to the risk level associated with the loan.

The book was a reaction to a chapter in Adam Smith's *Wealth of Nations*, which had been in print since 1776. Smith had defended the 5 per cent legal rate of interest in his *Wealth of Nations* (Smith et al. 1976, bk. 2, chap. 4, para. 14). It was not high enough to impede trade, Smith wrote. What gave Playfair pause for thought was Smith's "invisible hand" argument (see Bishop 1995), which did not fit with Smith's position on the legal rate of interest: "Every individual is continually exerting himself to find out the most advantageous employment for whatever capital he can command. It is his own advantage, indeed, and not that of society he has in view. But the study of his own advantage, naturally, or rather necessarily, leads him to prefer that employment which is most advantageous to the society" (bk. 4, chap. 2, para. 4). By "capital" in this quotation, Smith meant the individual's efforts and activities, as well as his money.

Playfair's words response to this general concept is that government incentives might help:

> Dr. Smith, and other writers on finance, are of opinion, that money finds its own way into proper channels, with greater readiness and effect, than would result from the application of the greatest human sagacity. That money will naturally flow into its proper channels, as water descends

> from higher grounds and rocks, and finds its natural level, is certainly true. But as it is possible, by means of sluices and reservoirs, to collect the waters of the heavens, which would otherwise be lost, and to introduce them by canals and rills, where their fertilizing power is most wanted; so it is also possible for the legislative power of a nation to give such direction to its currency, as shall in the most effectual manner produce the various fruits of industry and art. (Playfair 1785, 17–18)

The whole of Playfair's book is an argument for what he thought would be an appropriate sluice for the treatment of interest charges.

Playfair had made a reasonable proposal and it was fairly well received by reviewers of the book in the literary journals of the day (*Critical Review*, *Monthly Review*, *New Annual Register*, and *Political Herald*). The reviewers for *Critical Review*, the *New Annual Register* and the *Political Herald* were all positive about the proposal, with the writer for the *Critical Review* commenting that he was glad to see that the anonymous author had the interests of the public in mind. The *Monthly Review* was generally positive but, as previously mentioned, thought the book too wordy. A later criticism came in 1787 from Jeremy Bentham in a letter, written while he was in Russia, to his friend George Wilson in London (Electronic Enlightenment 2008). Bentham had just finished his own manuscript on interest rates, published in 1788 as *Defense of Usury*. Bentham was also against the stipulated legal rate of interest set at 5 per cent; what motived him was that Pitt's government was considering a decrease in the legal rate to 4 per cent. In his letter to Wilson, he called Playfair's book a trumpery book in which the proposal was vague, ill formed, and poorly argued. At the same time, Bentham wrote, "I understand it has been well enough spoken of by several people."

While Playfair claimed to have had the interests of the public in mind, his proposal might also have helped him return to business. He divided the business world into merchants and manufacturers – those who sold goods and those who produced them. He argued in the book that merchants had no need for this kind of bank; it was easier for them to obtain credit. It was the manufacturers who found it harder to obtain credit. Manufacturers would benefit the most and, in a reciprocal way, benefit the country. According to Playfair, since the larger established manufacturers had easy access to capital, it would be the smaller manufacturers who would really benefit from the proposal. Playfair had just failed at being one of those smaller manufacturers.

Just as he was conscious of the need to market his products when he was in industry, Playfair must have realized that, if he were to be successful as an author, he had to promote his new book. The printer for the book, George Robinson and two other family members, does not appear to have given it much help. The advertisement for the availability of *Regulations for the Interest of Money* was brief and may have appeared in only one issue of the *Morning Post and Daily Advertiser* (28 July 1785). Also, the Robinsons' bookshop was not where those who could change the usury laws necessarily met. Playfair likely began to promote his book through a network of family connections made through his brother John. Two nodes in the network included George Dempster and Sir Adam Ferguson.[1]

Between 1761 and 1790, George Dempster was the member of Parliament for Perth, the area of Scotland from which the Playfairs hailed. Sir Adam Ferguson was a close friend of Dempster and also served as an MP, mainly for Ayrshire, from 1774 to 1786. George Dempster and John Playfair crossed paths at least once before 1785; both attended the General Assembly of the Church of Scotland in 1783, a meeting that lasted thirteen days. In that same year, Sir Adam Ferguson and John Playfair became founding members of the Royal Society of Edinburgh. Dempster shared their scientific interests to the point that John Playfair was one of the Dempster's sponsors for fellowship in the Royal Society of Edinburgh in 1789.

Advertisements for *Regulations for the Interest of Money* came out as early as 28 July 1785. Within a month, Dempster knew of the book and mentioned it in a letter to Sir Adam Ferguson on 22 August 1785. When Dempster wrote to Ferguson on 8 September 1785 about William Playfair's next project, which turned out to be *The Commercial and Political Atlas*, Dempster said, "The author is brother to the Professor of Mathematics at Edinburgh, and manufacturer of plated ware here" (Dempster and Ferguson 1934, 145). From the context of the letter, it appears that George Dempster and William Playfair made the initial contact and later Dempster promoted Playfair's work to Ferguson.

Through Dempster, William Playfair's network expanded. On 30 January 1786, the prominent Glaswegian merchant Patrick Colquhoun wrote to Dempster on a number of topics, including William Playfair.[2] He thanked Dempster for introducing him to Playfair and commented that Playfair's "personal worth, ingenuity and abilities render his society a peculiar acquisition." Playfair and Colquhoun continued to have "much conversation" after the initial introduction. Prior to Colquhoun's letter to Dempster, Playfair had sent Colquhoun a copy of *Regulations for the Interest of Money*. Colquhoun found Playfair's ideas in the book "excellent." From the letter, it is evident that Dempster knew Playfair before Colquhoun did, and that Dempster was introducing Playfair to people who might be interested in the publications of this new author.

Colquhoun promised Dempster that he would give Playfair any assistance he could in Playfair's next project, which turned out to be *The Commercial and Political Atlas*.

In his letter to Adam Ferguson, what interested Dempster more than Playfair's book on interest rates was Playfair's work that developed into *The Commercial and Political Atlas*. In the letter to Ferguson, Dempster mentions "tables of trade by William Playfair" (Dempster and Ferguson 1934, 143). These "tables" were a new idea and "will put cyphers out of fashion in public accounts of this sort." To put it in other words, these are tables without numbers, or statistical graphics. Playfair later called his graphical technique "lineal arithmetic." As a postscript to the letter, Dempster writes, "I cannot part with my Tables of Trade till I can procure another copy and they are not yet published." In another letter to Ferguson less than three weeks later, Dempster did send a copy of the tables of trade and refers to the work as "chart accounts which I mentioned to you in my letter of some date." Dempster wanted Ferguson's opinion of the work, because "he [Playfair] means to go on with the work if properly encouraged and to give us a view of the whole trade of Great Britain and Ireland" (145).

As may be gathered from Dempster's letters to Ferguson, *The Commercial and Political Atlas* did not see light of day as a fully developed book. In fact, the intention was to publish the work in instalments. And in August, Dempster seems to have been in on an early copy of the first instalment. Advertisements for it appear in the newspapers beginning on 13 September 1785. Dempster's letter to Ferguson in which he encloses a copy of the chart accounts is dated 8 September 1785. The first instalment, which sold for five shillings, contains five charts with explanations. The five charts of the instalment are the first five charts that appear in the final product. They show, in order, English exports and imports to and from "all parts," Ireland, Germany, West Indies, and all North America.[3] John Ainslie, a well-established engraver working in Edinburgh, engraved the five charts. It was probably family connections, Playfair's two brothers working in Edinburgh, that resulted in the choice of Ainslie as the engraver. The full book, advertised in the newspapers as the second instalment, was available to the public by the end of May 1786 (the book is advertised in the *Morning Herald*, 30 May 1786). Samuel John Neale, a London engraver who had been in business since 1782, engraved almost all the remaining graphs in the final version.

William Playfair's connection to George Dempster may explain the choice of John Debrett as a lead publisher for *The Commercial and Political Atlas* and two other publications before Playfair left for France in 1787. Dempster was an independent Whig in politics. He voted according to his conscience, sometimes with Pitt and sometimes against him, and belonged to none of cliques among the Whigs. For example, in 1784 another politician, John Robinson, saw Dempster as an opponent of Pitt. On the other hand, Dempster supported Pitt's attempt to eliminate the national debt (Namier and Brooke 1964, vol. 2). During 1785–6, Dempster may have found Debrett's a more congenial lounging place than Stockdale's.

I have described Debrett as a "lead publisher" for *The Commercial and Political Atlas*. This requires a little explanation. The title page reads, "Printed for J. Debrett, Piccadilly; G.G. and J. Robinson, Pater-Noster Row; J. Sewell, Cornhill; The Engraver, S.J. Neele, No. 352 Strand; W. Creech and C. Elliot, Edinburgh; and L. White Dublin." For the first instalment of *The Commercial and Political Atlas*, Robinson came first on the list and an additional two booksellers were involved. It is likely that initially Robinson arranged for the book to be printed and then Debrett took over when the full book was ready for publication.

How William Playfair first came into contact with George Dempster is a matter of speculation. Most likely Playfair approached Dempster, perhaps in a chance meeting at Debrett's bookshop. It may have been a self-introduction through a letter using John Playfair's acquaintance with Dempster.

Playfair similarly promoted the project to Matthew Boulton and James Watt, but it was after the first instalment was publicly available. On 20 September 1785 Playfair wrote to Boulton, opening his letter with, "I have taken the liberty of asking will you do me the favour to accept of the first number of a work which I hope may be one day of considerable utility."[4] A week later, Playfair wrote to Watt with a similar opening sentence.[5] In both letters, Playfair said that he had offers of assistance in obtaining materials for his publication, presumably the data behind the graphs. And the offers came from well-placed people, in Playfair's words, "from those whose situation both in England and Ireland makes it easily in their power to." He asked both Boulton and Watt to provide him with any comments that would make improvements to the work in progress. If Boulton replied, there is no record of it. Watt commented that it would be good to have the data, on which the charts were constructed, published in the book, as Lord Sheffield had done in his book. Watt's reasoning was that "the charts now seem to rest on your authority, and it will be naturally enquired from whence you have derived your intelligence."[6] Playfair followed this advice, but in a limited way. The complete version of *The Commercial and Political Atlas* contains data, for the most part, for 1700 and every tenth year thereafter up to 1780.

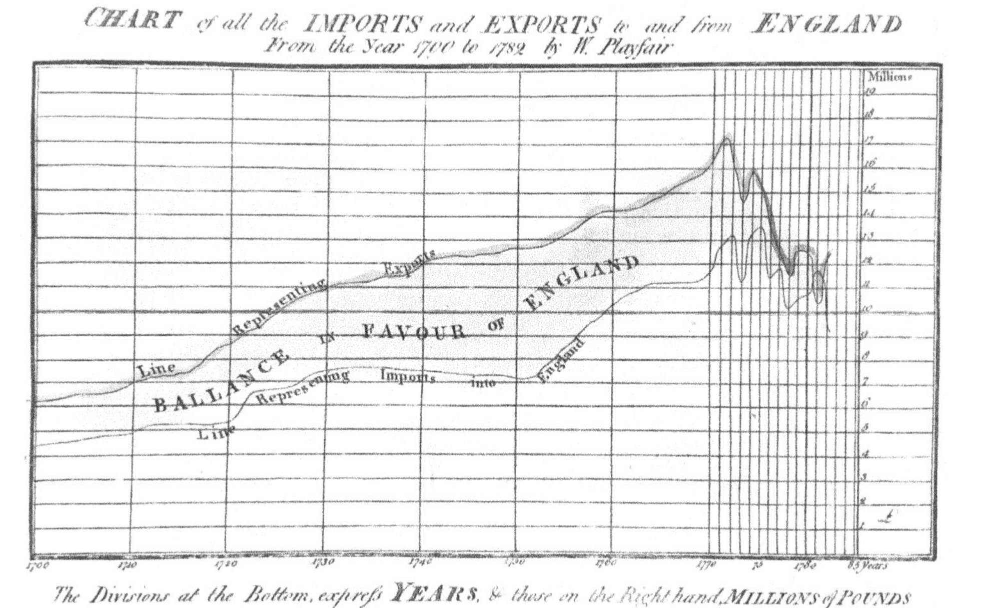

Figure 6.1. Playfair's first graph in *The Commercial and Political Atlas*. Courtesy of William L. Clements Library, University of Michigan.

It was a little later, perhaps in early 1786, that Playfair met John Holroyd, Lord Sheffield. It was Patrick Colquhoun who introduced the two when Colquhoun was in London.[7] Sheffield's book, mentioned by Watt, is *Observations on the Commerce of the American States*, published in 1784. The book contains about forty tables of statistics with extensive commentary. Playfair's first graph in *The Commercial and Political Atlas*, showing all imports and exports to and from England between 1700 and 1782 (figure 6.1), corresponds to Sheffield's table IX, showing the same thing in numbers. Sheffield's numbers for the same time period, leaving off 1781 and 1782, are averaged by decade. Had Sheffield provided yearly data for all eighty years, the effect would have been overwhelming and difficult to digest. Playfair's graph gives more easily interpretable information at a glance. The graph also shows the annual effect that the American Revolution had on imports and exports, an effect that is lost by Sheffield after he averaged his data for the decade 1770 to 1780. Sheffield only comments on this effect and the reasons for the general decline (Sheffield 1784, 221–3). Playfair's graph shows that the largest annual decline in that decade occurred in 1772 before military hostilities broke out. On examining figure 6.1, one can understand George Dempster's excitement over Playfair's project.

As shown in figure 6.1, Playfair used colour as an informational element of his graphs. This would have added to the cost of publication. The area coloured beige shows the size of the British trade surplus, with the small red area showing the trade deficit in the few years when it occurred. The red line accentuates the printed black line for the time series of British exports, while the yellow line is the same for imports. As with most of Playfair's graphs, the charts were coloured by hand (see the discussion by Spence and Wainer in Playfair 2005, 16). Some graphs later in Playfair's career appear to have been done using an aquatint technique.

Although still living in 1786 and presumably meeting Playfair's brother John regularly, Adam Smith made no known comments on William Playfair's graphs. Had he done so, he may have questioned the reliability of the data on which Playfair based his graphs. In 1776, Smith wrote, "There is no certain criterion by which we can determine on which side what is called the balance between any two countries lies, or which of them exports to the greatest value" (Smith et al. 1976, bk. 4, chap. 3, para. 4). One reason for this uncertainty was that valuations of goods in the custom house accounts (the data used by Playfair) were inaccurate.

What leaps off Playfair's graph (figure 6.1) for me is the shape of the trend lines. The imports and exports increase fairly smoothly over the century until 1770, at which point there is more volatility or bumpiness to the graph, which carries through the 1770s and 1780s. Was this really so, or did Playfair smooth his data prior to 1770? Or even after 1770? The first indication that Playfair manipulated the data to make them more presentable is from Sheffield's discussion of his table IX. Sheffield states that he took ten-year averages for the import and export data because "a balance [exports over imports] may be exhibited as very large or very small" (Sheffield 1784, 221). In other words, Sheffield was concerned about the volatility of the data and took averages to obtain a stable trend.

Before examining the issue of smoothing it might be useful to try to track down Playfair's data sources. Playfair begins *The Commercial and Political Atlas* by stating, "In this volume of Charts I have endeavoured to include all the different branches of our Trade, and such of the Public Accounts as I judged to be of the most importance." This is not much to go by, but the phrase "public accounts" probably refers to the

Table 6.1. Total English Imports and Exports in Millions of £

	Imports		Exports	
Year	Playfair	Other sources	Playfair	Other sources
1700	4.55	5.97	6.30	7.30
1710	4.90	4.01	7.00	6.69
1720	5.35	6.09	8.60	7.94
1730	7.50	7.78	10.90	11.97
1740	7.55	6.70	12.00	8.87
1750	7.25	7.77	12.65	15.13
1760	10.30	9.83	14.25	15.58
1770	11.65	12.22	16.30	14.27
1780	10.75	10.81	12.40	12.70

custom house accounts mentioned by Adam Smith. For the data behind figure 6.1, Playfair provides only part of the series of imports and exports. In the book, he includes data only for 1700 and every decade thereafter until 1780; further on in the book he gives yearly data for 1771 through 1782 (Playfair 1786, 20 and 44). All these values correspond to the values on his graph (figure 6.1). Original data do exist; some public accounts have been published. They do not correspond to Playfair's data by decade but do agree with his yearly data. The whole run of public accounts data can be extracted from other contemporary or near-contemporary publications: Charles Whitworth's *State of the Trade of Great Britain in Its Imports and Exports*, published in 1776, and David Macpherson's *Annals of Commerce*, published in 1805 (Whitworth 1776; Macpherson 1805). Sheffield's ten-year averages in his table IX can be obtained (with perhaps an odd typo and addition error) from Whitworth's data. A comparison of the total import-export numbers from Playfair and from Whitworth or Macpherson ("Other Sources" in the table) is given in table 6.1.

My conjecture, which at this point is not provable, is that Playfair smoothed the first part of his data, and it is the smoothed data that appear in table 6.1. For smoothing, Playfair did not use some fancy mathematical smoothing method, which would make him well ahead of his time, but instead used a draftsman's spline. This kind of spline is made with flexible wood, which can be bent into curved shapes that are held in place by weights. He probably learned to use the draftsman's spline at Boulton and Watt while doing engineering drawings there, or earlier during his apprenticeship at Andrew Meikle's. His brother James would naturally have used such splines in his architectural work.

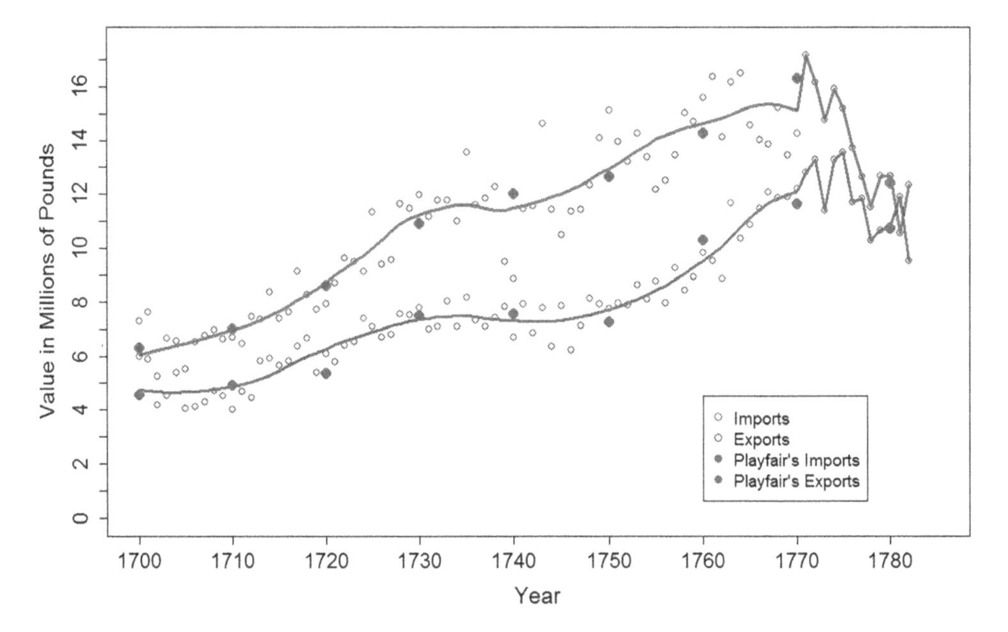

Figure 6.2. Smoothed import and export data using LOESS. Drawn by the author.

To support my conjecture, I smoothed the data extracted from Whitworth and Macpherson using a modern smoothing technique called LOESS. Whitworth has data for 1697 through 1699 and so I included them as well. Since these values are low, it reduced the smoothed values at the beginning of the series, just as Playfair's values for 1700 are well below Whitworth's reported values. For the years 1700 through 1770, I used a smoothing parameter (called the "span") set at 0.4, while the years 1771 through 1782 were left unsmoothed. The result is shown in figure 6.2. While my smoothing does not exactly replicate Playfair's, it comes close. I have also inserted Playfair's data into figure 6.2 to show how closely my smoothed data come to Playfair's. As a result of this exercise, I find it reasonable to conclude that Playfair used a draftsman's spline to smooth his data from 1697 through 1782, leaving out the first three years from his graph. For some reason, he decided to have his spline pass through, or pass very close to, the data points for 1771 through 1782, probably to show the effect of the American Revolution on British trade, which Playfair comments upon in his text accompanying the graph.

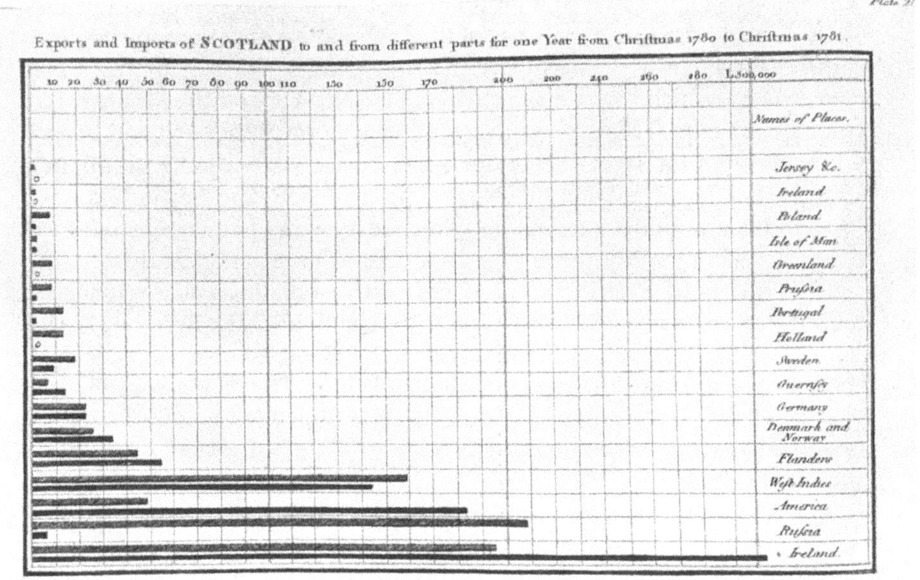

Figure 6.3. Import and export data for Scotland from *The Commercial and Political Atlas.* Courtesy of Stephen M. Stigler, University of Chicago.

The Commercial and Political Atlas contains forty coloured graphs along with commentary and explanations concerning the graphs. Most graphs are line graphs (like the one in figure 6.1) of economic time series related to imports and exports for a variety of countries and territories. Data up to 1773 for almost all of the first twenty-five graphs can be found in Whitworth's *State of the Trade of Great Britain in Its Imports and Exports.* There are a few exceptions to this line graph rule – a bar graph (figure 6.3) showing Scottish trade, and a chart (figure 6.4) showing how the national debt could be reduced and eliminated using a sinking fund.

For the completed book, James Corry contributed data on the revenue of Ireland; Corry provided the last five graphs in the book done in the same style as Playfair's graphs. His involvement in the publication of *The Commercial and Political Atlas* illustrates Playfair's growing network of people in high places. Corry was a clerk in the Irish House of

Commons.[8] He received his appointment on the recommendation of John Foster, who was a member of the Irish House of Commons beginning in the 1760s. It was also Foster who provided Corry with much of the data on which Corry's graphs were based. Apparently, Foster was the "go to" man for Irish financial statistics. For a decade, he had been chief baron of the Irish Exchequer. Also, Foster's grandfather, William Burgh, had been the accountant general for Ireland, and Foster had access to his grandfather's papers.[9] How Corry came to be contacted by Playfair was possibly through a roundabout connection with Foster. Foster and Lord Sheffield carried on a correspondence that spanned about forty-five years (Ireland, Oriel, and Sheffield 1976). Between 1784 and 1790, when he did not hold a seat in the House of Commons, Sheffield was in the anti-Pitt camp among the Whigs. The final and perhaps weakest link in this suggested chain is that the Glaswegian merchant Patrick Colquhoun introduced Playfair to Sheffield sometime in the first half of 1786.[10]

For the trade data for Scotland, Playfair constructed a bar graph showing the amount of imports and exports between Scotland and seventeen other countries or territories during 1781 (figure 6.3). The original data can be found in Macpherson's *Annals of Commerce*. As Playfair mentions in his book, he did not have an extended time series as he did for England. In Whitworth's *State of the Trade of Great Britain in Its Imports and Exports*, Scottish trade data are given only for the years 1697 to 1707, when Scotland and England were united to become Britain. In Macpherson's *Annals of Commerce*, annual trade data for Scotland begin in 1760. Playfair's graph agrees with Macpherson's data.[11] As with figure 6.1, the overall effect of the graph is better than looking at Macpherson's table of the original data (Macpherson 1805, 706). It is easy to see at a glance which are Scotland's major and minor trading partners and the general extent of the trade.

One graph (shown here in figure 6.4) in *The Commercial and Political Atlas* is not based on economic data. It shows how the British national debt could be eliminated through the establishment of a sinking fund. Playfair had the debt pegged at about £245 million and assumed that £1 million annually would be put into a sinking fund. This would eliminate the debt in about fifty-three years. The curved line running from 245 on the y-axis to 53 on the x-axis shows the decrease over the years in this £245 million debt. Playfair also considered the accumulation of new debt. This is shown by the line rising from no new debt (0 on the y-axis) in the current year (0 on the x-axis) to about £177 million when the original debt is eliminated after fifty-three years. The annual amount of £1 million is then put towards this debt, which is eliminated after seventeen years, or a total of seventy years after the current year.

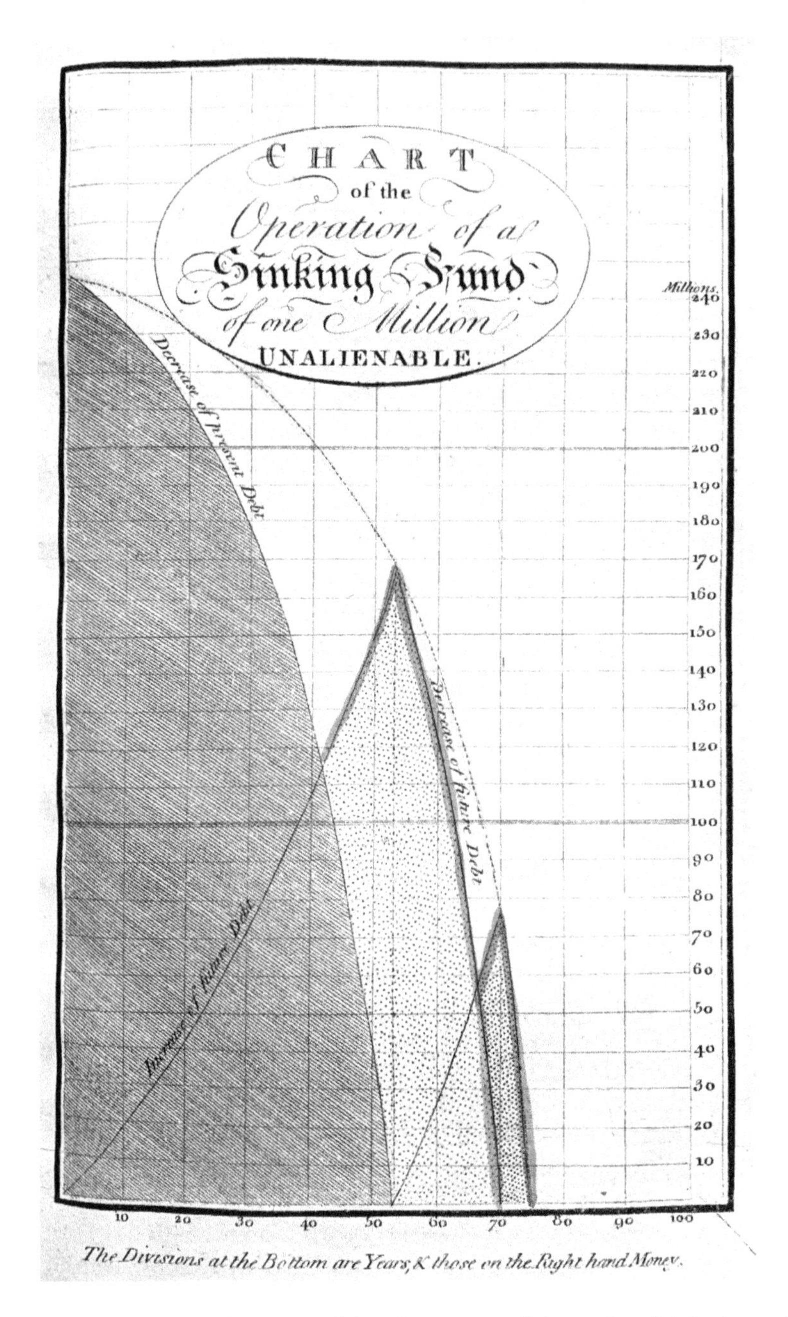

Figure 6.4. Playfair's depiction of the elimination of the national debt through a sinking fund. Courtesy of Stephen M. Stigler, University of Chicago.

At fifty-three years, new debt could again accumulate and Playfair shows graphically that this could be eliminated in another five years after applying the annual amount of £1 million to the sinking fund. I have chosen to show the graph from Playfair's 1798 *Lineal Arithmetic* (Playfair 1798b). With the exception of the added colour, it is the same graph that appears in the 1787 Stockdale edition of the *Commercial and Political Atlas* (Playfair 1787a). The graph shows Playfair's use of hachure and stippling illustrated in his early drawings at Boulton and Watt in figure 2.1.

At the most basic level, there are two ways to pay off a debt, the British national debt included. The first is to make regular payments that cover interest and principal until the debt is paid. The second is to pay only the interest as it comes due and then regularly invest money in an interest-bearing fund so that at some point in the future the fund accumulates to the value of the principal of the loan. At that point the loan is paid off. The technical name for the second method is a sinking fund. Monetarily, the two methods are equivalent, but on the surface they look quite different. Here is a simple example. Suppose a loan of £1,000 is made today at 5 per cent interest per annum, and the loan is to be repaid in, or over, twenty years. According to the first way to repay the loan, an annual payment of £80.24 is made on the anniversaries of the loan for twenty years, the final payment being made twenty years after the loan was made. According to the second way, the interest of £50 is paid on each anniversary of the loan for twenty years. On the twentieth anniversary the £1,000 is repaid in full. In order to make this final payment, £30.24 is invested on each anniversary for twenty years. At 5 per cent interest, this will accumulate to £1,000 in twenty years. Under both methods, an annual payment of £80.24 is made.

When it comes to paying off the national debt through a sinking fund, the idea is simple, but the execution is difficult. The government would set aside regular contributions and put them into a fund that would accumulate with interest to the value of the debt at some future date. For a government, the exceedingly difficult temptation to resist is to dip into the fund, or even exhaust it, when other unexpected expenses suddenly arise. By the mid-eighteenth century, such had been the case with a sinking fund established in Britain in 1716 to reduce the national debt (Ross 1892). The basic idea behind the British sinking fund in 1786 was to put the yearly fixed amount of money towards buying government bonds, or buying back its own debt. The bonds, which were instituted in the 1750s, came to be known as 3 per cent consols. Since the consols were redeemable at the option of the government, the interest that the government earned from its own bonds would go towards buying more bonds. The compounded interest that the government would earn on these purchased bonds would lead to an acceleration in the reduction of the national debt without a change in the annual payment made through taxes. The appeal of the sinking fund to British politicians was the "magic of compound interest" in the acceleration of the reduction of the debt.

It was the graph shown in figure 6.4 that led to an expansion of Playfair's network of contacts. The background to the expansion begins several years earlier. Deeply concerned about the national debt, which had ballooned three-fold between 1716 and 1771, Richard Price, the Nonconformist minister, philosopher, and actuary, proposed a version of a sinking fund in 1771 to reduce the debt (Price 1771, 163–5). Price continued to promote sinking funds throughout the 1770s and 1780s. During the short-lived Shelburne ministry (July 1782 to April 1783), headed by the Earl of Shelburne, the government once again considered a sinking fund to reduce the national debt and was in contact with Price about the scheme. William Pitt was the chancellor of the exchequer during this ministry. When Shelburne's ministry fell, it was replaced by an even shorter-lived Fox-North ministry that lasted until December 1783. At this point, Pitt formed his ministry, which lasted until the end of the century, and Shelburne was sidelined. Pitt resurrected the sinking fund through new legislation in 1786 (Cone 1951).

Enter once again George Dempster, one of several politicians who also had been deeply concerned about the national debt. On 23 June 1784, he spoke in the House of Commons about the necessity of reducing the debt and referred to calculations done by Richard Price that would eliminate £267 million of the debt in sixty years through a sinking fund. Pitt responded in the House by agreeing that without setting aside money for a sinking fund, it would be impossible to reduce the debt (Great Britain 1815b, 1014–17). By January 1786, Price was in communication with Pitt about the establishment of such a fund. He suggested three possible plans to Pitt, all variations on a sinking fund. What Pitt finally adopted was one closely related to one of Price's plans, but Price's least favourite one. Pitt presented his plan to the House on 29 March 1786. The next month Dempster was part of a group that discussed the sinking fund proposal with Price (Cone 1951). During debate in the House on the sinking fund that took place on 12 May 1786, Dempster spoke to the bill before the House. He praised some amendments that were made and, at the same time, claimed that more improvements to the bill needed to be made. He cited Playfair's discussion of sinking funds in *The Commercial and Political Arithmetic* in support of his argument, although the advertisement in the newspapers for the book did not appear for another two weeks (*Morning Post and Daily Advertiser*, 13 May 1786; *Morning Chronicle and London Advertiser*, 13 May 1786; and *Morning Herald*, 30 May 1786). Dempster suggested further amendments, but they were voted down (Great Britain 1803, 804).

It was on 22 November 1786 that Lord Shelburne wrote to Richard Price mentioning Playfair's "commercial tables ... as well as of one to shew the operation of a Sinking Fund," with Price replying three days later that he "had not yet seen any work of Mr. Playfair's" (Price 1994, 86 and 91). With Dempster's long association with Shelburne through their work in Parliament, with Dempster promoting Playfair's work, and with Dempster's and Shelburne's mutual interest in sinking funds and the national debt, it does not stretch the imagination to assume that Dempster introduced Playfair to Shelburne. From Shelburne's letter to Price, it is apparent that Playfair and Shelburne were well acquainted by November 1786 when the letter was written.

The second edition of *The Commercial and Political Atlas*, which appeared in 1787, was dedicated to George Dempster, with Playfair acknowledging Dempster's former patronage and his "assistance in procuring materials" for the publication. The major difference between the editions is that the graphs in the first edition were hand coloured, while those in the second edition were not (Playfair 1787b). Playfair replaced the colouring with hachure and stippling – drawing techniques that he had used at Boulton and Watt (see figure 2.1). The new edition was sold through Debrett's, Robinson's, and Sewell's bookstores. Another edition appeared in 1787. The latter edition was sold at John Stockdale's bookstore (Playfair 1787a). The only difference in the two 1787 editions is that the dedication to Dempster does not appear in the one sold by Stockdale, perhaps for political reasons.

When Shelburne wrote to Richard Price in November 1786, his main point about Playfair's work was that Playfair was writing a new book "to shew the different operation of annuities and perpetuities" (Price 1994, 86). Playfair was going to dedicate the book to Shelburne and, using Shelburne's connections, wanted Price's opinion of his new literary venture. Near the end of the year, Playfair visited Price and gave him a copy of his manuscript. It turned out not to be a book on the calculation of the values of annuities and perpetuities, a topic with which Price was very familiar, but instead was to be a book on the national debt and the sinking fund, as suggested by figure 6.4, a topic close to Price's heart. Price liked what Playfair had written. He found Playfair's graphs "agreeable" and "useful." He wrote positively to Playfair about the manuscript and then expressed his disappointment with the structure of the sinking fund that Pitt had set up to tackle the national debt.

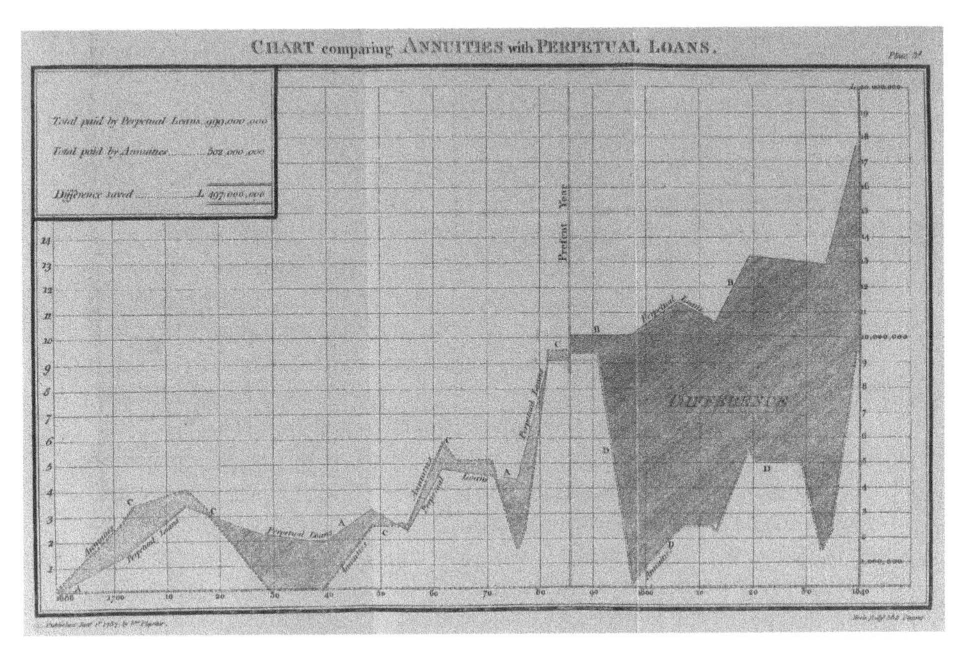

Figure 6.5. Playfair's graph of the growth of the national debt with projections. Courtesy of University of Edinburgh Library, United Kingdom.

The manuscript was published under the title *An Essay on the National Debt* with a publication date of 1787 (Playfair 1787c). In his *Essay*, Playfair wrote at length about the dangers of the growing national debt and of financing it through perpetuities, or the 3 per cent consols. He drew a set of graphs to illustrate his point. First, he graphed the size of the national debt from 1688 to the present day (1786) and projected what the debt would be between 1786 and 1840. Then he calculated what the national debt would look like annually had it been financed through fixed-term annuities of fifteen years in length and graphed these results. Figure 6.5 shows his third graph, which is an overlay of the first two. The dotted line shows the national debt based on fixed-term annuities and the solid line shows the actual debt and projections based on perpetual loans. The area between the lines shows the cost difference between the two methods. Playfair claims, and shows in the graph, that after 1786 (the vertical line showing the "present year" on the graph) the national debt would balloon out of control when funded by perpetual loans.

Playfair's projection was based on some assumptions, which in retrospect were not true. First, Playfair assumed that the yearly contribution to the sinking fund, which Pitt had put in place in 1786, would be carried through until 1840. Further, on the basis of the graphs and what can be gleaned from the text, he assumed that from 1791 to 1840 there would be three wars, each lasting about seven years followed by ten years of peace. His fifteen-year annuities would easily keep the national debt in check. Good projections come only from good assumptions. The French Revolutionary Wars began in 1792 and were followed by the Napoleonic Wars, which ended in 1815. For Britain, this was twenty-three years of almost uninterrupted war fought with a mixed chorus of changing allies on the Continent. It was accompanied by soaring costs, which led to a strain on the national finances.

There were evils, so Playfair claimed, attached to this ballooning debt. Preaching doom and gloom, Playfair predicted that these evils would lead to disaster. The greatest was that a proportion of the population, the part that was living off the income from the perpetuities or the 3 per cent consols, would become idle and unproductive. With an increasing debt because of the expense of war, funding the debt through perpetual loans would lead to increased idleness in the population. This would be followed by an eventual destruction of the nation's industry, the bedrock on which the wealth of the nation sat. A decline in industry would imperil the whole British political system, including the monarchy itself, Playfair concluded. This was a theme that would recur in 1805 in Playfair's *Decline and Fall of Powerful and Wealthy Nations*.

James Anderson, a Scottish agriculturalist and political economist, provided a very insightful review of *An Essay on the National Debt* in the *Monthly Review*.[12] Anderson liked the graphs. After describing them, he says, "These are the particulars which are delineated with accuracy in the work before us; and on the account that the varying size and dimensions of a geometrical figure are more palpable to the eye, and convey a clearer idea to the mind, of a change in the dimensions, than can be given by arithmetical notation, it will be deemed a happy invention for facilitating the attainment of political knowle[d]ge: and thus far does the work before us merit, in our opinion, the approbation of the public." Anderson then focussed on the naivety of some of Playfair's arguments, which are obvious, I think, from my description of the idleness caused by the use of perpetual loans to fund the national debt. After picking apart Playfair's arguments, Anderson concludes, "We might quote many other passages which alike discover the embarrassment of the author; but we decline the ungracious task. In short, Mr. Playfair is here evidently wading beyond his depth, and it might be well if he could be advised to enter, in future, on the *public* discussion of such questions with greater caution."

Anderson also reviewed *The Commercial and Political Atlas* (*Monthly Review* 1788b, 505–9). He began by pointing out that Playfair was not the first to draw such graphs, citing graphs of historical chronologies published earlier by Joseph Priestley and by John Playfair's colleague, Adam Ferguson of the University of Edinburgh. (The works of Priestley and Ferguson, and the predecessors, are reviewed in Rosenberg and Grafton 2010, chap. 4.) Despite Playfair's lack of priority in graphical methods, Anderson saw something new and exciting in Playfair's graphs, which were meant to aid in the interpretation of economic statistics:

> And it will be acknowle[d]ged by every unprejudiced person who looks at these charts, and compares them with the commercial tables of Sir Charles Whitworth, that, in matters of this nature, the mode of elucidation here adopted possesses some great advantages over that by arithmetical figures alone: in looking at a chart, the eye at one glance sees the encrease and decrease of any article there delineated, with the utmost facility, and we can trace its progress with much satisfaction and ease; whereas, from the arithmetical tables alone, it is obliged to draw its conclusions from separate, detached facts, which can only be compared together by a process far more tedious and painful, though equally accurate: We cannot, therefore, help thinking the public obliged to Mr. Playfair for this specimen of ingenuity; for whatever facilitates the acquisition of knowle[d]ge, we must always think, adds greatly to the sum total of human acquirements.

Anderson's sentiments ring true today. A graph can more easily convey information and is more easily interpretable than a table of numbers.

The Commercial and Political Atlas was Playfair's first published venture into statistical graphics. Several more publications with pictures, including *An Essay on the National Debt*, were to follow from Playfair's pen. Despite Playfair's lack of priority in the publication of graphs, today Playfair is seen as the father of modern statistical graphics. Howard Wainer, for example, has made a very strong case for why Playfair should carry this title (Wainer 2005, 9). Playfair invented or popularized many graphical forms, such as the time series plot and bar graph seen in figures 6.1 and 6.3. After Playfair, such graphs became more widely used over many disciplines. He was a very prolific author and had some insightful publications. Many have fallen by the wayside and have been forgotten today. It is his insights into data through graphs that have long stood the test of time.

Chapter Seven

Playfair Goes to Paris

By 1787, William Playfair's partnerships in his manufacturing interests had been dissolved for about a year. Working without partners, he had carried on his business as a button maker, apparently on a much smaller scale. At the same time, he had tried his hand at writing. In this latter profession, like his manufacturing activity, he was moderately successful. Of his five books published prior to his departure for France, the critics panned two of them (*Essay on the National Debt* and *Joseph and Benjamin*). His most successful book was *The Commercial and Political Atlas*, which found support in high places from the MP George Dempster, as well as from former prime minister Lord Shelburne, now Marquess of Lansdowne. As Playfair later admitted to a relative from Scotland, who had visited him in London, writing did not pay that well (Rogers 1872, 22).

Why did Playfair go to France in the late spring of 1787? It could not have been because he was escaping creditors pursuing him in England. He was not penniless; he seems to have spent the first several months in France at leisure without any kind of employment that would bring in any money. On his first Sunday in Paris, with his wife Mary at his side, he strolled through the Champs-Élysées gardens and through the nearby Tuileries Garden, with dinner afterward in Place Louis XV (now Place de la Concorde), which connected the two gardens.[1] I think that a desire to visit and write about France for an English audience can also be dismissed, since his command of the French language at the time was almost non-existent and he did not write about France until 1792. Most likely he went to France to reconnoitre in order to begin anew as a manufacturer. The climate was right for him. French industry was growing rapidly, although in terms of total output it paled in comparison to what was happening in Britain (Schama 1989, 190). His former employers Boulton and Watt may also have inspired Playfair. In

November 1786, the two partners arrived in France on the invitation of the French government under Louis XVI for an all-expenses-paid visit. They returned to England on 16 January 1787 (Jones 1999).

From the French point of view, the major reason for the invitation to Boulton and Watt was to obtain advice on the machine de Marly, the hydraulic system, completed in 1684, that brought water about seven kilometres from the Seine River to Versailles (Brandstetter 2005). Though an engineering marvel in its day, the machine was clumsy and inefficient and could not meet the increasing demand for water in Versailles, including both the water supply in the town of Versailles itself and the ornamental fountains at the palace. Over the years 1784–6, Comte d'Angivillier, director of the Bâtiments du roi (the king's buildings), the office in the French government responsible for the machine de Marly, ran a competition for proposals to refurbish or rebuild the machine. More than forty-five proposals were received. While considering them, d'Angivillier intended to invite Boulton and Watt to submit one, if he found that none were appropriate. Apparently none were acceptable; at the same time Boulton and Watt declined to make a submission, probably because of the tangled "delicate web of French administrative politics" (209).

There were other reasons behind Boulton and Watt's visit to France. On the French side, the French minister Charles-Alexandre de Calonne wanted advice on an iron-smelting factory in which he had a financial interest. For their part, Boulton and Watt had their own reasons for accepting the invitation. There was the possibility, unfulfilled in the long run, that they could obtain an exclusive licence to supply and erect steam engines in France. On the personal side, Boulton wanted his son to study in France.

Playfair arrived in France with letters of introduction in hand. As I have mentioned in chapter 1, he did not find them particularly useful. On the other hand, he found at least two interesting and entertaining. One was for André Morellet and the other for Alexander Gordon. Morellet, who held the ecclesiastical position of abbé, was a friend of Voltaire and Benjamin Franklin, and was the last surviving Enlightenment philosopher when Playfair arrived on the scene. Very influential in intellectual circles, he was secretary to the Académie française at the time when it was closed in 1793 by the French revolutionary government (Académie française n.d.). Alexander Gordon was the principal of Scots College, a college of the Sorbonne that was a centre for Scottish Catholics. Mainly as the result of disputes between Gordon and his bishop, but also illness (possibly a bout of tuberculosis that ran through the college earlier in the century), other theological disputes, and financial problems related to old Jacobites in exile (followers of James II and his descendants), the college was in serious decline by the 1780s (Halloran 1996, 320–42). Gordon left for Scotland in September 1792. The French revolutionary government closed the college the next year. The main theological dispute was over accusations of Jansenism in the college dating from earlier in the century when Gordon was a student at the college. Gordon may have been sympathetic to Jansenism. Less strict than the Jesuits who were key players in the Counter-Reformation, Jansenists held certain Calvinist views that would be compatible with Playfair's Scottish Presbyterian upbringing. Whoever wrote Playfair's letter of introduction to Gordon must have thought that the college would be a congenial place to be for a Presbyterian Scot who could speak very little French at the time.

The Marquess of Lansdowne wrote Playfair's letter of introduction to Morellet. Prior to Playfair's arrival in France, Lansdowne also sent Morellet a copy of Playfair's *Commercial and Political Atlas*. Several years before, Lansdowne and Morellet had met in London, and they had maintained a friendship since that time, at times in person and more often by letter. Both had similar interests in economics, undoubtedly the reason Lansdowne wanted to introduce Playfair to Morellet (Dziembowski 2011, 216).

There are two versions of the meeting Playfair had with Morellet, one from Morellet and the other from Playfair. They agree on substance but vary in tone. Playfair came to dislike Morellet,[2] and Morellet thought that Playfair's graphical work was useless (Morellet and Fitzmaurice 1898, 243–4). Playfair thought that Morellet was a grumpy old man. Some of the grumpiness went away when Playfair presented Morellet with his letter from Lansdowne. According to Playfair, the conversation was difficult because, except for a few words, neither spoke the other's language. Morellet offered to see Playfair again, but after the meeting Playfair decided to nip the relationship in the bud. Morellet wrote to Lansdowne, saying that he preferred numbers to pictures when discussing economic data. It was better to discuss these issues verbally or in writing. According to Morellet, the charts were useless for this purpose. He thought Playfair's business proposal of selling inexpensive buttons and shoe buckles (which Playfair had been doing in London) a much better idea. He offered to have Playfair bring his business plan to him in writing so that he could translate it into French. He also offered to introduce Playfair to some of his friends and acquaintances. As a result of Playfair's bad impression of Morellet – his first and only – nothing came of their interaction.

The letter of introduction to Alexander Gordon seems to have been more of a social introduction rather than one that could relate to business. Playfair's description of his interaction with Gordon tells only of the social aspect. He visited Gordon several times and enjoyed his conservation. He dined in the college refectory with Gordon and the only other cleric on the college staff, Alexander Innes, who was the college procurator. There were no students left in the college. The only other person around was an "old domestic." The college had several tables in the refectory capable of handling seventy to eighty students and so Gordon, Innes, and Playfair ate in cavernous silence at the head table that seated ten to twelve people. What impressed Playfair about the college was its coolness in the heat of the summer, and the library, which contained some valuable books and manuscripts. He was especially taken by letters that the college held, written by King James II and his wife, Mary of Modena. Playfair's *Memoirs*, in which these incidents are recorded, give a small hint of his Jacobite sympathies. When Gordon left the college in 1792, leaving Innes behind to guard the college and its property, he took with him some manuscripts concerning the college. They are now held by the University of Aberdeen.[3]

In early 1788, Playfair went into partnership with Joseph Gerentet, an old acquaintance from Birmingham.[4] They received letters patent from the King Louis XVI on 5 April 1788 for the Société Gerentet et Playfair. The purpose of their business was to operate a factory that would produce and polish laminated metals. This would require a rolling mill similar to what Playfair had previously operated. It more or less replicated what Playfair had been doing in London with his former partners there, James Keir and William Wilson. The new partnership formally went into operation on 5 June 1788.

At some later point, most likely in 1790 as the French Revolution took a firm hold, Playfair was required to post a bond, amounting to 3,200 French livres, as a guarantee for the Société Gerentet et Playfair. The money was deposited with an English banker in Paris, James Carrey. It is hard to determine exactly, but Playfair probably obtained the money from a French aristocrat who was involved in some way financially with Société Gerentet et Playfair, possibly as a purchaser of shares. Through a Paris notary, Charles-Louis Farmain, there is record that money to the tune of 3,200 livres was due to Playfair from Pierre-Louis Brachet, Vicomte de Brachet. The surviving records show only that Playfair was chasing Brachet through legal means to obtain his money. The first notice was sent to Brachet's wife on 1 September 1790. She was then living at her home in Bauzemont about 300 kilometres northeast of Paris. A second notice was sent on 20 September 1790. The vicomte had been closely connected with the French court. He was a military officer who held the court position of gentleman of the chamber to the Comte de Provence, who later became Louis XVIII (Ëataux 1885, 17). By 1791, Brachet had fallen into financial difficulty in France.[5] Playfair did get his money from Brachet. When the French revolutionary government seized Carrey's records in 1792, Playfair's account for 3,200 livres was still on Carrey's books.[6]

To put the money in context, a standard British arithmetic book from 1788 values £1 sterling at between 22.5 and 24 French livres (Bonnycastle 1788, 191). This puts the value of Playfair's bond at between £133.33 and £142.22 or between about £20,000 and £21,000 today.

In his *Memoirs*, Playfair described Gerentet as a maker of coloured foil in Birmingham and as a manufacturer of *ballons* in Paris. Playfair's description of Gerentet's Birmingham occupation corresponds to what appears in the entry for Gerentet in a Birmingham directory: a "foreign foil maker" (Bailey 1781, 14). Gerentet and Playfair's experience with running a rolling mill were probably similar. The Paris occupation is a little more open to interpretation. *Ballon* in French means "ball" or "balloon." I would go with the balloon interpretation. In 1794, Gerentet was working with Louis-Bernard Guyton de Morveau. Guyton had served briefly as the president of the Committee of Public Safety in 1793, but left to work in arms manufacturing with a concurrent interest in the military application of ballooning. Gerentet was involved with Guyton in the manufacture of hydrometers necessary for optimizing the production of saltpetre in the production of gunpowder (Bret 2017).

The rolling mill operated by the Société Gerentet et Playfair was located on Île Louviers (also spelled "Louvier" in some documents and on the map).[7] Shown on the map in figure 7.1, this was one of four islands in the Seine River running through Paris (only three are shown on the map). Today, it is no longer an island. The channel between the island and the Right Bank was filled in in the nineteenth century. In Playfair's time, the island was the principal depot for firewood, where several wood merchants operated (Galignani and Galignani 1838, 111 and 321). It was a convenient location for Société Gerentet et Playfair; a furnace to make the metal more malleable prior to the process or afterward for annealing, fuelled by the wood held on the island, was probably attached to the rolling mill.[8]

Gerentet and Playfair went to the Paris notary Toussaint-Nicolas Garnier to draw up the articles of association for their company.[9] Two other names appear on these documents besides Gerentet and Playfair: Étienne Jean François Dusoulier, Chevalier d'Azeux, and Alexander Bond. I cannot find any information on Bond. However, in his *Memoirs* Playfair has a good description of the chevalier, whose name

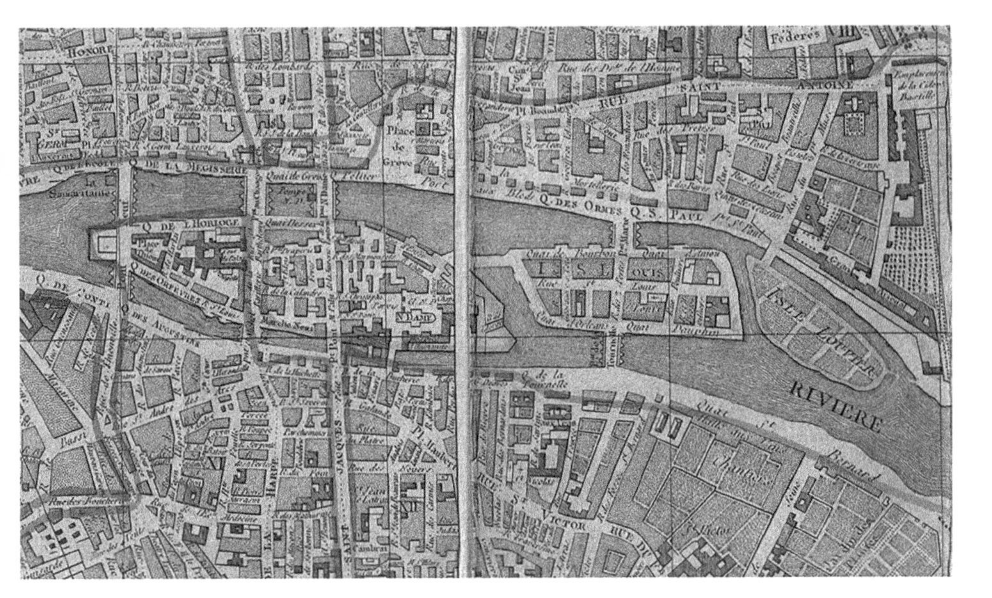

Figure 7.1. Central Paris and the Seine River from an 1802 map. Courtesy of Geographicus Rare Antique Maps, Brooklyn, New York.

he renders as "d'Azeuse" (one of many misspellings of the names of his acquaintances):

> I had not been two months in Paris when my conversation with one of my countrymen attracted the notice of a French Gentleman who spoke English perfectly well. His name was d'Azeuse and he had been travelling companion to a Mr. Cecil of the Exeter family.
>
> This chevalier, for he took that title, was of great service to me – not so much from friendship as interest for he learned from my conversation that I knew many things that might be established with advantage in France and he had connection and protection amongst the great.
>
> He was a man who had read a great deal and as his income was not considerable he made himself a welcome guest at the houses of a number of persons of consequence who like to have persons at their tables to entertain them and their guests by their conversation.
>
> D'Azeuse had some turn for wit – was not destitute of taste and had sufficient skill to pass himself for a man of much more knowledge than he possessed. With all these he had sufficient pride to avoid the appearance of a hanger on and he had just sufficient income to be independent. (Playfair n.d., Memoir 1)

According to Playfair, d'Azeux would be instrumental in getting Société Gerentet et Playfair off the ground.

As Playfair recalled in his *Memoirs* years later, there was much more to getting a business started than simply applying to the king or his minister for letters patent. It required breaking into a close circle of patronage and greasing palms along the way. And d'Azeux was the man to do it for him. Another letter of introduction for Playfair via Lansdowne was also of great help. This was not an actual letter, but instead the *Commercial and Political Atlas*. It arrived in Paris before Playfair did. In addition to Morellet, Lansdowne had sent several copies of the book to Charles Gravier, Comte de Vergennes, the French foreign minister. Lansdowne and Vergennes knew each other politically. When Lansdowne was Shelburne and prime minister, the two negotiated the end to the war that led to American independence from Britain. They did not negotiate face-to-face; Vergennes sent an emissary to London to work on his behalf. As a result of the positive negotiations and the treaties that followed in 1783, Shelburne and Vergennes had great respect for one another (Stockley 2001, 98–103). Knowing Louis XVI's interest in geography, Vergennes sent a copy of the book to Louis, and Louis liked it (Playfair 1796a, iv).

D'Azeux initiated a chain reaction of connections that ended at the feet of Louis XVI. In Playfair's early Paris days, he often dined with d'Azeux, either at a restaurant of d'Azeux's choosing or at Playfair's hotel. When Playfair made connection with Gerentet in Paris and told d'Azeux of it and the plans for a rolling mill, d'Azeux offered to help, which in due course had a price. As Playfair mentioned, the comfortable but less than wealthy d'Azeux was a bit of a social butterfly and often dined at others' tables. The first link in the chain of connections was Jean-Claude Richard, Abbé de Saint Non. The younger son of an aristocrat, Saint Non was a patron of artists and an artist himself. He was known as a "congenial figure in Paris society." D'Azeux, who probably met Saint Non socially, introduced Playfair to Saint Non. According to Playfair, after some bargaining, it was agreed to set up what became Société Gerentet et Playfair as a company with seventy-two shares. Half of the shares would be sold at 1,650 livres (or about £70) per share to raise capital for the new company. Of the other half, six would go to Saint Non to use for bribes of government officials. This settled, Saint Non arranged for Playfair to meet Jean Pierre de Batz, Baron de Sainte-Croix (typically known as Baron de Batz), who at the time was a thirty-four-year-old financier with a recently made fortune. Playfair met the baron a few times and on one occasion Playfair lent the baron a copy of the *Commercial and Political Atlas*. Batz was sufficiently impressed and invited Playfair to dine with him and Étienne Clavière, another French financier, who later became a major player in the French Revolution. It turned out that Clavière had his own copy of the *Commercial and Political Atlas*. Baron de Batz then took it upon himself to take Playfair's proposal to Louis Charles Auguste Le Tonnelier, Baron de Breteuil. Up until 24 July 1788, Breteuil held the position of Ministre de la Maison du roi and had the king's ear. When the proposal got to Louis, it was made evident to Louis that the proposer was the author of *Commercial and Political Atlas*, a book that Louis had enjoyed reading. On that, Playfair received his royal patent for the Société Gerentet et Playfair. At the conclusion of this business, Saint Non gave some of his shares in Société Gerentet et Playfair to d'Azeux and some to a lady friend. Once established, it cost about 10,000 livres to put the equipment and buildings in place on Île Louviers.[10] A month after Société Gerentet et Playfair began operations in July 1788, Clavière was sufficiently impressed with the Société's prospects that he invested 6,600 livres (four shares at 1,650 livres per share) in it (Antonetti 2007, 85).

Clavière and Playfair became friends and remained so until their political opinions diverged as a result of the French Revolution. At the beginning of the revolution, when Playfair was hoping for a constitutional monarchy for the French, Playfair visited Clavière at his Château de Bel-Air in Suresnes about six miles from central Paris (Playfair 1793c, 43–4).[11] During a stroll in the garden before dinner, the two discussed politics, with Clavière expounding on what he thought should be the disappearance of inequality in political power in a country because of the invention of printing and gunpowder.

Playfair was quick off the mark to try to generate business. Since how to deal with the machine de Marly was still up in the air, Playfair decided to submit his own proposal on 1 October 1788. On 31 July 1789, about two weeks after the storming of the Bastille, Playfair submitted a second proposal, essentially the same as the first, but adding Gerentet as a co-applicant. Both proposals were reviewed by the Intendant des eaux et fontaines (the administrator in charge of waters and fountains), working under the Comte d'Angivillier as director of the Bâtiments du roi. The scientist Charles-Augustin de Coulomb was the intendant at the time. Coulomb's main occupation until the French Revolution was that of military engineer. For the earlier proposal, Coulomb wrote two lengthy reports, one on 22 November and the other on 30 November 1788.[12] The report on the second proposal, essentially marginalia on the first page of the proposal, was short and negative. Coulomb's assessment of Playfair himself in the first submission was positive, but he had serious questions about the viability of the execution of the project.

Prior to the first proposal, Coulomb had never met Playfair. Coulomb had seen the name Playfair in the newspapers several times and thought that Playfair was a carpenter from London. He knew nothing of Playfair's abilities. By the time of the second report, Coulomb had gone to visit Playfair at his rolling mill. He was impressed by Playfair's engineering works and thought it should stand as an example to French manufacturers. He was easily able to understand Playfair's manufacturing process and saw that the goods from the mill were economically produced.

Playfair's proposals were both specific on promised results and vague on the details of how to get there. As for the promises, the replacement machine would be built at no cost to the French government; the only compensation Playfair required during the construction process was that he be given the materials recovered from the decommissioned old machine (presumably to sell; in his *Memoirs* he estimated that the materials were worth £38,000) (Playfair n.d., Memoir 11). He further promised that the amount of water raised by the replacement machine would be three-fold more than the current machine. And finally, he promised that there would be no interruption in service for the delivery of water to Versailles during the whole reconstruction. Playfair provided absolutely no information as to how he was going to achieve all this.

Playfair's vagueness led Coulomb to question Playfair's ability to deliver the final product. Coulomb could not see how the new machine could be built without disturbing the old one during construction, especially since Playfair was planning to use the pipes from the old machine. The problem that Coulomb saw was compounded by one of Playfair's conditions in his proposal. Playfair reserved the right to reject pipes from the old machine that did not suit his purpose. Not only would the construction of the new machine interfere with the working of the old one, it might also cost a considerable amount of money, perhaps as much as 200,000–300,000 livres, Coulomb speculated. Coulomb's final objection had nothing to do with engineering problems and Playfair's vagueness. For Coulomb, the machine de Marly should be restored by a Frenchman.

Playfair claimed in his *Memoirs* that he would have obtained the contract except that the French Revolution got in the way. As he had done with the Société Gerentet et Playfair, a bribe to a nobleman was required to seal the deal. The bribe was one third of the profits (Playfair n.d., Memoir 11) and the nobleman in question was Louis-Paul de Brancas, Duc de Céreste. In view of Coulomb's comments, despite the revolution, the bribe would never have resulted in obtaining the contract.

An examination of Playfair's background from working at Boulton and Watt and a reading between the lines of his proposal suggest what he was planning to do and how he would make his money. The current set of waterwheels powered pumps that lifted water from the Seine River to a reservoir about forty-nine metres above the river. The same waterwheels and a series of articulated connecting rods pumped the water from this reservoir to a higher reservoir fifty-seven metres above the first, and finally from the second reservoir to an aqueduct a further fifty-seven metres higher. A painting from 1722, shown in figure 7.2, shows the extent of the operation. The waterwheels were highly inefficient. It is estimated that these waterwheels expended only 70 horsepower, out of a potential 1,000 horsepower (Ermene 1956). Playfair intended that the waterwheels would be replaced by one or more steam engines. He had considerable experience at Boulton and Watt with steam engine design and with their erection for pumping water out of mines. Nothing was said about steam engines in his proposal, but the one detail that Playfair did let slip in his proposal was that the materials he needed would be built at his workshops in England, with no indication whether these workshops already existed or if they would be newly built when the proposal was accepted. For compensation, Playfair took a page from Boulton and Watt's playbook. Boulton and Watt's charges were based on the savings made on the reduction of coal needed to run their engines compared to other steam engines. Playfair's charges were going to be based on the savings made in the maintenance of the new machine. For an eighteen-year contract, Playfair was going to charge 38,000 livres per year to maintain the new machine, less than half the current maintenance cost. Since the machine de Marly was very expensive to maintain and a modern steam engine much less so, Playfair was probably counting on a very tidy yearly profit from this proposal. Other details were probably insignificant to him.

While in Paris, Playfair also met other scientists, though more in a social than a professional capacity. These included the Marquis de Condorcet, Gaspard Monge, and Alexandre Vandermonde, all mathematicians who had other major scientific interests, as had Playfair's brother John. Writing after his return to England following the French Revolution, Playfair showed his dislike for Condorcet, including him on a list of persons who "have propagated with such effect the system of riot, robbery and murder" during the revolution (Playfair 1798b, 1:92–3).[13] Joining Condorcet on the same list was Clavière, with whom Playfair had dined at Baron de Batz's residence when trying to get Société Gerentet et Playfair established. Among all the Jacobin politicians of the revolution, Playfair thought that Monge "was the

Figure 7.2. Machine de Marly circa 1700 by Pierre-Denis Martin. Public domain via Wikimedia Commons, original painting in Château de Versailles.

most unexceptionable of them all with regard to his past life" (Playfair 1798b, 2:472).[14] However, he had a different opinion of Vandermonde, although their relationship deteriorated as the revolution progressed. Playfair met Vandermonde soon after his arrival in Paris. The connection may have been mathematical, through Playfair's brother John. Or Playfair may have taken the initiative on the basis of a paper co-written by Vandermonde on the manufacture of steel, a topic of interest to Playfair. It was written in 1786 but did not appear in the publications of the Académie royale des sciences until 1788 (Vandermonde, Berthollet, and Monge 1786). However they met, Vandermonde took Playfair to meetings of the Académie (Playfair 1801a, ix). They often dined together and discussed politics, with Vandermonde supporting the revolution. As the revolution progressed, Playfair gradually withdrew himself from Vandermonde's company. Playfair last met Vandermonde on the street just prior to Playfair's departure from France. Playfair thought Vandermonde an honest, well-intentioned man who had been led astray.[15]

There were two Americans in Paris with whom Playfair made an acquaintance. Both were speculating in land in the United States and looking to attract French clients. One had a significant impact on Playfair's career and an eventual negative impact on Playfair's historical legacy, which has only begun to be corrected in the last twenty years (Moreau-Zanelli 2000). The first of the two to arrive, Samuel Blackden, had been a colonel in the American army during the American Revolutionary War. He arrived in France in 1786, initially interested in establishing trade connections with the port of Honfleur at the mouth of the Seine River. By the next year, 1787, he was selling land in Kentucky along the Ohio River to French emigrants. He also dabbled in trying to arrange for flour and wheat sales from America to France in response to crop failures in France.[16] Joel Barlow, the second to arrive, was operating as an agent in Paris for the Scioto Company. Organized in the United States in 1787 by William Duer, a British-born American speculator, and others, the company's original purpose was pure speculation. In a deal with the Ohio Company, the Scioto Company obtained an option to purchase land from the U.S. government in Ohio by the Scioto River. Theodore Belote describes the nature of the speculation:

> It was intended by the Scioto Company to make an immediate sale of its rights of preëmption in Holland and France. In both these countries large amounts of United States securities were held. These securities were then almost worthless. It was natural to suppose that their holders would gladly part with them in exchange for fertile lands in the west of the United States. The securities thus acquired would be used by the Scioto Associates to pay Congress for their lands. Since Congress would accept securities at par, while the Scioto Associates had secured them at a greatly depreciated value, the latter would soon be able to pay for their lands, and the sums derived thereafter would be clear profit. (Belote 1907a, 20–1)

About twenty-five years after Barlow first arrived in Paris to peddle this scheme to French investors, Playfair wrote in his *Political Portraits in This New Æra* a description of the extent of the potential profits projected from the proposed scheme, and his opinion of the likelihood of failure: "The title to the lands was merely one of preference, in case the persons contracting should pay for half a million of acres at a time, at the rate of about eight-pence: with such a title an American agent thought he might sell half a million of acres at a time, at five shillings an acre, that is, get about £200,000 for £18,000, without any kind of security for the delivery!!" (Playfair 1813, 1:138). It is unknown when and how Playfair met Blackden. What is known is that Blackden helped Barlow in his work as an agent by introducing Barlow to Playfair early in 1789 in Paris after Barlow was failing to provide any results for his handlers back in America.

It was odd for the Scioto Company to send Barlow to Paris. He had no connections in Paris and no knowledge of the language. This was where Playfair came in. He had both: his network of contacts appears to have been extensive and his language skills had gone in two years from almost nil to having reasonable competency in spoken and written French. After Blackden introduced the two to each other, in April 1789, Barlow asked Playfair for his help in disposing of three million acres of land in the newly opened territory in Ohio by the Scioto River. Playfair declined to be involved, citing his reasons in a 1791 letter to Alexander Hamilton: "This was certainly an Idea formed without Reflection as neither an Individual nor any Company could be supposed to pay 18 millions for lands uninhabited & unknown & which did not belong to the Persons who were to sell them."[17] Playfair's friend and sometime partner in scams Thomas Byerley provided some other reasons that he heard from Playfair. The land was situated too near the First Nations people living in the area and too far from the more inhabited parts of the country (Byerley 1817).

At the time that Playfair met Barlow, Playfair was living in the Hôtel Lamoignon (LeCocq 1881, 270). From the address, he was either prospering or living well beyond his means, more likely the former. Owned by the architect Jean-Baptiste Le Boursier from 1774 to 1794, the Hôtel Lamoignon was an elegant townhouse that still stands in the Marais district of Paris housing a library, the Bibliothèque historique de la ville de Paris. William's brother James, the architect, visited Paris in 1787 (Leach 2021). It is possible that it was through his brother's connections that William came to live in the Hôtel Lamoignon (shown in figure 7.3).

As Playfair was trying to re-establish his career in manufacturing, as well as rejecting Barlow's proposal, the storm clouds of the French Revolution were quickly gathering.[18] Just after Playfair got his business up and running, the French government faced a financial crisis. On 8 August 1788, the minister of finance in the French government, Étienne-Charles de Loménie de Brienne, declared that the royal treasury was empty; a week later the government suspended its debt payments. Brienne had tried to reform the Byzantine tax system of France by royal decree, but was blocked by the Parlement de Paris, one of thirteen sovereign legal courts across the country. The Parlement's veto was achieved on a procedural technicality. In order for a royal decree to take effect, it had to be registered with each Parlement. And this the

Figure 7.3. Hôtel Lamoignon. BHVP, CC BY-SA 3.0 https://creativecommons.org/licenses/by-sa/3.0, via Wikimedia Commons.

Parlement de Paris refused to do. Brienne was not able to raise any more money through loans. To break the logjam, Brienne agreed to call a meeting of the États-Généraux (Estates General), a legislative body comprising three different classes of deputies: nobility (First Estate), clergy (Second Estate), and commoners (Third Estate). The Estates General had not met since 1614. Brienne set the first meeting of the deputies of the Estates General for 5 May 1789. Then the situation changed rapidly. Brienne resigned on 25 August 1788 and was replaced by the Swiss banker Jacques Necker, who took the position on the condition that the Estates General reform the tax system.

While these financial and political troubles were unfolding in France, Playfair published in French an abbreviated version of his *Commercial and Political Atlas* and his 1787 *An Essay on the National Debt* (Playfair 1789). This was in early March 1789. It came out under the title *Tableaux d'arithmétique linéaire, du commerce, des finances, et de la dette nationale de l'Angleterre*. For the material on trade that appears in *Commercial and Political Atlas*, discussion and graphs were reduced to total imports and exports and then the same items specifically between Britain and the United States, and between Britain and India. This was followed by graphs of yearly expenditures on the military. As the French title suggests, Playfair also included a discussion of the British national debt and how it could be eliminated over several years by a sinking fund. The graph related to the national debt and sinking fund in *Tableaux d'arithmétique linéaire* is shown in figure 7.4. With the exception of French replacing English and the amount of money given in French livres at 24 livres to the British pound, the graph is virtually the same as figure 6.4, which appears in an English publication. Playfair used a Parisian engraver for his graphs, a man named Pillot. Playfair's French was not yet proficient and he had a man named Jensen translate his English to French. He dedicated the book to Breteuil, who had helped him obtain his letters patent for Société Gerentet et Playfair, although Breteuil had resigned his court office a few months after Playfair commenced his business. Official approval to publish was given on 5 March.

Playfair's then friend Alexandre Vandermonde took Playfair to a meeting of the Académie royale des sciences, where the newly published book was presented to the academy (Playfair 1801a, ix). At the presentation, Vandermonde had Playfair seated beside the president of the academy, Jean-Baptiste-Gaspard Bochart de Saron. Saron was a lawyer, physicist, and astronomer who financed the publication of some of Pierre-Simon de Laplace's work on the orbits of planets. Ever the one to promote himself, a few days after his book was published, Playfair sent a copy to Thomas Jefferson, the current American ambassador to France.[19]

It might seem odd to publish a book on British trade statistics and the British national debt for a French audience. With a hint of his ambition to establish himself in French society, Playfair addressed this possible conundrum in his preface. Two points are notable. First, Playfair wrote that the French were in better shape than the English to handle their debt, provided that France follow Playfair's advice on applying life annuities and a sinking fund to handle the debt. The British had followed that route but were doing it badly. It was now a propitious time to follow Playfair's advice as the Estates General would soon meet. Perhaps Playfair was aiming to achieve in France the same kind of status that Richard Price had earlier achieved in England. With his English version, he already had the ear of the king. Perhaps the French version would come to the attention of the new Estates General. Second, Playfair stressed the need to have good data and to present them well when making a political or economic argument. To support this argument, he referred to the letters of Junius. The pseudonymous Junius

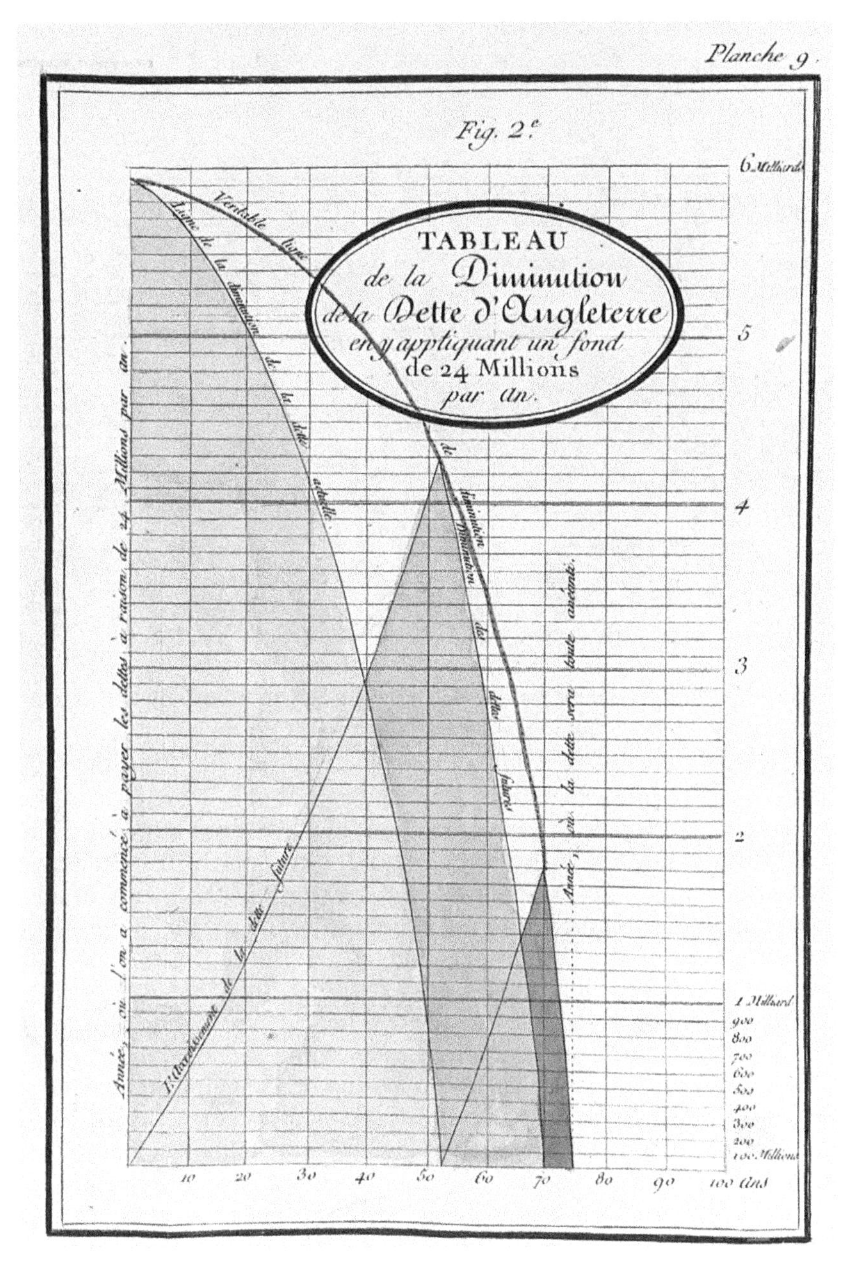

Figure 7.4. Elimination of the British national debt appearing in *Tableaux d'arithmétique linéaire*. Courtesy of Stephen M. Stigler, University of Chicago.

Figure 7.5. Suppression of the riots in Faubourg Saint-Antoine. Bibliothèque nationale de France, Public Domain via Wikemedia Commons.

was a political writer aligned with the Whigs. Without a specific citation, Playfair referred to the first letter of Junius from 1770. In it Junius had criticized the government of 1769, saying that the merits of any government should be tied to the condition of the people. Junius was negative about the people's condition, arguing in part that there had been a rapid decline in trade (Junius 1770, 2–3). Referring to figure 6.1, Playfair said that Junius's conclusion was faulty; the data and the graph showed British trade on the increase at that time.

The political changes unleashed in France in 1788 gained momentum in the next year. There was significant social unrest. On 27 April 1789, riots broke out in Paris among workers of the Réveillon wallpaper factory in Faubourg Saint-Antoine – the district of Paris where Playfair lived. Twenty-five workers were killed in skirmishes with police. An engraving of the event is shown in figure 7.5. The event would have been very disturbing to Playfair, who comes across in his writings as a law-and-order man who supported a form of constitutional monarchy.

The political process towards some form of reform continued. In January 1789, country-wide elections began for the deputies in the Estates General, which formally opened in Versailles on 5 May. Initially, Louis treated the Third Estate deputies as second-class citizens Significant politicking ensued. Necker continued to try to reform the tax system, but on 6 June the deputies of the nobility rejected a compromise proposal. During June and early July, the Third Estate rebelled against the whole legislative process and declared themselves the Assemblée nationale (National Assembly), which Louis did not recognize. By the end of June, Louis had reversed course and recognized the new National Assembly as the legitimate legislative body. On 6 July, the new National Assembly set about writing a new constitution. Unhappy with the situation, five days later, Louis abruptly dismissed Necker. As a result, there was further rioting in Paris and some of the military sided with the crowd. For Louis, the situation was out of control.

Having experienced the riots in April, the citizens of Faubourg Saint-Antoine met on 13 July. At the meeting they expressed their concerns about growing lawlessness in the city. There were armed and disorderly groups in the streets. The police either ignored them or joined them. Hôtel de la Force prison had been broken into and the prisoners left unattended. When the meeting concluded, the citizens decided to form their own militia (LeCocq 1881, 275–6). Playfair joined them (LeCocq 270). The next day, a large crowd stormed another prison, the Bastille, which held only seven prisoners but also a cache of gunpowder. After several hours of resistance, the governor surrendered and was killed on leaving the prison. Several members of the new Faubourg Saint-Antoine militia joined the crowd. Some have interpreted this as Playfair being present at the storming of the Bastille. In view of Playfair's belief in law and order, it is unlikely that he took part.

Chapter Eight

Playfair Tries to Take Advantage of the French Revolution

Following the storming of the Bastille, Playfair quickly embraced the French Revolution (Playfair 1792b, 2). On the basis of his experiences after his arrival in France, by 1789 he had come to believe that reform of the French government was necessary (5–9). He had had several reasons for his belief. Under the ancien régime, there was no rule of habeas corpus. Critics of the state could be imprisoned at the will of the Crown; Playfair had seen examples. The tax system was corrupt and unjust. The ruling aristocracy had "degenerated into mere luxury, dissipation and intrigue" (6). Consequently, "the most luxurious court in the world consisted of the most uninformed and the least deserving of courtiers" (6). The rich oppressed the poor, and there was no justice through the justice system. With the revolution, the road to necessary reform was now open. Playfair wanted to see a constitutional monarchy established in France based on the British model.

But by 4 August, three weeks after the fall of the Bastille, Playfair's enthusiasm for the revolution had waned, soon to disappear, although he did admit to enjoy seeing the demolition of the Bastille in the days after 14 July. The catalyst was the abolition of seigneurial feudalism during a lengthy meeting of the National Assembly on 4 August (Mavidal et al. 1893, tome 8, 350). Prior to leaving Britain, Playfair had advocated for more political power for manufacturers and merchants, sharing power in the House of Commons with the landed interests. But the 4 August decrees implied that all citizens could have a share in political power. In his later writings, Playfair was not in favour of universal male suffrage or what he called "levelling." The decrees also led to the eventual abolition of the aristocracy, a group that in Britain Playfair thought was essential to good governance.

As shown in his later writings, Playfair took a keen interest in French politics. Later in life, he recalled meeting with Étienne Clavière in 1790

(Playfair to *Morning Post*, 4 November 1819). He had met Clavière before, in 1788 at the home of the Baron de Batz when Playfair was trying to obtain letters patent for the Société Gerentet et Playfair. A member of the Jacobin club, Clavière had been active in the revolution, writing articles in newspapers, as well as speeches for Jacques Pierre Brissot, another revolutionary leader. In his meeting with Clavière, Playfair recalled,

> One day in the summer of 1790, conversing with him on the affairs of the time, I observed that "it was strange that men, professing to seek liberty, should be enemies to the Christian religion, since to that religion we owed the abolition of slavery, and that sort of equality that has hitherto prevailed amongst mankind." To this Claviere returned for answer: "That is true, but they care nothing about the Christian religion nor any other. The business is to make the people revolt, and the first step to revolution is to destroy every principle of religion."

Although by 1790 Playfair opposed the revolution, he did maintain contact with several revolutionary leaders. By 1792, he knew "directly or indirectly most of those persons who have figured in the present Revolution."[1]

Playfair attended at least one meeting of the National Assembly, sitting in the Suppléants' Gallery, which was reserved for alternate deputies. Probably he was able to take his seat there, rather than in the public gallery, courtesy of a deputy of the nobility in the National Assembly, possibly Louis Marthe de Gouy D'Arsy or Jean-Jacques Duval d'Eprémesnil. Both worked with Playfair in his dealings with the Scioto Company. The meeting on which Playfair reported was raucous. Those sitting in the public galleries hurled insults at the deputies of the National Assembly when their decisions seemed to favour the king and his court. The issue at this particular meeting was the Tuileries Garden, which had been shut to the public. On 6 October 1789, after a group of women from Paris marched on Versailles protesting the lack of bread and demanding that Louis XVI accompany them back to Paris, Louis acquiesced. In Paris, the royal family stayed in the Tuileries Palace. Part of the Tuileries Garden was shut to the public and given over to the queen, Marie Antoinette, and the dauphin for their private use. The meeting of the Assembly that Playfair attended was probably on 15 August 1790. In April the National Assembly had asked the king to declare which of his chateaux he needed for his personal use. The request was repeated on 15 August 1790, in the meeting that Playfair probably attended. The issue of royal residences seems to be one that could enrage the public galleries. Three days later, the queen, in well-crafted political language, agreed to open the whole of the Tuileries Garden to the public (Mavidal et al. 1893, tome 8, 139).

The storming of the Bastille brought Playfair back into contact with Joel Barlow. Realizing that many now wanted to leave France because of the current unrest and uncertainty, he suggested that emigration was the key to solving Barlow's land speculation problem in Ohio. The Scioto Company in America had pre-emption rights on three million acres of land. Rather than trying to sell this land in large blocks for the American Scioto Company, as was Barlow's original mandate from his handlers in America, Playfair argued that selling the land in smaller lots to French emigrants would more likely succeed (Byerley 1817). Barlow agreed. He brought Playfair on board, giving him one-thirtieth of a share in the contract he had made with his handlers in return for any services Playfair might give.[2] Playfair made good with his extensive connections in Paris. On 3 August 1789, Playfair brought together several French investors to form a partnership with Barlow called Compagnie du Scioto. The new company would purchase the lands in America from the American Scioto Company and sell them in smaller parcels to French emigrants. This set up a whole chain of connections, many of them weak, between the true owners of the land (the U.S. Congress) and the immigrants who wanted to settle the land: U.S. Congress (owners), Scioto Company in America (with only pre-emption rights to the land), Joel Barlow (fairly ineffective agent for the Scioto Company), Compagnie du Scioto in Paris (run by Playfair, who had problems running his businesses in England), and French settlers wanting to purchase the land in America (Moreau-Zanelli 2000, 96). An additional weak link in the chain was the physical distance between Barlow in Paris and William Duer, who ran the Scioto Company from New York. In his defence of Warren Hastings in 1786, the impeached governor-general of Bengal, Playfair had recognized the problem of distant communication for on-the-spot decision making and monitoring of business strategy ([Playfair] 1786, 54). Barlow's apparent lack of interest in the business compounded the problem (Moreau-Zanelli 2000, 87–8). In addition, Playfair was charged with keeping a register of the funds flowing through the business. From what we have seen of Playfair's dealings with James Keir in London and from what we shall see with the Original Security Bank that Playfair set up in London, keeping on top of the accounts was not Playfair's strong suit. All this was a recipe for the eventual disaster.

Whatever the problems were, Playfair assembled an impressive and varied group of investors ranging from middle-class merchants to aristocrats. Some had experience as speculators on the stock exchange, the Paris Bourse. On the basis of his experience with the Société Gerentet et Playfair, he also recruited investors with political connections, both in the royal court and in the emerging National Assembly. His initial contact likely was Claude Odille Joseph Baroud, a prominent speculator on the Paris Bourse. Playfair likely came into contact with Baroud through Étienne Clavière, whom he met earlier over the establishment of Société Gerentet et Playfair. Baroud and Clavière knew each other through financial dealings in the mid-1780s (Desan, Hunt, and Nelson 2013, 193). Baroud probably brought in at least one merchant, Antoine de Saint-Didier. The two were doing a business deal together with a third person,[3] François Troussier Guibert, who was running arms and supplies into Belgium to support the rebel army against Austrian rule during the Brabantine Revolution (Guibert 1790; *Le martyrologe Belgique*, 1791, 107–8). On the political side, Gouy D'Arsy was a deputy in the Estates General, and Jean François Noël Mahéas held the position of Contrôleur du bureau des traitements des domaines du Roi (comptroller of the pay office of the king's estates). The final partner, Guillaume Louis Joseph Chevalier de Caquelon was a lower order aristocrat. Playfair and Barlow completed the roster of partners.

Playfair soon changed his residence. By early November, his new address was N° 162 Rue Neuve des Petits Champs (now only Rue des Petits Champs), close to the Louvre (bill of sale reprinted in Belote 1907b, 55). He had been living in the Hôtel de Lamoignon, an upscale townhouse about 2.5 kilometres away. The impetus for the move may have come from Baroud, who was living at that time in Rue Neuve des Petits Champs.[4] N° 162 Rue Neuve des Petits Champs became the business office for the Compagnie du Scioto and remained Playfair's residence until he left France.

According to the first two articles of agreement in the formation of the Compagnie du Scioto, there were two ways to raise money for the company: capital and income.[5] To raise capital, 8,000 shares were to be sold in the company at 1,000 livres per share (Article 2). Using the capital raised, the partners would then buy the 3 million acres of land in Ohio from the Scioto Company and send out immigrants to settle the land (Article 1). Income would be produced by the immigrants, who would purchase the land at 6 livres per acre. The purchase from the Scioto Company would occur when 2 million livres were raised from the shares (Article 4). Most or all of the money must have been raised, for the sale agreement for the Compagnie du Scioto to obtain the land from the Scioto Company was completed on 3 November 1789. The 3 million acres were sold at 6 livres per acre. At that time, a new name appears on the Compagnie du Scioto roster – Jean Antoine Chais de Soisson, an avocat in the Parlement de Paris, who was given the power to sell land for the company (Belote (1907b, 57). He and Playfair became the main sales agents in Paris for the Compagnie du Scioto. Playfair later described Chais as his clerk.[6] The sales agreement sets out a schedule of payments for the 18 million livres: four payments of 1.5 million livres each due on 31 December 1789, 30 April 1790, 31 December 1790, and 30 April 1791, followed by four payments of 3 million livres each due on 30 September 1791, and on 30 April in each of 1792, 1793, and 1794. The Compagnie du Scioto missed the deadline for their first payment. According to Joel Barlow, the first payment was made early the next year in 1790 (Belote 1907b, 63 and 73), but it is uncertain exactly when the payment was actually made. In a letter on 25 January 1790, Barlow wrote William Duer that the first payment had not yet been made, but in a separate letter to Benjamin Walker on 21 December 1790, he said that the first payment was made in January.

One small number in the whole transaction stands out: the French immigrants were to pay six livres per acre to the Compagnie du Scioto, and the Compagnie du Scioto was to pay six livres per acre to the Scioto Company for the same acres. Where is the profit in this transaction? Later I will provide the answer, which shows the high level of risk that Playfair and his associates were taking.

Two other issues go hand in hand with the question of profit. The first comes from Playfair's friend Thomas Byerley, whose account of the Scioto venture mentions additional compensation for Playfair (Byerley 1817, 378). Playfair was to get a 10 per cent commission on the land sales that he made. Out of this commission, he was to pay the Compagnie du Scioto's expenses in France. Since the immigrants were to pay for their land with a 50 per cent down payment in cash and the remainder on credit, the 10 per cent was to be taken from the cash payment. In 1791, Playfair gave a slightly different account of the commission – it was 15 per cent of sales with no other conditions.[7] The second issue pertains to the abilities of Playfair and Barlow as businessmen. Although no business records for any of Playfair's ventures survive, I have the distinct impression that Playfair was not a very good businessman – long on ideas, deal-making, and promotion, but short on delivery and the minutiae of actually running a business. Nevertheless, despite his shortcomings, Playfair appears to be a better businessman than Barlow. As an example, Playfair negotiated a sweet deal with Barlow: one-thirtieth of the share in Barlow's deal with the Scioto Company and 10 per cent of sales made by the Compagnie du Scioto. Again, where is the profit for the Compagnie du Scioto, if Playfair is taking 10 per cent of the sales revenue, or essentially 20 per cent of the cash flow? All of the sales revenue appears to be earmarked to buy the land from the Scioto Company.

The new company soon issued a prospectus and a brochure, both written in French, the former mostly likely from Playfair and the latter translated by Playfair from a twenty-four-page pamphlet, published in English in 1787, describing the lands in Ohio ([Cutler] 1787). Although Playfair's French had become passable in two years, it still had a long way to go. Jocelyne Moreau-Zanelli, who has thoroughly researched the Compagnie du Scioto and the immigration scheme to Ohio, describes the French translation in the brochure as "assez malhabile et illisible" (quite clumsy and unreadable) (Moreau-Zanelli 2000, 106).

The 1787 pamphlet was written by Manasseh Cutler, a partner in the Ohio Company. It was a piece of hyperbolic advertising: "No part of the Federal territory unites so many advantages, in point of health, fertility, variety of production, and foreign intercourse" ([Cutler] 1787, 9). Cutler went on to write, "This country may, from a proper knowledge be affirmed to be the most healthy, the most pleasant, the most commodious and the most fertile spot on earth known to the European people" ([Cutler] 1787, 9). Knowing little or nothing about this part of America, Playfair followed suit in his translation. Cutler's pamphlet skirted around some important information. For example, the title of the pamphlet includes the phrase "confirmed to the United States by sundry tribes of Indians in the Treaties of 1784 and 1786 and now ready for settlement." The Iroquois had renounced the 1784 treaty and the Shawnee renounced the 1786 one. The situation was not completely rosy with the Indigenous people.

The prospectus for the Compagnie du Scioto came with a map. The map in figure 8.1 is a later version from 1791.

The prospectus was directed at two kinds of people: settlers and investors. For the settlers, land was offered at 6 livres per acre, again 50 per cent down and the rest on credit. The cost to transport the settlers to their new land was set out and itemized. Anticipating 4,000 settlers, Playfair estimated that it would cost 200 livres per settler. The press questioned this amount (*Révolutions de Paris, dédiée à la Nation*, 20–7 February 1790). Later, Playfair substantially revised his estimate. The 200 livres covered only transportation to America. An additional cost of 930 livres would include transportation within the United States, purchase of livestock, living in the United States for six months before the land would produce income, purchase of guns and ammunition for hunting, and a miscellaneous amount in case of emergency (Compagnie du Scioto 1790, 11). Like Cutler, Playfair left out some important bits of

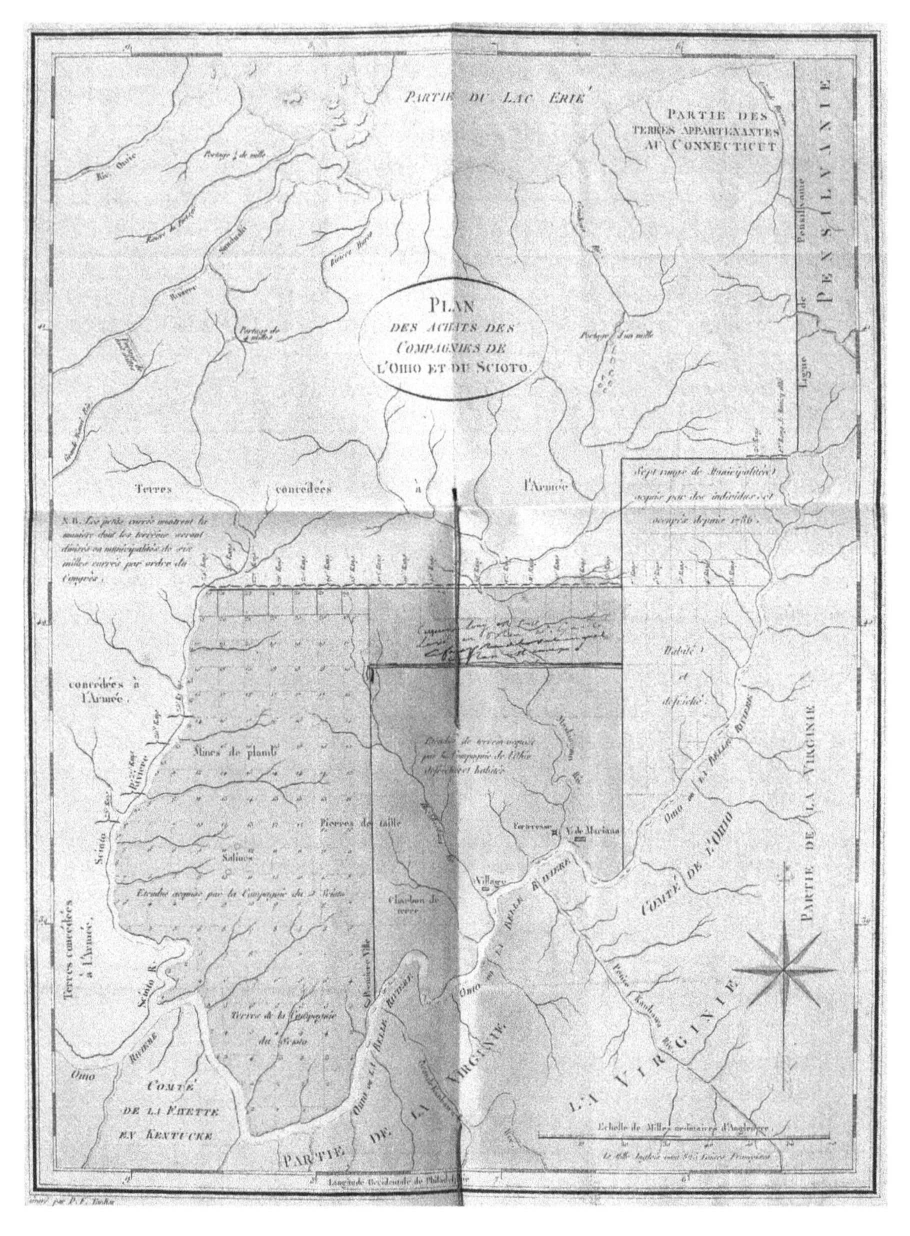

Figure 8.1. The Scioto Company pre-emption is the shape coloured blue to the left of the vertical line and the Ohio Company land is the shape coloured red immediately to the right of the Scioto land. Courtesy of Archives nationales de France, Paris.

information. Neither the Compagnie du Scioto nor the Scioto Company actually owned the land they were selling, only pre-emption rights on the part of the Scioto Company.

For the investors, Playfair laid out in reasonable detail the nature of their investment. It was an enticing speculation with a potentially large return on investment. The Compagnie du Scioto could pay the Scioto Company in American bonds held in France at 90 per cent of their face value. But, according to Playfair, these bonds could be obtained at 70 per cent of their face value (Compagnie du Scioto 1789, 9–10). In a letter to William Duer, Barlow wrote, "It is probable that the greater part will be paid in American French debt – or those bonds given for money borrowed of the government of France" (Belote 1907b, 58). This is where Mahéas would undoubtedly become useful with his court connections to those in the royal treasury. Since the French treasury was in a cash crunch and with Mahéas's connections, the Compagnie du Scioto would have easy access to buying these bonds. This was the same method, on a smaller scale, that Barlow and Duer were using in trying to sell large tracts of the Scioto purchase (which had not been officially purchased). And there was another twist to the speculation. With a very successful initial settlement, the rest of the land, to be settled later, could be sold at a price higher than 6 livres per acre, the purchase price the Compagnie du Scioto had paid (Compagnie du Scioto 1790, 14).

The problem with Playfair's plan was in his anticipation of the level of risk the Compagnie du Scioto faced. There was the obvious problem that neither the Compagnie du Scioto nor the Scioto Company owned the land outright. There was either lack of communication or incompetence, or both, on both sides of the Atlantic. Even if everything had gone well with land ownership, there were also high upfront costs with an uncertain return, since the return was based on a high level of immigration and an appreciation of land values.

There was an immediate public reaction to the scheme run by the Compagnie du Scioto, which picked up steam in the early months of 1790. On the positive side, sales began on the day the Compagnie du Scioto closed the deal on the land sale with the Scioto Company. On 3 November, five 100-acre purchases were made by individuals whose occupations ranged from coach driver to stone mason to bourgeois.[8] Thereafter, sales were brisk, and the first boat departed for America in January with about sixty settlers (Belote 1907b, 62). Of course, they did not have land to go to. Barlow wrote frantically to Duer on 25 January 1790, "Dont for God's sake fail to raise money enough to put the people in possession – make any sacrifice rather than fail in this essential object. If it fails we are ruined" (63). On the negative side, the French revolutionary press soon lined up to oppose the Scioto scheme and were often acerbic in their attacks. Such negative press could not have been good for sales.

From the contracts that survive, it appears that Playfair and Chais handled most of the sales. Barlow's signature appears on some early contracts, but soon disappears. Sales were brisk enough, and the contracts repetitive enough, that forms were printed for easy processing in late December.

The Compagnie du Scioto soon faced a problem. Playfair's group of investors was quickly falling apart. Mahéas fled by the end of 1789 to join the royalist army that was gathering in Coblenz and died while with the army (Viton de Saint-Allais 1872, tome 6, 18–19). Within the same time frame, Troussier Guibert, the arms dealer, was spending time in a Belgian jail because of his dealings. He walked off with fifty shares in the Compagnie du Scioto (Belote 1907b, 73).

Other problems emerged with negative press reports. Soon after the bill of sale was completed, the attacks began with a letter to the editor of *Révolutions de Paris, dédiée á la Nation* dated 16 November 1789. *Révolutions de Paris* was the most popular revolutionary paper of the day. One of its hallmarks was to focus on a topic of public concern, follow it closely for one or two issues, and then move on (Censer (1994, 30). Such was the case with the Compagnie du Scioto and its mission. The initial writer, a lawyer named Médard Louis Couturier Duhalton, warned of the Scioto scheme. He began by pointing out that the current problem in France was the lack of cold hard cash. He blamed the balance of trade in favour of the British, and aristocrats fleeing France and taking their money with them. Buying land in the United States and settling there would remove more specie from France. He went on to describe the plains of the United States as a generally poor and unhealthy place to live. The soil was unforgiving and coins so hard to come by that the people had to rely on paper money, which was an "extreme resource." He finished by saying that two of his friends had been tempted to fall into the trap of immigrating and he convinced them otherwise. He wanted these immigration schemes exposed so that the general public would not fall into the trap that had tempted his friends.

Couturier Duhalton was perhaps a little too quick in expressing his disdain for paper money. It would soon be a way of life in France. By November 1789, through decrees of the National Assembly, all Crown lands and lands owned by the Catholic Church became the property of the state. To pay its debts, which were enormous and mounting, the government decided to sell the land it now owned. In December 1789, it decided to issue assignats, government securities that paid 5 per cent interest (Levasseur 1894; Brezis and Crouzet n.d.). Creditors were paid in assignats. It was hoped that the bearers of the assignats would use them to purchase the land that was for sale. This did not happen to any great extent. In April 1790, the National Assembly decided to lower the interest rate to 3 per cent, to issue assignats in smaller denominations, and to make the assignat legal tender. By September 1790 the interest rate was abolished, more assignats were printed, and they were issued in even smaller denominations.

In late February 1790, three months after Couturier Duhalton's letter, writers for *Révolutions de Paris* followed up with investigative journalism (*Révolutions de Paris, dédiée à la Nation*, 20–7 February 1790). They came to the same conclusion as Couturier Duhalton: the Scioto scheme would drain specie from France and the scheme would drain people from France. A brief description of the Scioto scheme followed the opening paragraph. Then came a list, with commentary, of the disadvantages immigrants would face in America: having to get food from hunting while the sod is broken and crops are planted; having no bakeries to make bread; building your own home; having insufficient resources to bring in clergymen; separation from friends; and having to communicate in a foreign language. The writers believed that the new French government would make land available in France to many at a modest cost after the confiscated clergy property was sold. There was no reason to be duped by the Compagnie du Scioto into immigrating. A final letter appeared in the next issue expressing the hope that the National Assembly would issue a decree that would discourage immigration to America (*Révolutions de Paris, dédiée à la Nation*, 27 February–6 March 1790).

Le Spectateur national was a few weeks ahead of *Révolutions de Paris* in condemning the Scioto scheme. Over four issues of the paper in late January 1790, the writers pointed out problems that the immigrants would face. Not only were they going to farm virgin soil, the Indigenous people were disputing ownership of it. Further, the Compagnie du Scioto was overstating the fertility of the soil. Should the immigrant be successful in farming the land, the settlement would be far from the sea so that getting the produce to a European market would be difficult.

Several newspapers joined the chorus. In early March the *Moniteur*, which in 1799 became the official journal of the French government, appealed to fear, as well as to patriotism and the benefits of the revolution. It told its readers that it was no longer necessary to leave France to find happiness and to enjoy liberty, security, and property (*Gazette nationale ou le Moniteur universel*, 6 March 1790). Readers were also reminded that thousands of French settlers earlier had died in Mississippi, Louisiana, Acadia, and along the banks of the Saint Lawrence. In addition to the newspapers, several negative pamphlets were written about Scioto and even a play appeared. For several months, the Scioto scheme was very much in the public eye.

The issue reached the diplomatic level with letters from Louis-Guillaume Otto, a French diplomat in the United States, to Armand Marc, Comte de Montmorin Saint-Hérem, the French foreign minister in Paris (Wolfe and Wolfe 1994). The letters show the concern about the emigration from France that appears in the French newspapers and the conviction that the new colony along the Scioto River would fail. Otto wrote, "Among already settled colonists there are mainly all sorts of workers, servants, wigmakers, confectioners, musicians, all ill-suited to the sort of life they must embrace" (translation by Wolfe and Wolfe 1994, 51). Otto thought only two types of immigrants would do well in America: farmers and tradesmen. They were also the two groups that would benefit most from the revolution back in France. Otto's letters also show a division of opinion among the immigrants. Many were unhappy and wished to return to France. Others were able to obtain land and remained in America, living with their decision to emigrate.

Finally, the issue reached the floor of the National Assembly (Mavidal et al. 1893, tome 17, 505–6). On 2 August 1790, the chairman of the National Assembly read a letter from Elénor-François-Elie, Comte de Moustier, French ambassador to the United States: the efforts of the Compagnie du Scioto were now "frighteningly successful." Moustier wanted the National Assembly to find the most suitable means of putting an end to this emigration evil. The only response was from Alexandre-Théodore-Victor, Comte de Lameth, a moderate among the deputies. He expressed the belief that the National Assembly could not prevent emigrants from leaving France. He summed up the emigration problem as "good riddance to bad rubbish." It was not until 9 February 1792 that the National Assembly decided to confiscate the property of émigrés (304–5).

These articles and returning immigrants who had found no land for them at the end of their voyage put a distinct damper on sales. By going through Paris notaries' records, Jocelyn Moreau-Zanelli has given a virtually complete list of the land sales of the Compagnie du Scioto (Moreau-Zanelli 2000, 427–36).

The company needed to sell 250,000 acres to make their first payment without dipping into capital and additional tracts of 250,000 acres for each of their next three payments to the Scioto Company for the land. In total, according to Moreau-Zanelli's list, the company sold just under

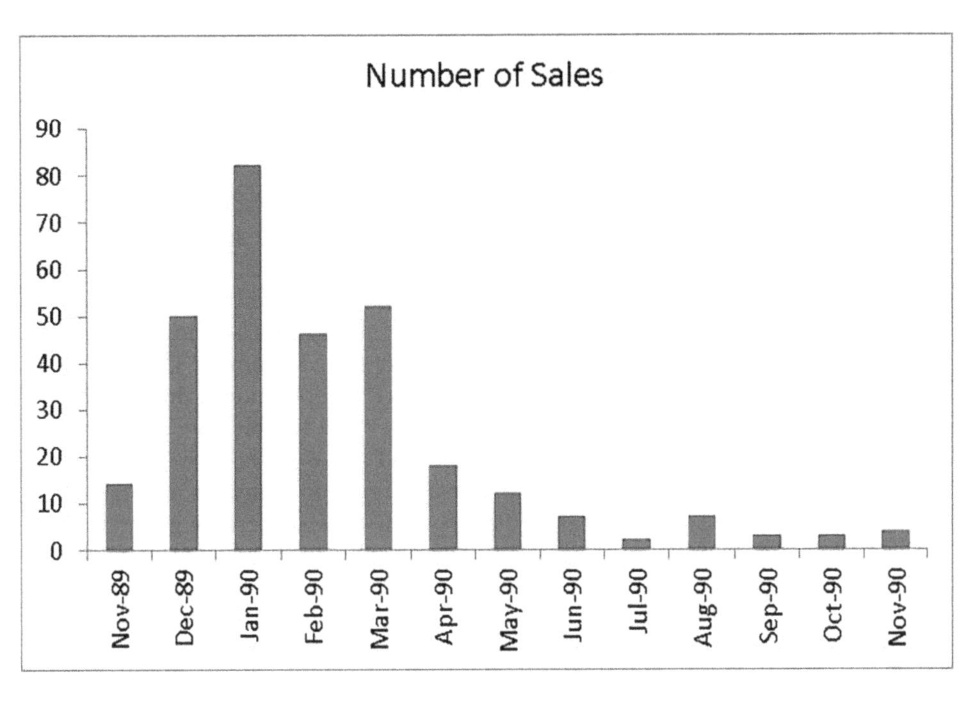

Figure 8.2. Monthly sales of the Compagnie du Scioto. Drawn by the author.

157,750 acres, less than the two-thirds of their target for the first payment, and nothing for future payments. In terms of the money raised, rather than the 1.5 million livres they needed, they raised only 954,682 livres, about 52 per cent of it in direct payments and the remainder on credit. If Byerley is the credible source that he appears to be, Playfair took a little in excess of 49,000 livres for his expenses. Achieving their target would require that the company send a large number of immigrants (which they had not done) or have a few people buy up large tracts of land. Only three purchasers accounted for slightly more than one-third of the land purchased. The remainder was in smaller plots requiring numerous purchasers to fill the land; 70 per cent of the plots purchased were 250 acres in size or less. The notary who handled most of the sales was Charles-Louis Farmain. He listed his daily work in "Répertoire chronologique pour la période du 25 avril 1785 au 29 novembre 1790."[9] Monthly sales of the Compagnie du Scioto from November 1789 through November 1790 appear in figure 8.2. The graph shows slumping sales after January 1790, with a distinct fall-off by April 1790, due to negative press and dissatisfied returning immigrants.

The speculation in buying American paper at 70 per cent face value and redeeming it at 90 per cent face value seems to have gone nowhere. Perhaps speculative investment was drying up because of the revolution or perhaps it was the loss of Mahéas and his contacts with the treasury. Whatever happened, Playfair seems to have speculated on improving the land in Ohio, thereby raising the value for future immigrants to buy at a higher price. He and Barlow contracted with Charles Félix Bué de Boulogne to go with the initial immigrants, staying on for a year if necessary (Joel Barlow to M. Boulogne, 31 December 1789 and 1 January 1790, in Belote 1907b, 63–5). He was a good choice. Boulogne had been in the United States since the American Revolution, serving with Lafayette and eventually became a land agent for French émigrés (Tassin 2007, 74–5). This was an expensive route for Playfair and Barlow to follow. Barlow increased the appeal for the settlers by granting each a house lot within a planned city (probably unplanned at the time and on property that the Scioto Company did not own). The lots would be situated around the city common on which the settlers would grow their first crop in common. In addition, Playfair hired labourers to help the settlers clear the land (Barlow to Benjamin, 21 December 1790, in Belote 1907b, 76). By November 1790, Playfair owed 45,000 livres in bills, 30,000 payable in sixty days and the rest due in ninety days (Livingston 1790).

Of course, when the settlers arrived in the United States there was confusion and lack of organization, coupled with the fact that the Scioto Company in America had no land to offer the settlers since they did not own any land. There was dissatisfaction for many and some returned to France. Those returning demanded refunds on what they had paid, and Playfair had to oblige, an obvious drain on finances. Duer and his associates scrambled in America to obtain land from the Ohio Company on which to settle the immigrants (see, for example, Jones 1975, 407).

Soon after the establishment of the Compagnie du Scioto and while the situation was still appealing, a number of aristocrats, along with Barlow and Playfair, met in February 1790 to form the Société des Vingt-Quatre. Their object was to purchase 24,000 contiguous acres of land from the Compagnie du Scioto (Moreau-Zanelli 2000, 184–6). To this land, they would bring settlers who would form a new community. They would also bring Catholic priests to serve the new community. It would be a reconstruction of their estates in France with the exception that they would be subject to the laws of the United States, not France. There were also plans to build a city within their newly acquired tract of land. Among the aristocrats in this group were Claude-François de Lezay-Marnésia, Jean-Joseph de Barth, and Jean-Jacques Duval d'Eprémesnil. They had all been deputies in the Estates General. And they were all royalists in the sense that they wanted a constitutional monarchy in France along the lines of Britain. Lezay-Marnésia left for the United States in May 1790, a few months after the departure of the initial group of immigrants. At about the same time, Barth embarked for America to oversee the new settlement. D'Eprémesnil remained in France and became one of Playfair's good friends (Playfair n.d., Memoir 5).

By July 1790 land sales had been steadily declining. In May, June, and July, the sales numbers were twelve, seven, and two respectively. At this point, Barth's son stepped in with an associate, Marc Antoine Coquet de Trazayle. After lengthy negotiations, it was agreed to wind up the current Compagnie du Scioto and begin another on 22 July 1790. In at least one source, the new company is called the Maison du Scioto.[10] Having large bills coming due very soon and with no money to pay them, Barlow was in a very weak bargaining position. Consequently, the new agreement was highly advantageous to the new partners, of whom Playfair was one and Barlow was not. Under the new agreement, the Scioto Company would receive only fifteen sols per acre (three-quarters of a livre, rather than six). From each sale, 10 per cent would be taken for local expenses, and the remainder would be held in escrow in order to pay the American Congress directly for title to the land (Moreau-Zanelli 2000, 239–40). All promises to immigrants who had already signed on with the former Compagnie du Scioto would be upheld. Farmain's notebook shows that after 22 July 1790 the Compagnie du Scioto made only two more sales. Beginning on 22 August 1790, the sales were made in the name of William Playfair. The entry for 22 August has Compagnie du Scioto crossed out and William Playfair inserted, with only Playfair's name appearing in the remaining entries.

Despite the setback in sales, shown in figure 8.2, but with a new Compagnie du Scioto in place, plans for the Société des Vingt-Quatre continued through the summer of 1790. Playfair's brother James, living in London, worked on plans for an "American city" over the three days of 15–17 August.[11] James Playfair's efforts are probably connected to a map of the planned town of Gallipolis with 619 numbered and contiguous plots of land, shown in figure 8.3, which survives in d'Eprémesnil's papers.[12]

During the spring and summer some immigrants returned home, understandably very disgruntled. The situation was so serious that Barlow received threats on his life. Late in the fall, a friend told Barlow to

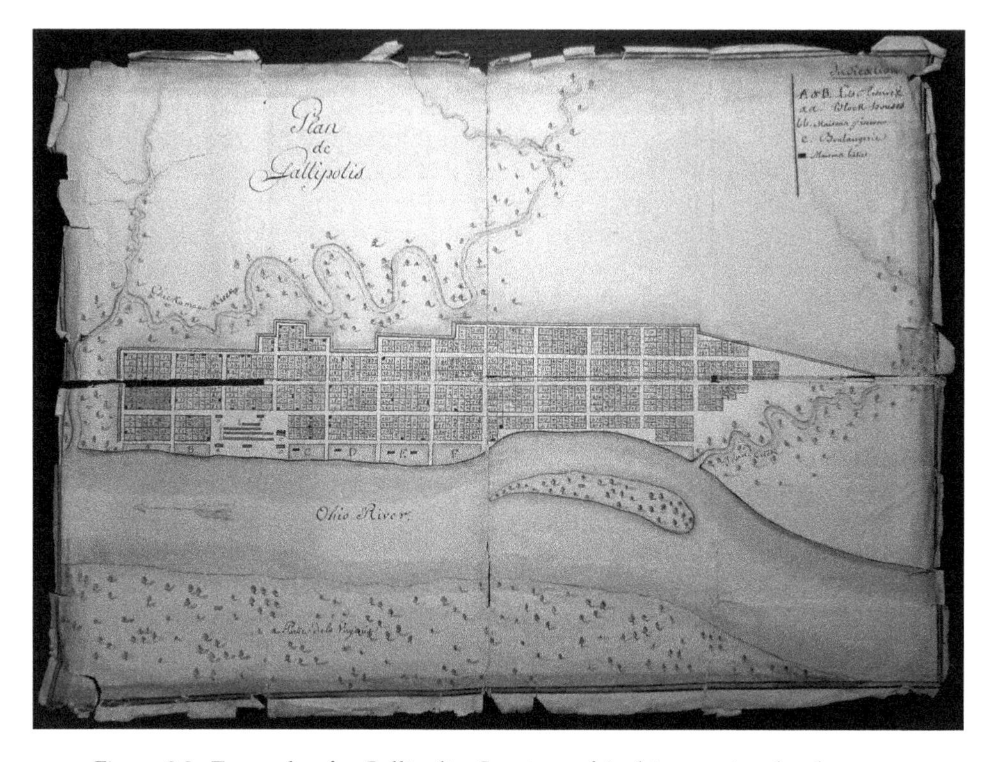

Figure 8.3. Town plan for Gallipolis. Courtesy of Archives nationales de France, Pierrefitte-sur-Seine.

leave Paris (Belote 1907b, 77). It was Playfair who seems to have been the one to deal with the returnees and it was not pretty. On 23 January 1790, a M. Caplazy purchased 200 acres from the Compagnie du Scioto and made his 50 per cent down payment of 600 livres. It is unclear if he sailed for America and came back, or if, seeing the negative publicity in France, he decided not to go – more likely the former. Whatever the case, he asked for a refund.[13] Playfair obliged by giving him 315 livres with the rest payable on 17 May 1791. When the balance did not materialize, Caplazy and his wife went to see Playfair on 12 July. The encounter became violent. Caplazy claimed that Playfair struck him on the arm with a stick and that Playfair's servant struck Caplazy's wife. Playfair told the police that he intended to settle the debt just as he was settling other outstanding debts. The encounter is described briefly in Bouchary (1940, 147) and Moreau-Zanelli (2000, 397).

Duer must have been shocked when he heard of the new agreement between Barlow on the one hand and Barth and Coquet on the other. Or, if he had not been given the details of the new agreement, he must have been wondering what was happening with the sales in France. He had paid money in America to settle some of the French immigrants and had not received much, if anything, in return. In September 1790, Duer appointed his friend and associate Benjamin Walker in New York to go to France to investigate (Belote 1907b, 66).

In a letter to William Duer dated 20 November 1790, Playfair wrote about how the situation had unfolded in the past few months (Sibley 1901, 43–53). He got Barlow's agreement on the details, as Barlow had written at the end of the letter, "I have read this letter & approve generally of the facts & the reasoning therein contained. J. Barlow." In an appeal to the Americans, Playfair was highly optimistic about the possibility of more land sales, although the experience shown in figure 8.2 proves otherwise. Playfair thought that all he needed were some good testimonial letters from immigrants. With the government abolishing the ancien régime's corrupt patronage positions and sinecures, while providing compensation through assignats, Playfair claimed that there should be many with money wanting to emigrate. To make the immigration viable, Playfair underlined the necessity of having land available for the immigrants when they arrive. At this point, Playfair has every appearance of a desperate gambler facing large losses and wanting to gamble more to recoup them.

Playfair followed up with a letter to Duer dated 27 December 1790 (Belote 1907b, 67–70). Barlow wrote to Walker on 21 December 1790 (71–8). By that time, Walker had arrived in Paris. As things were falling apart, both Playfair and Barlow were trying to excuse themselves from any blame or wrongdoing. After giving at length his version of how events unfolded, Barlow put some of the blame on Duer for his lack of interest in what was going on and for his not keeping Barlow informed about what was happening in Ohio. For his part, Playfair tried to put some of the blame on Barlow for making the deal with Barth and Coquet on 22 July. Playfair stated in his letter that nothing had been sold since that time. Literally, this was true. The Compagnie du Scioto no longer officially existed. What Playfair neglected to say was that since July 22 he had been selling land on behalf of the new company. It is difficult to say whether the Americans made any distinction between the two companies or if they saw either of the two French companies as a "branch plant" of the Scioto Company in America.

Still trying to recoup his losses, Playfair published another prospectus in December 1790 (Compagnie du Scioto 1790). Once again, he glowingly described the land the Compagnie du Scioto was selling, but did not own, to the French émigrés. At length, he tried to refute the many criticisms of the scheme that appeared in the French press. The price remained the same: six livres per acre with a 50 per cent down payment in cash and the remainder on credit to be paid in two years. To calm the fears of potential clients, he advertised that in the new agreement 90 per cent of the cash down payment would be deposited in a Paris bank until the proposed immigrant got title to his land; the remaining 10 per cent would be used for local expenses. After a statement of more realistic costs to immigrate to America and then more advertising hype, Playfair inserted a testimonial letter that he had received from Lezay-Marnésia. Of course, Lezay-Marnésia had his own vested interest in the success of the emigration scheme, so he was very positive.

When Walker made his report, he exonerated Barlow but was heavily critical of Playfair. He looked at Playfair's accounts but found no money. The accounts have a long list of cash flow entries up to 22 July 1790, the day that the Compagnie du Scioto was terminated and replaced by another company that kept the purchasers' money in escrow. The accounts show more than 850,000 livres flowing through the books, all meticulously itemized.[14] The difference between this amount and the approximately 950,000 livres obtained from Moreau-Zanelli's list of sales might be attributed to business carried out after 22 July. Despite Playfair's accounting, Walker concluded that Playfair had somehow pilfered funds (Moreau-Zanelli 2000, 255). Until Jocelyn Moreau-Zanelli made her careful study, many historians have assumed that Playfair was an embezzler. Despite the severe shortcomings of both Playfair and Barlow, she rightly concludes that Playfair was made the scapegoat for the badly managed and badly executed speculation of the partners in the Scioto Company in America (256).

Well into 1791, Playfair continued to try to keep the scheme afloat. On 20 March 1791, Playfair wrote a letter to Thomas Jefferson, at the time the American secretary of state,[15] and ten days later, he wrote a similar letter to Alexander Hamilton, secretary of the treasury.[16] In both letters, Playfair defended what he had done, complained of his treatment by Walker, and hoped that the American government would find a way to allow the settlers he had waiting in Paris to obtain title to land in Ohio. It all came to naught.

Playfair was embittered by the whole episode. More than twenty years after the Scioto debacle, Playfair wrote about the plan that he had devised for French émigrés: "This plan set Mr. Barlow afloat, though his avarice, and that of his employers, stopped the sales when 150,000 acres were sold." Despite all that had happened, and his own shortcomings, it appears that he still believed that the Scioto venture could have been successful (Playfair 1813, 1:138).

While the business of the Compagnie du Scioto sputtered on, Playfair had other things on his mind during 1790. The first was that his second son, Andrew William Playfair, was born that year. The second was the National Assembly decided to issue assignats in December 1789 paying 5 per cent interest.

Playfair responded to the new French paper money with a pamphlet entitled *Qu'est-ce que le papier-monnoie? Lettre d'un Anglois à un François* probably published in January 1790 (Playfair 1790b). In the pamphlet, Playfair described two kinds of paper money: money issued by institutions like the Bank of England where it was backed by gold, and government paper such as the Navy bills in England, which had a fixed redemption time and carried 5 per cent interest. In the latter case the paper was supported by the government's ability to pay its creditors through taxes raised and thus maintain confidence in the value of the paper. Playfair held up the United States as an example of the value of paper money when backed by gold. It was well known that the value of the revolutionary currency in the United States had collapsed. After the establishment of the Bank of North America, which issued paper currency backed by gold (loaned by France), the value of the money had remained stable for the past two years. Without mentioning the word "assignat," Playfair stated that if France issued any other form of paper money it must pay interest.

Probably in September 1790 or later, Playfair wrote a short pamphlet for his friend d'Eprémesnil[17] in which he discussed the workings of the Bank of England and its issue of paper money. The assignats issued by the French government could not be seen, Playfair wrote, as the same kind of money issued by the Bank of England. In England the holder of a Bank of England note has the option of exchanging it for gold. In France, the assignat is forced on the public as legal tender without gold backing its value – only property. It was in September that the National Assembly abolished the interest rate attached to the assignat.

Playfair also responded to the National Assembly's September discussion of, or decree for, removing the interest rate on the assignat. The response was a new pamphlet, *Lettre II. D'un Anglais à un Français sur les assignats* (Playfair 1790a). In it Playfair made his own proposal for issuing paper money in France. He distinguished between what he called an assignat *forcé* and an assignat *non forcé*. Like a Bank of England banknote, the assignat *non forcé* would be payable on demand in specie. The assignat *forcé* would be like the English Navy bills, payable at a fixed date with interest. Playfair suggested that the government issue three-quarters of its assignats as payable on demand and one-quarter as notes due with interest. He went on to suggest how this should be done. The government should sell all the land it owned (*domaines nationaux*) in stages. Beginning in December 1790 the sales would be made to the value of 100 million livres a month rising to 200 million a month in May 1791. Then the assignats should be issued in staged amounts. Should the assignats printed exceed the amount of land sold, they should be printed as assignats *forcés*. At the end of the pamphlet Playfair set out a schedule of what type and how many of the assignats should be in circulation. Despite Playfair's protestations that he had soured on the revolution as of 4 August 1789, his pamphlets on paper money, especially the second one, are an attempt to provide sound advice to the revolutionary government.

In the first few months of 1791, Playfair was winding up the affairs of the Compagnie du Scioto or its successor. It was a slow process; Playfair had to track down all the outstanding debts, many originating in America. It seems that Playfair was also cleaning up the mess caused by returning immigrants and by Barlow's dealings with Barth. On 18 March 1791, Playfair wrote to Walker mentioning outstanding debts to "Vibert, Lalemond & the notaire."[18] Vibert and Lallement (or Lalemond, Playfair sometimes mangled names when spelling what he heard) bought 150 and 100 acres respectively in Scioto. They were both well-trained professionals – Vibert, a writer and writing teacher, and Lallement, a master chandelier maker (Moreau-Zanelli 2000, 432 and 436). Playfair told Walker that he hoped to retire these debts soon.

There are other indications that Playfair was in financial difficulty, one more indication of his inability to keep on top of money.[19] In June 1792, he was in the offices of the commissioner of police responding to a debt of 2,400 livres to an engraver of cut glass, with Playfair's forte-piano (valued at 3,000 livres) held as collateral (the fortepiano was an early form of piano played by Haydn, Mozart and Beethoven). On the same day, he was in the same office with complaints about his sales agent, Chais de Soisson. On 1 May 1791, Playfair gave Chais two notes payable to the recipient for his work as sales agent. One was for 1,500 livres payable on 25 December 1791 and the other was for 1,800 livres payable on 31 January 1792. According to Playfair, the work was not done so Playfair did not want to pay. He went to see Chais at his house to retrieve the notes, but could not find him. Meanwhile, Chais had signed the notes over to a third person, a married woman named Colot. Since, at the time, it was illegal to pay a married woman directly and since Chais had not done the required work, Playfair filed a complaint with the commissioner of police. It appears that Playfair and Chais de Soisson's relationship did not end happily.

With the Compagnie du Scioto defunct, Playfair's third business venture in Paris soon took shape in the form of a bank. Known as the Compagnie de commission operating out of Rue des Bons Enfants near the Louvre, it began operations in September or October 1791 (Bouchary 1941, 137–51). On 18 October, a dissatisfied customer of the Compagnie du Scioto denounced the new bank as part of a wider complaint to the commissioner of police.[20] He questioned the solidity of the bank in view of the fact that Playfair was in charge, but the denunciation had little effect. By December 1791, the bank was being put through some bureaucratic hoops as part of its start-up. On 10 December, the bankers applied to have the government appoint auditors to oversee its operation.

In addition to the typical financial transactions of a bank, the activities of the Compagnie de commission included purchasing luxury items, paying for them half in cash and half in banknotes with a maturity date. Among these purchases were porcelain vases, gold watches and other jewellery, furniture, and several paintings by prominent artists such as Claude-Joseph Vernet (d. 1789) and Jean-Baptiste Le Prince (d. 1781). With the exception of the watches, the whole lot was valued at about 52,140 livres in 1792 or about £340,000 in today's money.[21]

Once in business, the bank was known popularly as the Caisse de billet de parchemin because its banknotes were printed on parchment. Initially, it issued banknotes in denominations of five, ten, fifteen and twenty sols, equal to one-quarter, half, three-quarters, and 1 French livre respectively. With its banknotes, the Compagnie de commission became caught up in a bit of revolutionary politics. Near the end of January 1792, the company began issuing banknotes of intermediate denominations: six, seven, eight, and nine sols. The Société de Sainte-Geneviève, a political club, objected to this move, fearing that this would lead to the hoarding of copper coins. The club denounced this issue of notes in several revolutionary newspapers, but nothing seems to have come of it.

Very soon after its inception, the bank had to deal with new government taxes. By a decree of 25 May 1791, banknotes issued at a value of twenty-five livres and below were exempt from a stamp duty, but that changed in the next year. As of 1 April 1792, the exemption was changed to notes of value ten sols (half a livre) up to sixty sols (three livres) (Mavidal et al. 1893, tome 40, 495–6). This meant that the bank's low-denomination notes were subject to the new duty. The Compagnie de commission had to scramble to meet the demands of the new decree.

They withdrew their low-denomination notes and reimbursed those who held them. As they were planning new issues when the decree came down, they also had to cancel their contract with the note maker. Parisian municipal representatives came to inspect the bank's operation and came away satisfied with the bank's actions, as well as its good management. The municipality recommended that the bank advertise that it withdrew its notes only to comply with the law and that the bank would continue to reimburse its notes (Bouchary 1941, 137–51).

Playfair's residence, and former office for the Compagnie du Scioto, at N° 162 Rue Neuve des Petits Champs is near his bank in Rue des Bons Enfants. His neighbourhood in 1790–2 is part of historic central Paris today. Another resident in Playfair's building, and probably the owner, had been Jean Antoine Comte d'Agoult, a royalist military officer (Courcelles 1826, 71–2). At the calling of the Estates General in 1789, he was a deputy for the nobility. He left France in 1791 to join the royalist troops mustering in Coblenz. The street appears on a 1797 map of Paris, shown in figure 8.4. Playfair's bank, with its offices in Rue des Bons Enfants, was nearby, down a cross street. Across the street from Playfair's residence was the garden of the Palais Royal with the palace itself below the garden. Playfair said he could see the Palais Royal and its adjoining garden from the window of his home. On the map, Palais Royal has been renamed Palais Égalité; it reverted to Palais Royal after the Bourbon restoration. The palace was the residence of the Duc d'Orléans. In the early 1780s the duke decided to commercialize his property, living in only part of it, and the garden was made public. The public parts of it opened in 1784 as a shopping and entertainment complex. In 1786, Antoine Beauvilliers, who had been in royal service, opened an upscale restaurant in the Palais Royal under the name La Grande Taverne de Londres (Fierro 1996, 1137). This was Paris's first fine-dining establishment and at the same time a meeting place for conservative political factions. The fact that Playfair dined there at least once shows his seemingly good financial situation at the time and his political sympathies.

Now disgusted with the course of the revolution, in May 1792 Playfair published a pamphlet highly critical of the French revolutionary government. *A Letter to the People of England on the Revolution in France* was published in London (Playfair 1792b). It covered three topics: (1) why the revolution in France was necessary, while in England it would be "highly dangerous" to change the British constitution; (2) French propaganda about its revolution, carried out "under the specious appearance of Liberty and philosophy," had stirred unrest in other countries and should be counteracted; and (3) the people of France had been

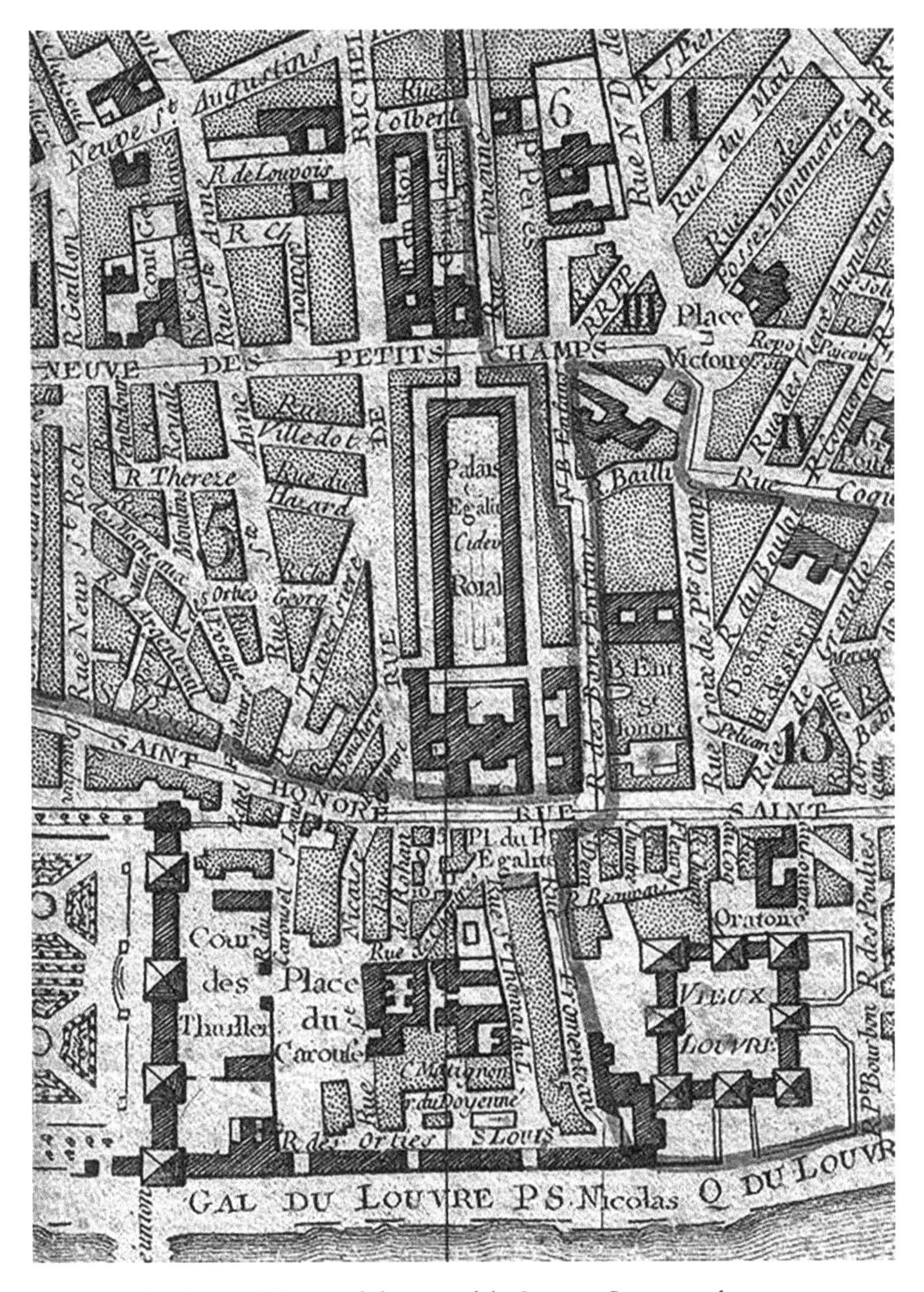

Figure 8.4. Paris 1797 around the area of the Louvre. Courtesy of Geographicus Rare Antique Maps, Brooklyn, New York.

deceived by their revolutionary leaders and in the long run this would lead to "a long series of poverty, discontent and crimes in this unfortunate country" (3–4).

In the pamphlet Playfair used words like "ridiculous" to describe the new form of government in France. He claimed that the new leaders of France were reducing the country to a second-rate nation and they knew it. Consequently, these same leaders conducted propaganda campaigns to spread revolution to other nations in order to reduce them to the same level as France. This would make a new level playing field for France and the rest of Europe. Playfair called these leaders "traitors to their country." He went on to say that the revolutionary leaders "must succeed in exciting Revolutions abroad or abandon their system at home, therefore, corruption, bribery, nothing will be spared to succeed" (Playfair 1792b, 40–5). None of these words endeared Playfair to the hearts of the revolutionary politicians in France. In particular, he offended one of the radical Jacobins, Betrand Barère. This would have some repercussions for Playfair further down the road.

By the summer of 1792, the French Revolution was becoming increasingly violent. During the previous summer, Louis XVI had tried to flee France with his family, but they were captured in Varennes and brought back to Paris, to be held under house arrest in the Tuileries Palace. The escalating violence in 1792 was due, in part, to riots caused by food shortages in France and by the threat of invasion by an Austro-Prussian alliance in aid of Louis XVI. When formal hostilities began in April 1792, the French invaded the Austrian Netherlands, but were eventually repulsed. Joined by French royalist émigré troops, the Prussian army, under the command of Charles William Ferdinand, Duke of Brunswick, invaded northeastern France in August 1792 with the intention of marching on Paris. Brunswick's army was defeated at the Battle of Valmy on 20 September 1792 and retreated beyond the French border. Prior to the invasion, on 25 July 1792 Brunswick issued a manifesto declaring that any harm that came to the French royal family would be met with harm to the French population. When the manifesto circulated in Paris, rather than striking fear into the French, it had opposite effect of turning the Parisian population against Louis XVI.

It is within this atmosphere of riot and mayhem that Playfair came across his friend d'Eprémesnil being beaten in the streets of Paris (Playfair n.d., Memoir 5). Playfair had just come from dining at La Grande Taverne de Londres in the Palais Royal. After dinner, as Playfair walked north towards his residence, he passed the garden of the Palais Royal. Seeing a large crowd on the other side of the garden, he quickly found that the crowd was threatening to kill d'Eprémesnil, who was naked and bloody. Despite the angry mob, Playfair was able to get help from a member of the National Guard. They took d'Eprémesnil towards Playfair's residence, but soon the mob began to follow. They were able to duck into a building that housed some of the National Guardsmen and shut the gate behind them with the mob clamouring outside. The guards wanted to put Playfair back in the street, but were persuaded to hide him. Soon Jérôme Pétion, the mayor of Paris, arrived with a force of about 1,200 cavalry, and eventually the mob was dispersed. Pétion left in his carriage with d'Eprémesnil, leaving Playfair to fend for himself. Away from the scene, Pétion placed d'Eprémesnil in prison for safeguarding. Ever resourceful, Playfair got a messenger boy of the guard to go out and buy several bottles of wine for the guardsmen. After drinking a toast to the king of England, but not to Louis XVI, one of the guardsmen escorted Playfair home.

Within a month of this episode, Playfair left France. His last ten days in Paris were eventful and tense (Playfair n.d., Memoir 6). The prelude starts on 4 August 1792 with the British ambassador in Paris, George Leveson-Gower, Earl Gower, writing to his superior in London, the foreign secretary, William Wyndham Grenville, 1st Baron Grenville. Gower expressed concern about the dangerous situation that was developing in Paris.[22] He was particularly concerned for the safety of the French royal family, especially the queen, Marie Antoinette. He awaited instructions from London, which did not arrive until 12 August.

By 9 August, Playfair had heard that soon there would be a confrontation between the revolutionary government on one side, which mainly comprised Jacobins, and supporters of the monarchy on the other. That night, he called a meeting of the four partners in the bank, the Compagnie de commission, at about 9:30 (Playfair n.d., Memoir 6). Playfair proposed that they divide the bank's assets and each leave with his share. He reasoned that since the call on their banknotes had declined to about 200 livres per day, probably only 5 per cent of them would ever be redeemed. After expressing fear that the bank would be ransacked by the expected mob the next day, Playfair suggested leaving the fixed assets, estimated at 250,000 livres – the vases, the paintings, the furniture, etc. – and take the money. Two partners balked at the proposal, but soon they came on board and the division began. Probably to protect himself and Playfair from any dissent from the two wavering partners, a third partner placed his two pistols on the table prior to the division. The total amount divided was 60,000 livres in assignats, 800 Louis d'or (gold coins, one Louis d'or equalling 24 livres), and 100,000 livres in gold and silver bullion. This was a considerable amount of money, about £7,700 in 1792 or well over a million in today's money.

Clearly, when Playfair left Paris he was fairly flush with cash. When the authorities did an accounting on 17 September 1792 of what was left, there were still 23,680 livres in assignats, 5,800 livres in banknotes of other banks, and nearly 80,000 livres in other financial instruments, as well as the fixed assets (Bouchary 1941, 140).

As Playfair had expected, there was violence the next day, 10 August. Several armed revolutionaries, including members of the National Guard, stormed the Tuileries Palace and were met by the king's Swiss Guards. At the end of the day, about 800 on the king's side, including about 600 Swiss Guards, died, as did more than 370 on the other side. It was a turning point in the French Revolution. Louis and his family survived after seeking refuge in the chamber of the Legislative Assembly. But the monarchy soon came to an end. On 21 September 1792, France officially abolished the monarchy and declared itself a republic.

Playfair experienced the preamble to the major events of 10 August, rising early in the morning and venturing into the streets. As he walked past the Palais Royal into Rue Saint-Honoré he came across "a mob of the lowest rabble preceded by two ragged fellows carrying heads still bleeding and recently cut off." Playfair later recounted that there were seven beheadings that morning ([Playfair] 1792, 23). He immediately returned home, then went to his bank in Rue des Bons Enfants, where he met with his partners, who informed him that they were glad of the advice he had given them the night before. The partners told the two remaining bank clerks to continue business as usual and left enough money to cover expenses and calls on their banknotes for the next ten days (perhaps most of the 23,680 livres in assignats left in the bank later in September when the authorities came). When they heard cannon and musket fire coming from the Tuileries Palace, Playfair returned home, but with some difficulty. He was caught up with some of the crowd fleeing the Tuileries Palace after the Swiss Guard repulsed the revolutionaries' first attack. When he did eventually get home, he stayed until he heard the shouts of victory from the revolutionaries who had won the day.

Playfair no longer felt safe where he lived in Rue Neuve des Petits Champs, as the building also housed an aristocrat, Jean Antoine d'Agoult. From his experience with d'Eprémesnil, any aristocrat who was sympathetic to the monarchy was now a target. D'Agoult was off with royalist troops, so it was probably only his wife and children who were left in their residence. A few days before, Playfair had sent some large trunks to a hotel (he calls it "hotel garni"), and on the evening of 10 August, he left his residence to stay there. Playfair says that "old Mother Antoine," a possible reference to d'Agoult's wife, accompanied him. Once in the hotel, he sent her back to the lodgings in Rue Neuve des Petits Champs.

A couple of days after 10 August, Playfair was out and about in Paris. He had breakfast in a coffeehouse near the Tuileries Palace. He saw little outward damage to the palace – a couple of musket ball holes in the windows and the marks of a couple of cannon shots high on the walls. But despite the lack of apparent damage, he had decided to leave France. In the street, Playfair met Auguste Camille Arnould, a member of the Jacobin Club.[23] Arnould was the illegitimate son of Louis Léon Félicité de Brancas, Duc de Lauraguais, and the opera singer Sophie Arnould. The Jacobin warned Playfair of the danger he was in, because of the help Playfair had given to d'Eprémesnil and because of a quarrel between Playfair and Barère, probably over the publication of *A Letter to the People of England on the Revolution in France*. As a result of the events of August, the Jacobins, Barère among them, now had the upper hand in government. It was really time for Playfair to leave France.

Although he was not a direct witness to the major events of the day, within a month Playfair wrote a detailed account of the events of 10 August along with the background leading up to it ([Playfair] 1792) and published it anonymously in London. The earliest newspaper advertisement for his pamphlet, *Short Account of the Revolt and Massacre Which Took Place in Paris, on the 10th of August 1792*, appeared on 8 September 1792 (*World*, 8 September 1792). It was popular enough that a second edition was issued in 1793 ([Playfair] 1793). Both editions open with "It is of the utmost importance that all Europe should know, not only the horrid crimes which have lately been perpetrated in Paris, but also the crooked and infamous labyrinth by which the principal movers of those crimes have arrived at their ends." Over the next forty pages, there follows a scathing indictment of the National Assembly, the revolutionaries, and the mob. This pamphlet and *A Letter to the People of England* launched Playfair into a lifelong literary career against Jacobinism and the French Revolution.

On 16 August, Playfair went to Gower to obtain a passport for him and two others (one of his partners and another acquaintance) so that they could leave France. His request was denied. Playfair then discussed with Gower whether Brissot, whom Playfair knew, could issue the passport. Brissot now held a powerful position related to foreign affairs in the French government.

On 17 August, Henry Dundas, secretary for war, wrote to Gower from England. The letter was not received until 23 August. However, Gower had already suspected what its contents would be. Gower's suspicions, which were confirmed, were the reasons for denying Playfair his passport. The despatch from London read,

> Under the present circumstances as it appears that the exercise of the executive power has been withdrawn from His Most Christian Majesty [Louis XVI], the credential under which your Excellency has hitherto acted can be no longer available: and His Majesty [George III] judges it proper, on this account, as well as most conformable to the principles of neutrality which His Majesty has hitherto observed, that you should no longer remain in Paris. It is therefore His Majesty's pleasure that you should quit it, and repair to England, as soon as you conveniently can, after procuring the necessary passports.

Gower was unable to leave until 28 August, at which time he was given a passport by the French government.

The same day that Dundas wrote his letter, Playfair went to see Brissot at 8:00 a.m. Brissot was ensconced in the offices of an insurance company run by Étienne Clavière, Brissot's close associate in politics. Playfair flattered Brissot enough that he was able to get the three passports he required. Later that day, Brissot's wife contacted Playfair for a favour. Since he was leaving for England, could he accompany two of her female relatives to Boulogne? The two French ports of embarkation for England were Dieppe and Calais. The road to Calais went through Boulogne-sur-Mer. Playfair agreed and contacted the two women, who were happy to join him on the journey.

On the next day, Saturday, 18 August, Playfair took one last stroll through his old neighbourhood before leaving Paris. By the Palais Royal, he met Jacques-Constantin Périer, an engineer who had visited Boulton and Watt while Playfair was in their employ. During the course of their conversation, Périer told Playfair that his workers were breaking up a magnificent equestrian bronze statue of Louis XV that had stood in Place Louis XV, now Place de la Concorde (McClellan 2000, 4–5). The mob had torn down the statue on 11 August, the day after the major rioting that Playfair witnessed. A few days later, the National Assembly passed a decree allowing the melting down of such statues, and similar items, to eliminate the last vestiges of feudalism in the country and, at the same time, to make cannon. Périer was taking full advantage of the situation. Playfair expressed concern over the destruction of a beautiful piece of art. Périer's reply was that "cannon are more useful than statues of kings."

According to Playfair, there was a major problem with his flight plan to England. Despite Madame Brissot's request and her husband's power, it was illegal, so Playfair said, for women to travel without their spouses in France. If he or his male friends were caught masquerading as the women's spouses, they were liable to be thrown in prison for six months. Consequently, on Sunday, 19 August, Playfair took a post chaise (a fast carriage, fast eighteenth-century style) to Dieppe, which went through Rouen. He left a note for Madame Brissot indicating that he could not carry through on the favour, as he was obliged to alter his travel arrangements to go through Rouen to Dieppe and not through Boulogne to Calais. He left instructions not to deliver the note until forty-eight hours after his departure.

By September, customers of the Compagnie de commission were now abandoned. Like bailiffs in a bankruptcy, the authorities moved in and closed the bank on 12 September. They appointed two commissioners to report on the assets of the bank, which they did on 17 September. There were 163,881 livres and 14 sols left to distribute in some way, much of it in fixed assets.

Playfair left before the Reign of Terror took hold. Of those names mentioned in this chapter, Louis XVI, Marie Antoinette, Duc d'Orléans (who renamed himself Philippe Égalité), Brissot, d'Eprémesnil, Marthe, and Pétion all went to the guillotine. Clavière and Baroud were both arrested with the expectation of the guillotine; Clavière committed suicide the day before he was to go to trial and Baroud somehow escaped execution through the protection of his friends. Montmorin was imprisoned and died during the September Massacres in which 1,100 to 1,600 people in Paris prisons were murdered over five days in early September 1792. Lezay-Marnésia returned to France, was imprisoned, and later released.

Chapter Nine

Playfair Escapes from France and Returns to England

The French Revolution had a profound effect on Playfair's political thinking, and he was quick to write about it. In France, he had seen much violence, rioting, and general civil unrest. Prior to leaving England for France, he had wanted to see parliamentary reform with more political power given to manufacturers and merchants. Now he took a much more reactionary stance, which is summed up in a single sentence in a twenty-seven-page pamphlet that he published through John Stockdale's bookshop: "To preserve the constitution and liberties of England untouched is what I wish" (Playfair 1792a, 4).

The pamphlet, entitled *Inevitable Consequences of a Reform in Parliament*, was first advertised on 19 November 1792 at the price of one shilling (*Star and Evening Advertiser*, 19 November 1792). Within a month, Stockdale promoted the pamphlet by tying it to two other popular pamphlets that had each gone through at least six editions. A purchaser could pay either the shilling or buy each of the pamphlets in bulk at forty-four shillings per hundred copies (*Diary or Woodfall's Register*, 14 December 1792).

In *Inevitable Consequences of a Reform in Parliament*, Playfair's main political premise was that representation should be based on property. Any move to universal male suffrage would lead to leadership of the country devolving to the lower classes, which in Playfair's eyes was highly undesirable. He recognized that representation by property was unequal across the country. But to make a change would require an arbitrary cut-off point that would not completely fix the problem. Therefore, he recommended staying with the present imperfect system; it had shown itself to be successful. He concluded that partial reform was ridiculous. Complete reform, he argued, would lead to revolution; and revolution would lead to civil war and bankruptcy of the nation. France was such an example. There was too much risk associated with reform. Looking to America would not provide a good case to argue for reform. With such vast amounts of land available in America, and land ownership widely spread, the property qualification would be automatically satisfied. The purpose of government, Playfair contended, was to punish crime, carry out public works, and protect its citizens from foreign enemies. America, Playfair naively stated, had none of these issues. After arguing that France and America should not be used as models, he concluded, "It appears to me that the Parliament consists of nearly the fittest persons in the nation. I do not think we can reasonably expect any better laws by means of any reform" (Playfair 1792a, 25–6). Wide representation was unnecessary. The king and Parliament already responded to public opinion when it decidedly went in a particular direction. Nearly two decades later Playfair would go to great lengths in his *British Family Antiquity* to prove that the landed classes were the fittest people to sit in Parliament and run the country.

On the surface, Playfair's views on the working of government are very similar to those of Edmund Burke. While Burke is famous for his political philosophy, Playfair is not. Why? First, Playfair was not a great writer; wordiness was only one of his problems. Second, while he was on the mark on several occasions, at other times he expressed highly eccentric opinions. Here is a sampling of what a few of the critics wrote about his work:

Review of *General View of the Actual Force and Resources of France in January 1793*: "He neither is, nor affects to be a fine writer" (*British Critic* 1793a, 107).

Review of *Letter to the Right Hon. the Earl Fitzwilliam*: "Mr. Playfair attacks Lord F. with too much flippancy and sarcasm" (*Monthly Review* 1795, 99).

Review of *Peace with the Jacobins Impossible*: "As true friends to the constitution of this country, we sincerely lamented that the defence of government should (by any chance), have fallen into hands so extremely incompetent as those of Mr. Playfair" (*Critical Review* 1794, 466).

Playfair's eccentric or extreme opinions on government show up eighteen months after he published *Inevitable Consequences of a Reform in Parliament*. In April 1794, Playfair put out a forty-two-page pamphlet, published anonymously through Stockdale ([Playfair] 1794). Taken from characters in Greek mythology, the pamphlet carried the title *Scylla More Dangerous Than Charybdis*. The two are mythical sea monsters on either side of a narrow strait, Scylla representing a shoal and Charybdis a whirlpool. Getting too close to either would be fatal.

In Playfair's world, Scylla represents absolute democracy and Charybdis represents absolute monarchy. Playfair argued that in the current circumstances, because of the French Revolution, there should be a move towards Charybdis, a move featuring the relaxation of habeas corpus and the suppression of even a hint of sedition. He wanted juries to convict those who stirred up any discontent against the government. It was a very broad brush. The critics were divided in their assessment of the pamphlet. One thought that Playfair's suggested suppression of freedom of opinions and freedom of speech, if instituted, would make Charybdis more dangerous that Scylla (*Analytical Review* 10 [1794]: 205). A more sympathetic critic, while not agreeing with all that Playfair wrote, thought that Playfair meant well and considered the whole pamphlet "an able performance" (*Monthly Review; or Literary Journal* 14 [1794]: 441–6).

When we last caught sight of Playfair in 1792, he was in Dieppe on 19 or 20 August. The usual route to England from Dieppe was to land at Brighton.[1] He may have sent his family much earlier; his report in his *Memoirs* of leaving Paris gives the distinct impression that he travelled with male colleagues only. On arriving at Brighton, since there was no real harbour there, passengers were rowed ashore. Soon Playfair was in London submitting the manuscript to John Stockdale that described the events of 10 August in Paris. He may have prepared the manuscript in France, as it went to press very quickly in England as a forty-page pamphlet ([Playfair] 1792). An advertisement for a print copy first appears in the press on 8 September 1792 (*World*, 8 September 1792). The pamphlet was popular enough that it went to a second edition the next year ([Playfair] 1793).

Playfair's activities in England from 1792 to 1796 were spent sniffing around the corridors of power while developing as a propagandist in John Stockdale's stable of writers. The period might be divided roughly into two parts: when Frederick Augustus, Duke of York, was campaigning in Flanders and just before (1792–4), and afterward when the British army had been pushed out of Flanders and beyond the Rhine River (1795–6). This division is based on Playfair's own statement: "Having been on the Continent during the campaigns of Flanders, in 1793 and 1794, and since have been much acquainted with military men" (Playfair 1808, 6). Playfair was a great supporter of the Duke of York as a military leader, writing in York's favour for the post of commander for the Peninsular War that began in 1808. But Arthur Wellesley, later Duke of Wellington, was appointed to this post. During this time, Playfair's most productive literary output was in 1793 and 1795. This chapter will concentrate on the period 1792–4.

In Britain, though there was some inkling the year before, the clouds of war rolled in early in 1793. On 21 January 1793, Louis XVI went to the guillotine, convicted of high treason and crimes against the state. Three days later, Britain broke diplomatic relations with France. France reciprocated on 1 February by declaring war on Britain. At the same time, France also declared war on the Dutch Republic.

Playfair was quick to respond to the declaration of war in a patriotic way with his purse. On 26 February 1793, he attended a meeting of a newly created and well-intentioned society with a lengthy name: The Society for the Relief of the Widows and Children of Seamen and Soldiers Who May Die or Be Killed in His Majesty's Service during the War. If enough money were raised, the money would also be used to support the wounded. Playfair pledged three guineas yearly (£3 3s.) for the duration of the war (*Diary or Woodfall's Register*, 28 February 1793).

Campaigning season in the eighteenth century usually started in the spring and ended in the fall. Over the winter, roads were too difficult to use for military supply vehicles and artillery. The season would start when there was enough green forage to feed the horses (Rogers 1977, 82). During the 1794 season, Playfair said he spent the five summer months (approximately May to September inclusive) following the progress of the war (Playfair 1795b, 2). For the 1793 season, George III appointed his son, the Duke of York, as commander of the British forces. Not receiving his formal order until April 1793, York left for the Continent the month before, arriving in Holland on 4 March.[2]

Hostilities in Europe had actually begun the year before when France declared war on Austria. Prussia soon allied with Austria. Commanded by the Duke of Brunswick, the allied troops were mustered at Coblenz, situated on the Rhine River where the Moselle River flows into the Rhine. Brunswick was set to invade France in July 1792. The powers behind Brunswick made a serious error. Since the revolution began, there had been talk in European courts and among French émigrés about issuing a threatening proclamation against the revolutionaries, and it finally came to fruition when the king of Prussia and the emperor of Austria agreed on the contents in mid-July. Brunswick signed it, and it was sent to Paris on 28 July (Barton 1967). The manifesto threatened the Parisian populace with violence if the French royal family came to harm. But it had the opposite of its intended effect. It coalesced support in France around the revolution.

Playfair was also quick off the mark with his pen. By 6 February 1793, less than a week after the French declaration of war on Britain, he had a pamphlet in print, *A General View of the Actual Force and Resources of France, in January, M.DCC.XCIII* (*World*, 6 February 1793). He dismissed the defeat of Brunswick's forces the year before, attributing it to the late start in the campaign, the disastrous manifesto, and other errors committed by Brunswick (Playfair 1793b, 29). This time, in Playfair's words, "It seems very clear that from a war with France we have, at present, very little to fear; we may rest assured that it will be a short one" (43). He had several reasons. The government was printing too many assignats without proper support, such as specie or property, as it had done before, thus putting the French economy on the verge of collapse. Even prior to hostilities, Playfair claimed that France could not even properly clothe its army. Further, France's army and navy would consist mainly of inexperienced raw recruits unable to properly do battle. It was all a recipe for defeat for the French. How wrong Playfair was. But what he did foresee accurately was one of the strategies of attack against the French government. As he expressed it, "Had some celebrated emigrants taken my advice in the year 1791, in making war on the credit of France, instead of combating her troops, we should not have had now to arm England" (24) The attack on credit was through the wholesale forging of assignats. It has been claimed that Playfair was central to the forging operation that occurred in England in the first half of the 1790s. That issue will be dealt with in chapter 11.

Confident in his prediction that the war with France would be a short one, he prematurely turned his thoughts to how post-war France should look. He put down his ideas in a hastily written pamphlet called *Thoughts on the Present State of French Politics* (Playfair 1793c), which appeared in mid-April 1793.[3] Playfair said that he would address five points. But by the time he had finished, he had covered only three, which could be summed up in the statement that the French were unfit to rule themselves, so it was necessary for other nations to step in to reduce the power of France and to make, in modern parlance, a regime change. Playfair's idea of regime change was to restore the Bourbon monarchy and to give amnesty to everyone in France except certain leaders of the revolution. His idea of reducing the power of France was to carve off certain French territories to cover the cost of war reparations. Playfair estimated that the total amount owed in reparations was £76.5 million. In a magnanimous gesture to relieve the French of paying such a large sum, territory would be swapped for money. Some areas in the north of France would be given to Austria and Prussia, while some southern parts would be given to Spain and Sardinia. Corsica would become an independent country and Holland would obtain the French West Indies, but British would control the trade of the West Indies. Taxes earned in these newly acquired territories from the swap would pay for the reparations.

With one exception, the critics were generally unfavourable to Playfair's ideas. The most negative one got to the point very bluntly: "We have thus detailed the whole of this very *humane, mild,* and *equitable* plan, that in addition to the superlative folly of disposing of the lion's skin, before he has been hunted into the toils, the reader may perceive the full extent of a scheme of fraud, rapine, and violence, unequalled by aught, except the ravings of some moon-struck and insatiate despot" (*Analytical Review* 1793, 197). The one positive critic claimed that the pamphlet contained, "as usual, many original thoughts, and useful hints" and called it "a curious piece of speculation" (*British Critic* 1793b, 215).

Playfair finished his trilogy of thoughts on the French Revolution with a pamphlet entitled *Better Prospects to the Merchants and Manufacturers of Great Britain*, once again published by Stockdale (Playfair 1793a), in mid-May 1793.[4] Playfair took the confidence he expressed in the earlier pamphlets about a short and successful war with France to predict a rosier future for Britain once the war was over. He began by dealing with the negativity that was "in the air." There was a certain anxiety among the public about the shortage of coinage and credit. Playfair attributed the coinage problem to British investors speculating on the assignat, which caused gold to flow from England to France, which led to the coin shortage. The government was currently dealing with the credit problem, Playfair assured his readers, by floating £5 million in Exchequer bills, short-term loans to the government. At the same time, Playfair took the opportunity to promote his earlier idea of removing regulation of the annual interest rate, which stood at 5 per cent. Better times were ahead, he trumpeted. When the French were defeated, he claimed that the radical elements that ran the country would immigrate to the United States; there was nowhere else for them to go. Because of the weakness of French manufacturing, whose origins Playfair attributed to the flight of the Huguenots in the late seventeenth century with the revocation of the Edict of Nantes, the demand for British goods would rise, especially in the United States. With the French loss of their West Indies colonies, the British naturally would come to take over this trade as well. With the end of what Playfair called the "confusion" in Europe, British trade would rise there as well. But it was all pie in the sky; the war did not end soon.

In the back of his mind, and even at the front of it, Playfair had thoughts of returning to business. He was an ideas man whose ideas often did not come to fruition; or, when they did, they withered or exploded. One that withered was a follow-up to his Scioto venture. The French press had severely criticized the scheme of sending French émigrés to settle in Ohio. It was too far inland. Farmers would have difficulty getting their produce to European markets. On the flip side, it would be expensive to import goods into the area. Playfair had a solution (Playfair 1793a, 17–18). He suggested building a series of wholesale warehouses by the mouths of rivers stretching perhaps 400 miles from Ohio to Kentucky to deal in the import-export trade. The syndicate operating the warehouses would import British manufactured goods, buy agricultural produce from American farmers, and ship it down the Ohio and Mississippi Rivers for export to the West Indies. He said that he had some interest from "some commercial people" but the start-up cost was prohibitive – £120,000.

The historical record is silent on the nature of Playfair's activities as a camp follower during the 1793 campaigning season, but it can be guessed with some accuracy that he was collecting information for articles and books to rally public opinion against the French Revolution. Some, perhaps many, in England were sympathetic to the revolution. Richard Price, who had liked Playfair's early work on the British national debt, was an early supporter of the French Revolution, though he did not live to see the excesses of it (Price 1789, 49). Even Playfair's old patron George Dempster, along with the Whig Club of Dundee, passed a motion in support of the French Revolution and conveyed it to the French National Assembly on 31 July 1790 (Mavidal et al. 1893, tome 17, 451–2). More generally, there was widespread popular unrest fuelled by the ideals of the French Revolution. Playfair wanted to reverse what he considered a pernicious trend.

Playfair sent a proposal to Henry Dundas,[5] one of whose duties as home secretary was to supress radical dissent and unrest in the country. It is probably no coincidence that Playfair's pamphlet on supressing dissent, *Scylla More Dangerous Than Charybdis*, is dated 20 April 1794. Playfair's proposal to Dundas is dated four days later, written while the issue was fresh in his mind and before his pamphlet came off the press. In the proposal, Playfair expressed concern about the direction he thought that public opinion was going in England. He went to great lengths to express how radical opinion had become successful and was now dangerous. After a wordy preamble, Playfair suggested an antidote: written propaganda that would appeal to the general public. Such propaganda would be widely advertised by the government, costing £8,000 to £12,000, he estimated. Specifically, Playfair wrote of the project, "I would beg leave to recommend employing some person to compile a short history of the *crimes & cruelties of the Revolution in France*, as details of that sort are read with great avidity." With a touch of modesty, feigned or otherwise, he made no hint that he should write the history, to the point that he did not even mention his own literary work. Rather, he suggested a French writer in London, Jean-Gabriel Peltier, who published several émigré journals that were anti-republican. Playfair's proposal was filed away and ignored. Somehow the clerks had mistakenly separated the proposal into two different sets of paper and had reassembled them a couple of days later, attaching a brief note that the correspondence was incomplete without the second set.

Perhaps this was Playfair's first venture into scams that he tried to pull in times of desperation, usually financial desperation. This time it was not financial. He had already written a history of the French Revolution. Offered by Stockdale, it was advertised in newspapers at the beginning of January 1794 as "in press" with the title *A Philosophical and Political History of the French Revolution from the Accession of Louis the 16th to the Present Time* (*Morning Chronicle*, 4 January 1794). But it seems that it never came off the press. It does not appear in the list of books Playfair submitted to the Literary Fund in 1807.[6] Nor does it appear in Thomas Byerley's obituary of Playfair in 1823, although it contains a lengthy list of Playfair's publications ([Byerley] 1823, 171–2). For some reason, Stockdale withdrew the book. Perhaps it was too great a financial risk for Stockdale, too expensive to produce with unknown prospects for sales. Also Playfair was better known for his publications on economic data such as the *Commercial and Political Atlas*, not for political history and commentary. Or, perhaps it was because Stockdale was already publishing what he considered a high-profile history of the French Revolution: a translation of *J.P. Brissot, Deputy of Eure and Loire, to His Constituents, on the Situation of the National Convention*. In the advertisements, it was featured as *History of the French Revolution with a Secret History of the Parties Acting Therein* (*Public Advertiser or Political and Literary Diary*, 27 January 1794). Brissot was a well-known early leader in the French Revolution, as well as a recent martyr to the more radical factions that were taking power in France. All in all, it was a better bet for the publisher than Playfair, whose credentials on the revolution were relatively unknown. Playfair's proposal to Dundas was a roundabout way to get his own book on the French Revolution published – and, of course, to make some money.

Playfair may have been the translator for Brissot's book and the writer of the forty-page preface to it. No hint of a name is given for the translator anywhere in the book. In what could be anonymous self-aggrandizement, Playfair described the preface as "incomparable and masterly" (Playfair 1794, 5). The style and general language of the preface are completely in sync with Playfair's sentiments.

For his own publication, Playfair was left with the crumbs, a thirty-two-page pamphlet written in January 1794 entitled *Peace with the Jacobins Impossible*.[7] It is a loud and strident description of the excesses of the French Revolution with descriptions of events listing executions and atrocities, in language that would rival any gothic novel of the day. It was the type of work that Playfair was recommending to Dundas.

Soon after Playfair sent his letter to Dundas, he was off to the Continent for the 1794 campaigning season. The previous year, during the summer campaign of 1793, the Duke of York and the British army had been relatively successful. The French were driven from Holland and Flanders, and then pushed back beyond their own borders. The one negative aspect of the campaign that year was a military blunder made during the autumn by a non-military man. The British government, on the urging of Dundas, who had been made secretary of war in July, ordered the Duke of York to take Dunkirk, at the time a relatively insignificant military target. York followed orders and laid siege to Dunkirk. The French were able to send reinforcements. York's troops were defeated in battle and he had to withdraw.

During the campaigning season of 1794, the tide turned against the British and their allies. Because of continuing French success throughout the summer, hostilities continued into December, well after the end of the usual campaigning season. The British and their allies were driven back beyond the Rhine River and the French took Flanders. The Batavian Revolution in the Netherlands had been brewing for years and came to a head in the winter of 1794–5. The Dutch Republic under William V, Prince of Orange and stadtholder, was overthrown and replaced by the Batavian Republic. The Austrian Netherlands ceased to be part of the Habsburg Empire and became part of France. William V fled to Britain in 1795. Over the fall of 1794, the Duke of York retreated to Bremen where the army wintered and then returned to England in the spring.

The only solid information that we have about Playfair on his European journeys is that he reached Frankfurt some time during the 1794 campaign, probably later in the summer as the military reverses were taking effect. There he met a member of the Parlement de Bordeaux, who described to Playfair the newly invented French telegraph. Playfair's story is described in a few near-contemporary publications, with Playfair's own version appearing in 1814 in his *Political Portraits*.[8]

Invented by Claude Chappe in 1792, the telegraph was a series of towers, spaced so that pairs of towers were in sight of one another. Each contained a mechanical device with a semaphore system to communicate messages between the towers. According to the *Oxford English*

Dictionary, it was "a signalling apparatus consisting of an upright post with one or more arms that can be moved in a vertical plane, different positions of the arm or arms representing different letters or messages."

After the Bordelais gave Playfair his description, Playfair spent two days constructing two working models. He also composed an alphabet for his telegraph. Then he set up two telegraph towers and showed that the system worked by sending messages through them. He took his models to James Ross, private secretary to Lord Malmesbury, a British diplomat in Prussia, then showed his model to Colonel George William Ramsay of the York Rangers, a British military regiment made up mainly of French émigrés and German soldiers. Ramsay sent the models on to his commander-in-chief, the Duke of York. On his return to England, Playfair tried to take a model of his telegraph to the British Admiralty, with arguments about how his model was superior to the French one, but the Admiralty refused to look at it. They already had a telegraph and were satisfied with it.

In his book *Playfair: The True Story of the British Secret Agent Who Changed How We See the World* Bruce Berkowitz has interpreted this story as evidence that William Playfair was a spy for the British (Berkowitz 2018, 178–81). But there are severe flaws in Berkowitz's argument:

- The Parlement of Bordeaux was a law court so that a Bordeaux parlementaire was a magistrate. During the French Revolution, a few of these parlementaires became émigrés. Among those who did not leave, a few died in the Terror and several were imprisoned (Doyle 1967, 397–402). The person whom Playfair met was an émigré. Thus, Playfair's information came from someone who was probably a royalist and not sympathetic to the French Revolution.
- In 1793, the French government published a report on their telegraph. Written by Joseph Lakanal, it contained a physical description of the telegraph (Lakanal 1793, 1794). The report was reprinted the following year. The only thing that was missing from the report in both printings was a description of the correspondence between the signal given and the associated letter of the alphabet. Lakanal refused to give the key, stating that he did not want to rob the author (Chappe) of his property. News of the publication of Lakanal's 1794 report reached London in September 1794 (*London Packet or New Lloyd's Evening Post*, 19 September 1794).
- The form of Chappe's telegraph was already known in England by September 1794, about, or just after, the time Playfair was learning of it from the Bordelais émigré.[9] By November 1794, a British newspaper had published a diagram of Chappe's telegraph (*True Briton*,

15 November 1794) (see figure 9.1). All that was left to the British authorities was to know how the semaphore signals translated to letters of the alphabet.

- By November 1794 telegraphs were already being tested and constructed in England and elsewhere. Near the beginning of November there were news reports of a telegraph in Manchester and another one in Sweden (*General Evening Post*, 6 November 1794). The latter was based on the same principles as the French one. By the end of 1794, the British Admiralty was in the process of completing a series of telegraphs on the coast of England (*Star*, 16 December 1794).
- In his *A Mathematical and Philosophical Dictionary*, published in 1795, Charles Hutton provides an alphabet for the semaphore signals of the French telegraph (Hutton 1795, 564–6 and plate 28). It is unlikely that the signals correspond to what the French were using. Hutton's key to the semaphore signals includes the letter *W* which was not in use in French in the eighteenth century (see, for example Abel Boyer's 1797 *The Royal Dictionary Abridged* where there are no entries under *W* in French) and rarely used today (see, for example, Ormal-Grenon and Pomier's 2004 *Concise Oxford-Hachette French Dictionary*, p. 631 where the entries under *W* take up only one-quarter of a page) (Boyer 1797, 631).
- The Rev. John Gamble was chaplain to the Duke of York in 1794 (Snape 2008, 28). The next year he wrote a twenty-page pamphlet, dedicated to the Duke of York, containing a history of signalling, from the use of beacon lights by the Romans to the modern telegraph (Gamble 1795[?]). He finished the pamphlet with a description of his own development of a telegraph. No mention is made of Playfair and his work.
- In the Royal Archives in Windsor the Secret Service accounts for the years 1779 to 1801[10] contain no mention of William Playfair. He does not appear to have been part of the British Secret Service.

In view of all these developments, it seems unlikely that Playfair was being paid by the British government to obtain information to which it had already had access. It rather seems that Playfair was a little too late to gain advantage from the model he constructed.

However, another possibility is that Playfair was collecting information for articles and books to rally public opinion in England against the French Revolution – a guess that is supported by much circumstantial evidence. His activities culminated in 1795 with a book carrying the lengthy title *The History of Jacobinism, Its Crimes, Cruelties and Perfidies: Comprising an Inquiry into the Manner of Disseminating, under the*

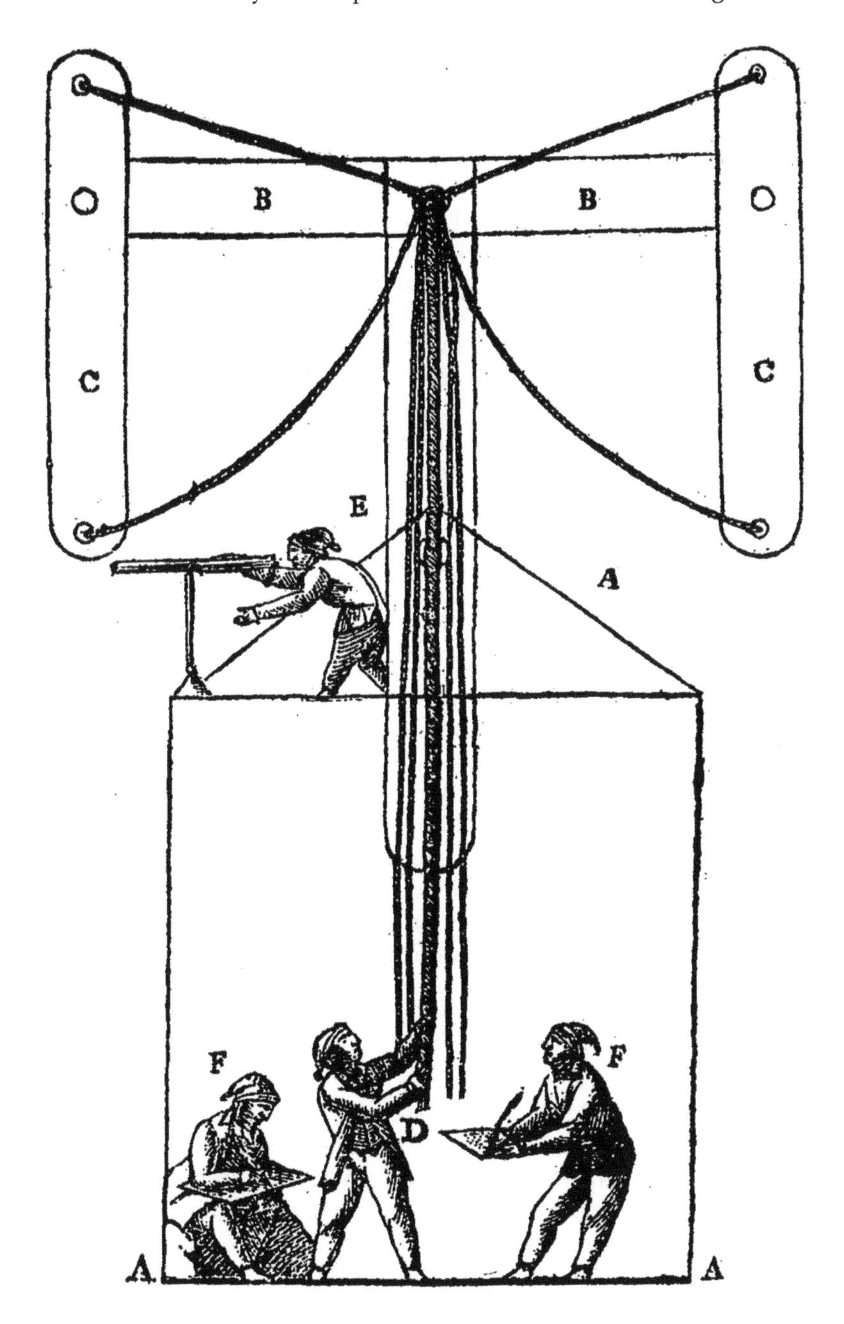

Figure 9.1. Depiction of the French telegraph by the *True Briton* Newspaper. Burney Collection of newspapers, British Library.

Appearance of Philosophy and Virtue, Principles Which Are Equally Subversive of Order, Virtue, Religion, Liberty and Happiness (Playfair 1795a). The title has all the elements of the propaganda campaign that Playfair put to Dundas. It was a major book of 821 pages, much longer (and more expensive to produce) than the 120-page book of Brissot with its 40-page preface. In a prospectus for *The History of Jacobinism*, also published in 1795, Playfair wrote, "During five months travelling in Holland, Flanders and Germany last summer, I found the general notion of all ranks of people was, that we fought for opinions, and not for any real object" (Playfair 1795b). The quotation suggest that Playfair was collecting information, as well as inspiration, for the book that did not make it into print early in 1794, but did by the end of May 1795, after revisions and alterations.[11] Further, in support of the thought that money was the real issue for the aborted 1794 book, *The History of Jacobinism* was published by subscription, giving Stockdale some up-front money.

This possibility gathers further support from Playfair's probable movements during the summer of 1794. His first move was from England to Flanders, where the troops were stationed. Since he left for home in September and wrote that he was on the Continent for five months, Playfair likely arrived in May at Ostend. On 10 March, a British naval officer described the port as full of troops. And there were many English in Ostend who were sympathetic to the "democrats" (French republicans) (*Oracle*, 10 March 1794). News from the battles was sent back to England from Ostend well into June. Further, non-military British passengers could travel to and from Ostend by packet boat, run by the British Navy to deliver mail (see, for example, *Sun*, 21 April 1794). Ostend was a good starting place for Playfair to collect any additional information he needed for *The History of Jacobinism*.

If Playfair tried to follow the troops that summer, it would have been at a distance. During 1794, the British troops were in almost constant retreat. To see Playfair's approximate movements that summer, through what Playfair would have called Holland, Flanders, and Germany, and the path of his return, it is useful to follow the progress of the 1794 campaign. At its beginning, the Duke of York came out of his winter quarters at Courtrai in the western part of Flanders. Key locations of the towns in the Flanders campaign are underlined in red in figure 9.2. The British and allied troops were strung out along a line in Flanders beginning in Nieuwpoort on the coast and then proceeding inland to Charleroi, about 200 kilometres or 125 miles in length. The line was close to the border with France. Fresh troops and supplies came by ship through Ostend up the coast from Nieuwpoort. Colonel Ramsay and his York Rangers were stationed at Menin nearly halfway between

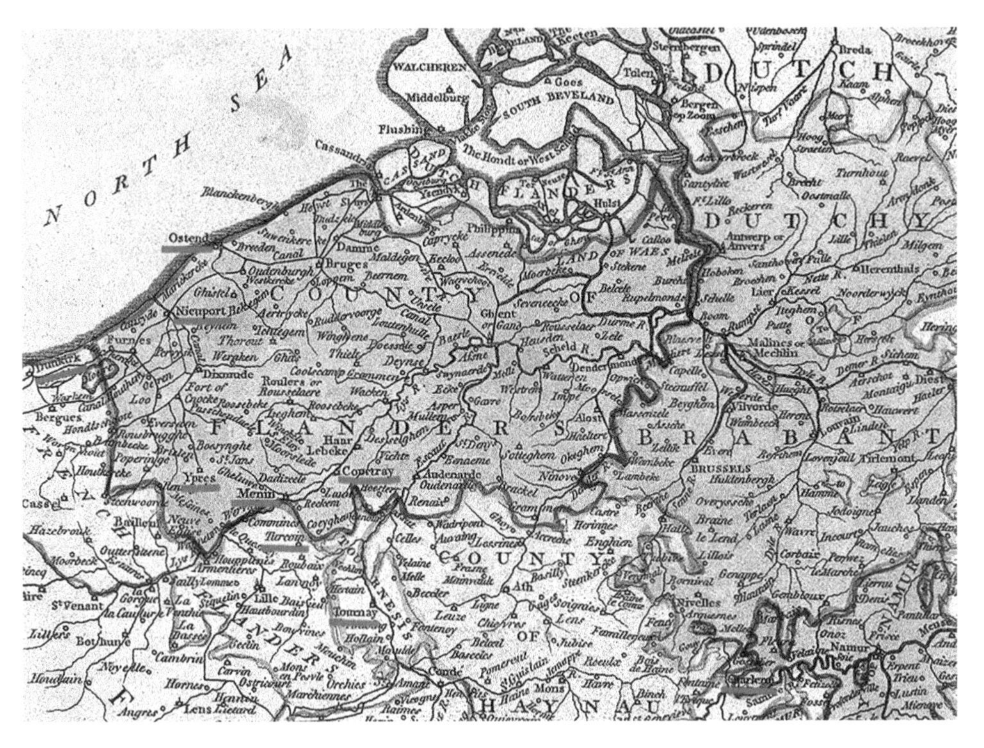

Figure 9.2. Map of Flanders prior to the French victories during 1794. Courtesy of Geographicus Rare Antique Maps, Brooklyn, New York.

Courtrai and Ypres on the western part of this line. In May, the British lost battles to the French in nearby areas, Tourcoing and Tournai. The British and allied troops were pushed back from their positions near the French border. By July, Antwerp had fallen and Belgium – or the Austrian Netherlands as it was then known – was essentially in French hands. In the late summer and early fall, the French began to make inroads across the French border with German territories. Key towns in the latter part of the campaign are underlined in red in figure 9.3. French forces moved towards the Rhine River, taking Trier (Treves) and Aachen (Aix-la-Chapelle) in August and September respectively. During October, after Playfair began making his way home, the French took Cologne, Bonn, and Coblenz in that order. The York Rangers were part of the British retreat from the Dutch Republic in the fall of 1794 and were nowhere near the military action along the Rhine. By August, the Rangers were in Dommel southeast of Amsterdam, and by the end

Figure 9.3. Germany from 1794 after losses to the French early in the year. Courtesy of Geographicus Rare Antique Maps, Brooklyn, New York.

of the year were defending Oldenzaal in the east of the Dutch Republic near the border with Germany. They wound up in winter quarters in Hanover, south of Bremen, where the Duke of York had his winter quarters (Chartrand 1999, 41). Figure 9.3 also shows the French territorial gains in Flanders during the early part of 1794.

Allied troops along the Rhine were mainly Austrian with some French royalist troops, known in 1794 as the Armée de Condé after its commander Louis Joseph de Bourbon, Prince de Condé, a cousin of Louis XVI. From 1791, their base of operation had been in Coblenz, sheltered by the prince-archbishop of Trier, Clemens Wenzeslaus. French émigrés in Coblenz remained under Wenzeslaus's protection until the city fell to the French republicans in the fall of 1794.

The military situation and Playfair's *Memoirs* provide hints about Playfair's handover of his model telegraph that enhance what Playfair wrote in his *Political Portraits*. Playfair was in Frankfurt at the time of the initial handover. He first took his telegraph to Lord Malmesbury's secretary. This was a reasonable thing to do. Malmesbury was a high-ranking British diplomat who was in Frankfurt from 15 July 1794 to 7 November 1794 trying to convince the Prussians to commit more troops in support of Britain and her allies (Harris 1844, 121–50). He was also dealing with the partition of Poland, whose second phase had been completed in 1793 (between Prussia and Russia) and whose third phase took place in 1795 (between Austria, Prussia, and Russia). Playfair's telegraph is not mentioned in Malmsebury's published diaries and correspondence.

Playfair next took his model telegraph to Colonel Ramsay of the York Rangers. Ramsay was nowhere near Frankfurt; he and his troops were retreating through central Holland at the time. A clue to how Playfair met Ramsay at this time is in Playfair's *Memoirs* (Playfair n.d., Memoir 7). Once Playfair decided to return home because of the advancing French troops, he was having trouble hiring horses and a carriage to get to Mainz from Frankfurt. He decided instead to hire a boat to go from Frankfurt to Cologne. Initially, he was accompanied by Ferdinand d'Esterno, but soon joined by Jacques Anne Joseph Le Prestre, Comte de Vauban, and his brother's family. D'Esterno had served in the Condé's army (Bonvarlet 1868, 228–9). According to Playfair, d'Esterno was on his way to join the York Rangers (Playfair n.d., Memoir 7). Vauban had been in Russia in the service of Catherine the Great. Recruited by the Comte d'Artois (later to become King Charles X of France) during his mission to Russia to obtain Russian support for the French royalists, Vauban was on his way to England to join the French forces (Vauban and Comte de Le Prestre 1806, 57). He returned to France and eventually was part of the French émigré ill-fated invasion of Quiberon in Brittany in June and July 1795. Playfair's group travelling on the Rhine did not have comfortable lodgings along the way. When they stopped for the night at a village on the right bank of the Rhine (on the other side of the river from the French republicans), they slept on straw and, as Playfair described it, "The bread was bad and the wine detestable." Since they were all going in the same general direction, it is likely that the entire group travelled together beyond Cologne. Perhaps Playfair and d'Esterno caught up with Ramsay and his regiment during their retreat through Holland, at which point Playfair showed his model telegraph to Ramsay.

Playfair's mention of Peltier in his proposal to Dundas is also relevant to Playfair's likely activities on the Continent. Peltier had a network of correspondents who contributed letters and information about what was happening in France, which included émigrés from Belgium and Germany, as well as several other places (Burrows 2000, 23–4). At the end of *The History of Jacobinism*, Playfair mentions several newspapers, books, and individuals "to be consulted in ascertaining facts relative to the revolution." One individual mentioned is Peltier. Earlier in *The History of Jacobinism*, Playfair quotes and translates an excerpt from Peltier's *Dernier tableau de Paris*, which is Peltier's history of the French Revolution up to about 10 August 1792 (Peltier 1793). It is quite possible, even strongly possible, that Playfair met with Peltier and got from him names of people to visit while Playfair was doing his rounds as a camp follower on the Continent. It is a reasonable alternative to Bruce Berkowitz's claim that Playfair was a spy.

Once back home, Playfair continued to have no luck catching the attention of the government of the day. In November, Playfair wrote from his home in Highgate Hill in north London to William Pitt, the prime minister, with a very vague proposal to change "the present system of gunnery" in the war with France.[12] In the letter, there is also a roundabout suggestion of a meeting with Pitt to discuss the proposal. According to Playfair, he had written to Pitt twice before on the subject and had been ignored. Either that, or the earlier letters did not get delivered. Nothing appears to have come of Playfair's proposal.

Chapter Ten

Playfair Becomes an Avid Anti-Jacobin Propagandist

A notice in the newspapers in early April 1795 informed the public that Playfair would publish a history of Jacobinism.[1] The printed proposal was available gratis from Stockdale's bookshop. Copies of the history would be sold to subscribers at 8s. 6d. Sitting at Stockdale's was the completed manuscript waiting to go to press, which would happen as soon as enough subscriptions were sold to defray expenses. Enough subscriptions were sold within about a month and the book was available at the end of May.[2]

As mentioned in chapter 4, 142 books were sold by subscription. Of that number, 89 went to booksellers, 10 booksellers in London, and 1 each in Bristol and Newcastle. The advertised price to nonsubscribers was 10s. 6d. so that a two-shilling profit could be had by the subscribing booksellers (*St. James's Chronicle or the British Evening Post*, 21–3 May 1795). Among the remaining 52 subscribers, 9 were clergymen, 8 were members of Parliament, and 8 were titled noblemen. The MPs are an interesting group. For modern readers, they are all near non-entities. Even in their day, for the most part they were quite quiet in Parliament and they all generally supported Pitt. The one who stands out is Charles Pierrepont, MP for Nottinghamshire from 1778 to 1796. Until 1793, Pierrepont sided with the opposition Whigs. In 1793, he joined William Windham, who had organized a group among the Whigs, known as the "third party" who favoured war against the French revolutionaries and were increasingly in opposition to the French Revolution.

While waiting for the *History of Jacobinism* to come off the press, Playfair had time to toss off a twenty-four-page pamphlet with all guns blazing against William Wentworth Fitzwilliam, 2nd Earl Fitzwilliam (Playfair 1795c). Fitzwilliam had meant well on Catholic emancipation in Ireland, but had completely flubbed it. The verbal daggers came out, one of them Playfair's. Appointed lord lieutenant of Ireland in January

1795, Fitzwilliam began poorly in his new office with some attempted patronage appointments in Dublin that went against Pitt's wishes back in Westminster (Wilkinson 2004). Then Fitzwilliam began to promote a Catholic relief measure in Ireland that the government in Westminster thought Fitzwilliam had agreed to defer. After some miscommunications, toing and froing, and subterfuge along the way, a bill came forward in the Irish Parliament for Catholic relief. Fitzwilliam was recalled. Later, a defence of Fitzwilliam's conduct appeared in print in the form of two letters addressed to Lord Carlisle, who in the early 1780s had been lord lieutenant of Ireland and sympathetic to Catholic relief (Geoghegan 2008).

It was the publication of these letters to which Playfair responded with his pamphlet. Many others reacted in print to the two letters, so Playfair was not alone (*Monthly Review* 1795, 98–102). Playfair attacked Fitzwilliam and questioned his motives on everything he did in Ireland from the patronage appointments to his handling of Catholic emancipation. With the French Revolution always in the back of his mind, Playfair compared Fitzwilliam to Jacques Necker, finance minister to Louis XVI from 1777 to 1781 and again from 1788 to 1789. By currying the favour of the French people, Playfair saw Necker as "one of the causes of the [French] revolution breaking out when it did" (Playfair 1795a, 67). In Playfair's opinion, Necker was a "quack doctor" (88). One reviewer of Playfair's pamphlet on Fitzwilliam best describes the contents of it, and adds his own opinion of Playfair: "This author accuses Lord Fitzwilliam of personal views in his administration, and of vanity and conceit: and if his lordship possessed half as much of the latter qualities as William Playfair, we should have thought the charge sufficiently heavy, without the scurrilous petulance which here accompanies it" (*Critical Review* 1795, 446).

Underlying Playfair's stand on Catholic emancipation may have been a streak of anti-Catholicism that hardened during his experiences in France. What he saw as abuses of the clergy in France would never occur in a Protestant country, so he claimed (Playfair (1795a, 53–5).

The History of Jacobinism arrived in bookstalls on 30 May 1795. If the critics are anything to go by, it was very successful. Where there were complaints, the most common one was of his writing style. Here is a review from the *English Review*:

> We have already delivered it as our opinion, that the present is incomparably the most philosophical and satisfactory account of the rise, progress, and actuating principles, of the French revolution, and existing government or anarchy, or whatever it may be called, that has yet been published.

> Mr. Playfair seems to unite a fine genius, and a turn for speculation, with a knowledge of the world; and, could he arrange and express his matter as well as he can investigate the truth, his writings would have double effect; though, in their present unpolished, and, to say truth, slovenly state, they abound with solid sense, profound reflection, and not a little of genuine pleasantry and humour. (*English Review* 1796, 252–3)

With the exception of the *Commercial and Political Atlas*, where Playfair showed much originality and creativity, *History of Jacobinism* was Playfair's best publication to date.

Perhaps his background in the Scottish Enlightenment looked to historical precedents to make an argument. To his credit, Playfair opened his *History of Jacobinism* by looking back for the causes of the French Revolution. He saw the emergence of feudalism after the fall of Rome as a development that gave protection to the people in return for service to the lord. The bond was broken between lord and vassal when standing armies were introduced, so Playfair claimed, so that the bond became merely mercenary. The feudal systems in England and in France evolved in different ways. In France, the evolution allowed for the wealth and extravagance of the court of Louis XIV and later courts, as well as the nobility in general. Likewise, the church in France had become wealthy with clergy appearing "in days of ceremony before the public in a blaze of gold, but ill imitating the poverty, and simplicity of their Divine Master" (Playfair 1795a, 52). Such activity by the church, aristocracy, and state created unrest in the population.

After Playfair had spent one chapter on the causes of the French Revolution and another three on the events leading to the fall of the Bastille, he made four assertions about the nature of the French Revolution, which appear to be responses to statements in Thomas Paine's *Rights of Man* (1792a). Playfair begins with a broad swipe at the new French political system (Playfair 1795a, 189):

> The foundation of the system of *anarchy*, *pillage*, and *murder* was laid on the following principles:
>
> 1 That insurrection is one of the rights of man.
> 2 That the good of the public is the supreme law, before which all others give way.
> 3 That all men are born and remain equal in rights.
> 4 That men are never bound by what their ancestors have done; this last is only a kind of repetition of the perpetual right of insurrection.

What Paine (1792a) actually wrote was:

1 "The end of all political associations is the preservation of the natural and imprescriptible rights of man; and these rights are liberty, property, security, and resistance of oppression" [quoted from the National Assembly's Declaration] (49).
2 "The whole of the Declaration of Rights is of more value to the world, and will do more good, than all the laws and statutes that have yet been promulgated" (51).
3 "Men are born, and always continue, free, and equal in respect of their rights" [quoted from the National Assembly's Declaration] (49).
4 "Every age and generation must be as free to act for itself, in all cases, as the age and generations which preceded it. The vanity and presumption of governing beyond the grave, is the most insolent and ridiculous of all tyrannies. Man has no property in man; neither has any generation a property in the generations which are to follow" (7).

In his statement (2), Paine wrote at a later date nearly what Playfair wrote (Paine 1792b, 60). Playfair claimed to have read only part 1 of Paine's *Rights of Man* (Playfair 1793c, 16). The problem with the French Revolution, Playfair thought, was that it was based on philosophy, not practical experience. This philosophy, as Paine espoused it, was false and pernicious. People had been duped by the philosophers offering terms like "liberty," "equality," "rights of man," "friends and brothers," and "universal benevolence." What they got instead was anarchy. Throughout the *History of Jacobinism*, Playfair spilt a lot of ink cataloguing incidents of mob violence and state-sponsored violence, and he placed the inspiration for that violence at the feet of the Jacobins.

Playfair had a fixation on the propertied class as the only group fit to run a country, and it shows up years later in his *British Family Antiquity*, where he expressed it with regard to the French Revolution:

> Had the leaders of the assembly been men of property and of plain good sense, as were our English barons at Runnymede, they would have been contented with laying a solid foundation for liberty; but they were mostly men of no property, smatterers in metaphysics and philosophy, who, thinking themselves equal to any task, would not be content with laying the foundations of a better order of things; they must destroy the old order to establish a new one in its place, and risk the safety and welfare of their country for the sake of wild theories which they had invented, and which were totally impracticable. (Playfair 1795a, 163)

The situation could not be compared to what had happened in America, Playfair claimed. There almost everyone was a proprietor of property, unlike in France or the rest of Europe, where 90 per cent were not (190–1). Though they had some traction in his own day, Playfair's political beliefs seem completely out of date today.

Beyond the issue of the landed class, Playfair found fault with the new legislative system. Louis XVI had called together the Estates General, consisting of three groups: clergy, nobility, and commoners. In preparation for meeting, the elected deputies had put together cahiers in which they collected grievances of their constituents. The cahiers had been ignored and set aside so that the deputies were no longer representing their people, so Playfair claimed. The Estates General quickly changed into the National Assembly and met as a single group. According to Playfair, the new National Assembly passed many hasty decrees but had no executive power. That still rested with the king and his councillors. Unlike Britain with its House of Lords, no second chamber provided a counterbalance to the work of the National Assembly. Further, they set aside old laws before introducing new laws to replace them, resulting in an interregnum of anarchy. Of course, Playfair's opinion worsened further after the uprising of 10 August 1792, after which the National Convention replaced the National Assembly, and after the Republic was declared. Playfair carefully catalogued all these events, along with his own commentary.

Reflecting his attempts to have Henry Dundas fund a propaganda campaign the previous year, Playfair recognized the need to have a positive effect on the hearts and minds of the people. Insightfully, he argued that one of the reasons the reformers in France, later revolutionaries whom he called "messengers of confusion," were so successful was the use propaganda (Playfair 1795a, 108). Generously funded, the reformers produced a lot of it. The king and his supporters had the money but failed to respond to it. Playfair argued that power is found not by force but by controlling public opinion. Oppression of opponents leads only to greater vigour among the oppressed.

Since his return to England there had not been any graphs in Playfair's publications, and *History of Jacobinism* is no exception. He does sprinkle data here and there in his publications to make a point; and again, *History of Jacobinism* is no exception. His strongest use of data is in his description of the end of the Reign of Terror (Playfair 1795a, 697–9). His only comment is that "the execution of accused persons had gone on as usual." And then he writes, "It may not be improper to give the list of executions for Paris only," in order to show the severity of the Terror. Playfair follows up by providing a table of the daily number

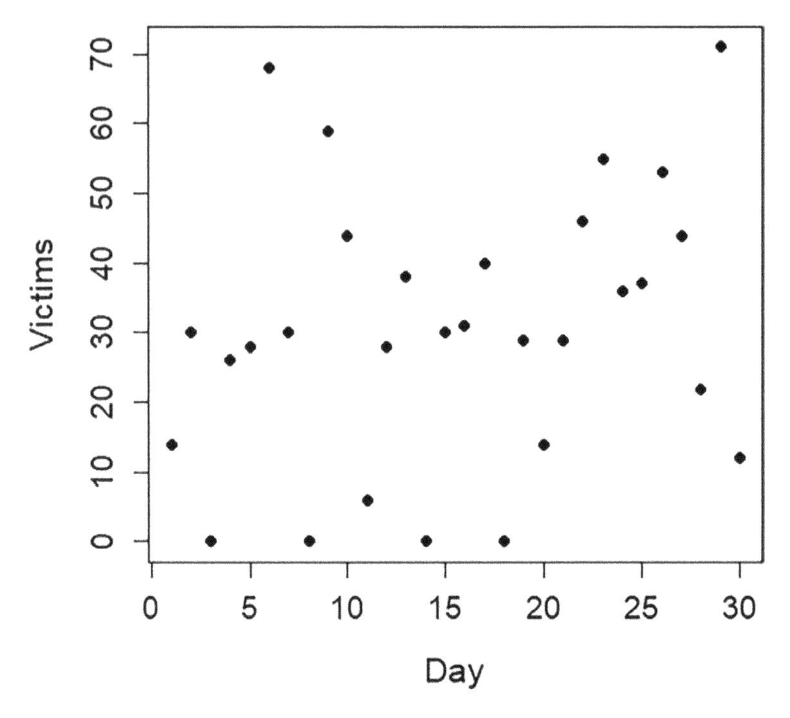

Figure 10.1. Daily victims of the guillotine in Paris, July 1794. Drawn by the author.

of executions in Paris during July 1794, categorized by type of victim: clergy, nobles, and common people. Maximilien Robespierre, the embodiment of the Reign of Terror, was executed on 28 July after his rule was overthrown. There were two more days of executions, and then the Terror ended. The last three days do not appear in the table, but Playfair gives the totals for those three days. Following Playfair's graphical inspiration, the number of executions by day for July 1794 is plotted in figure 10.1). The total number of executions shows no real trend in the data; as to any possible trends, or lack of them, by total or by category, Playfair made no comment on the data.

While modern-day Americans revere the Marquis de Lafayette as a hero of the American Revolution, Playfair had a completely different opinion (Playfair 1795a, 159–61). It is an example of beauty or the opposite being in the eye of the beholder. Playfair thought that Lafayette was vain and ambitious. At the beginning of the revolution, Lafayette was made commander-in-chief of the new National Guard – a militia instituted by the National Assembly to keep order in Paris. Lafayette's initial (and to Playfair, unforgivable) sin was that he allowed atrocities to happen early under his watch, so Playfair claimed. Having crossed that line, everything that Lafayette subsequently did during the revolution was questionable in Playfair's mind.

Perhaps the *History of Jacobinism* was not bringing in the money or recognition that Playfair hoped for. In July, he initiated a full-blown scam, to get money or recognition or both. On 11 July 1795, Playfair wrote to fellow Scot, Sir John Sinclair, author of *The Statistical Account of Scotland*, about a proposal he had received from the Comte de Vinezac, a French aristocratic émigré at the time purportedly in London.[3] Vinezac wanted to return to France to fight against the republican forces and had a detailed written plan to mount a rebellion in the southeast of France that would split the forces of the Republic. Sinclair, MP for a Scottish constituency, sent Playfair's letter and the plan to Sir William Pulteney, another Scot who was an MP, but in an English constituency. After Pulteney interviewed the alleged Vinezac, he wrote on 22 July to Henry Dundas enclosing Playfair's letter and the plan. Vinezac told Pulteney that he needed £100 to return to France. Dundas, another Scot, was home secretary at the time. The scam had a fairly good set-up, for many of the facts mentioned in the plan are correct, but with one glaring exception. At the time the letters were written, the Comte de Vinezac was already in the west of France fighting the republican forces with troops organized by someone else in London, Joseph-Geneviève de Puisaye (Gauer 2011, 15). Puisaye had already recruited Vinezac to serve in royalist armies in France and had recently convinced the British to support an ill-fated expedition of French royalist émigrés to take Quiberon, a fort on the south coast of Brittany. So the plan, purportedly from Vinezac, was a forgery, and the person who went to see Pulteney an imposter.

Since Vinezac's plan was a scam, the written proposal, in a fine copperplate hand, was almost certainly a forgery. There are signs that William Playfair was behind the forgery. After presenting the plan in fine copperplate writing, Vinezac signs his name:

This signature can be compared to other Vinezac signatures. In 1796, Vinezac wrote reports on the military situation at Clos Poulet in Brittany (northwest France, not southeast).[4] From these reports, here are five examples of his signature:

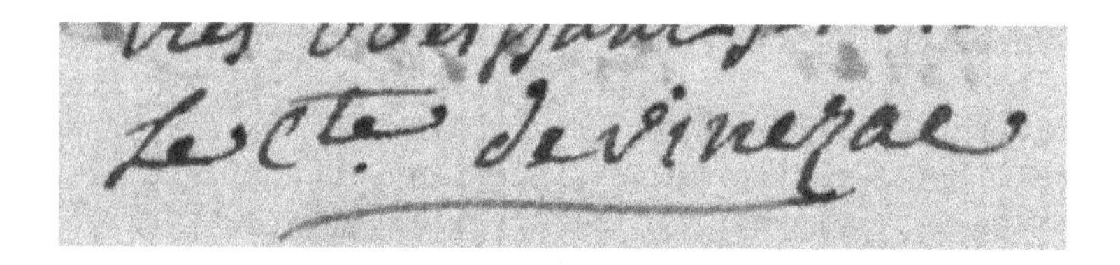

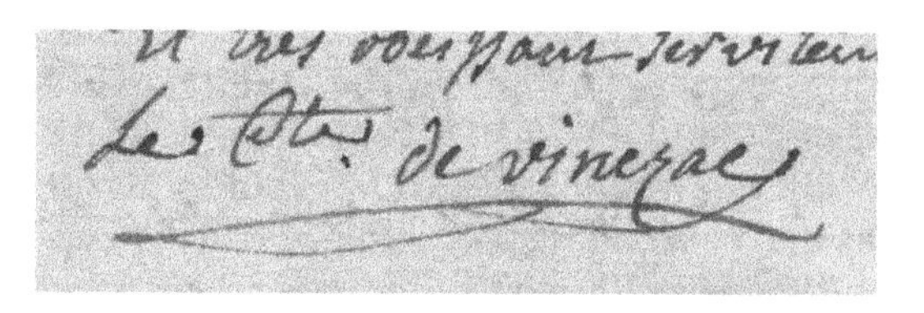

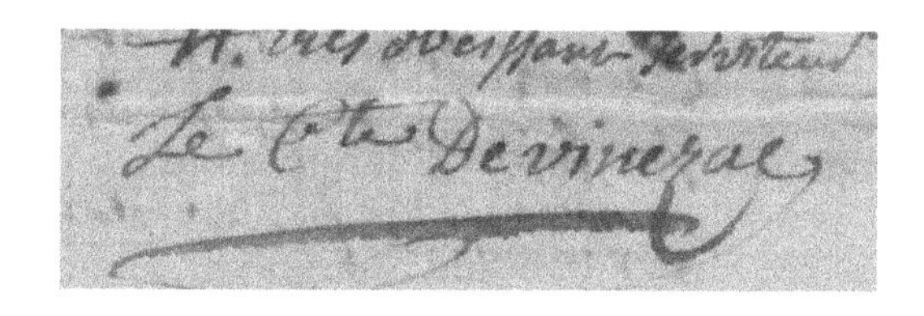

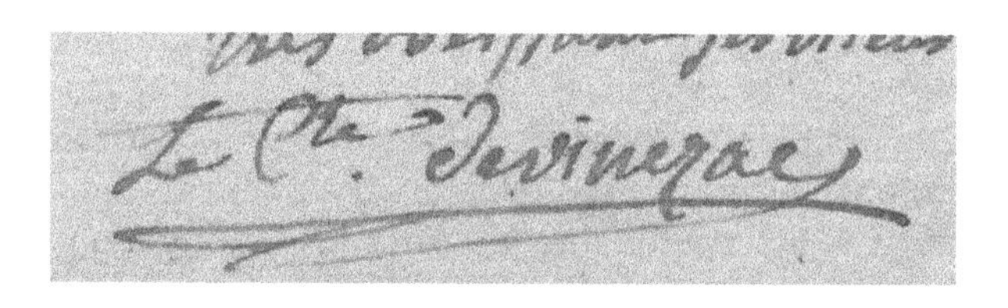

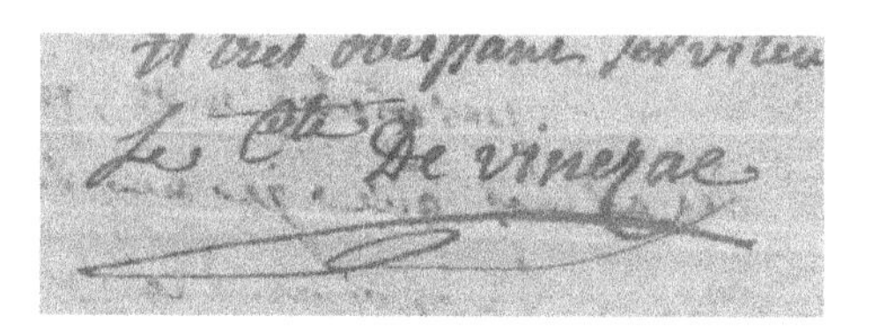

The signatures are similar to the one in Playfair's document. However, there are differences. The *L* in the article "Le" is below the *e* in the five signatures, while in the Playfair document it is on the same line. In addition, the *L* in the purported plan is written in a slightly different way from the five true signatures. The *C* in "Cte" is more rounded in the five signatures compared to the one in the Playfair document. In the genuine signatures "Le Cte" has an upward slant, whereas in the Playfair document, there is no slant. When he signs his name with a capital *D*, there is a break between "De" and "vinezac," unlike "Devinezac" in the Playfair document. Finally, the formation of the *z* in "vinezac" or "Vinezac" is different between the five signatures and the Playfair document. It may be concluded that Playfair or one of his associates had a copy of the genuine signature and tried to reproduce it. The reason for the copperplate writing in the proposal is that it would be difficult to copy Vinezac's writing for the whole document. Vinezac's reports in the National Archives are all written in his own hand.

Why can the forgery be attributed to Playfair? Every so often in the copperplate writing of the plan, Playfair breaks down and uses his distinctive rendition of the letter *p* instead of the more usual one. For example:

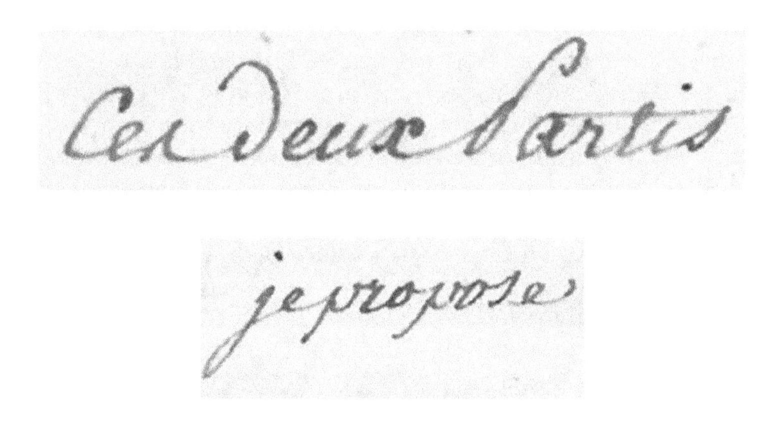

This unusual way to write *p* can be compared to Playfair's signature in a document from 1795:[5]

By August 1795, Playfair had another idea that went in an old direction. It was built on his proposal to Henry Dundas the year before, which Dundas had ignored.[6] Entitled *Revolutionary Magazine*, it was to be published weekly, beginning on Saturday, 29 August. It was promised that the magazine would contain anecdotes of the French Revolution and the French revolutionary army, including details of the treatment of prisoners as well as the prisoners' wives and daughters. There would be reviews of the main players and major publications "on both sides of the question." The idea came to reality under a different magazine name later in the year.

Playfair, the fiercely anti-Jacobin writer, finally crossed paths with a fiercely anti-Jacobin politician, William Windham. Windham's views are succinctly expressed in a 1793 letter to the Whig political hostess Frances Anne Crewe (Windham 1866, 291–2): "My hostility to Jacobinism and all its works, and all its supporters, weak or wicked, is more steady and strong that ever. If Pitt is the man by whom this must be stopped, Pitt is the man whom I shall stand by." Originally allied with the opposition Whigs under Charles James Fox, Windham initially supported the French Revolution (Thorne 1986). Subsequently, Windham broke with Fox and became increasingly influenced by Edmund Burke, joining with Burke in opposition to parliamentary reform and to the French Revolution, both issues dear to Playfair's heart. Windham joined Pitt's government as secretary at war in July 1794, while Pitt's right-hand man, Henry Dundas, became secretary for war. In his new position, Windham supported French royalist émigrés in their military ventures against the French republican government. One of Playfair's attractions was that he had contacts within the émigré community.

Playfair most likely came into contact with Windham through Charles Pierrepont, who earlier in 1795 had subscribed to the *History of Jacobinism*. Shortly before 24 October, Playfair met with Windham and discussed his proposal for a magazine that he had put together the previous August.[7] Encouraged by Windham's response, on 24 October Playfair sent Windham a letter enclosing a typeset copy of the proposal, along with a handwritten description of his plans. Arguing that the government needed to win the minds of the people, Playfair suggested that the magazine should be cheap and entertaining. In August, Playfair had thought that the magazine should be sent free to every curate in England with the hope that each parish would attract as many as ten subscribers. Now he suggested specifically to Windham that he needed support in the amount of £500 to £600 to print 2,000 free copies, with £250 of it going for start-up costs and general expenses. However, he did not have the money himself to carry out his proposal.

In his financial support, Windham came short of the proposed start-up costs, giving Playfair £100 in November 1795 and another £125 over four instalments (one of £50 and three of £25) by March 1796.[8]

The strength of Playfair's anti-Jacobin feelings also provides us with a window into Playfair's possible contradictory views on slavery. He fell into conversation with Gordon Turnbull, who lived on Grenada in the Caribbean. While it was originally a French colony from the seventeenth century, the British took Grenada in 1763. It was taken back by the French during the American Revolution, and then returned to British hands in 1783. In the middle of 1795 there was a pro-French uprising on the island, often interpreted as a slave revolt. Turnbull witnessed the uprising and wrote about it (Turnbull 1795). In his conversations with Turnbull, Playfair was impressed by Turnbull's ideas of how to prevent slaves from revolting when encouraged by the French government. Playfair sent Windham a copy of Turnbull's book and, in a letter of 22 January 1796, offered to bring Turnbull to meet Windham.[9] This is a bit at variance with Playfair's comments on a conversation he had with Étienne Clavière in 1790 where he expressed an abolitionist viewpoint (chapter 8). The circle might be squared by noting that Playfair wrote about his interaction with Clavière in 1819 after the British had abolished the slave trade in the British colonies a dozen years earlier and sentiment against slavery was growing in Britain.

Under Windham's patronage, Playfair published *The Tomahawk! or, Censor General*, an anti-Jacobin newspaper, whose first issue came out on 27 October 1795. In his unpublished *Memoirs*, Playfair stated that he was in partnership with "Dr. Arnold" and two unnamed others for this publication (Playfair n.d., Memoir 8). Dr. Arnold appears in the debit column of the ledgers for Charles Whittingham, the printer for the *Tomahawk* (Warren 1896, 24–6). Apparently Arnold handled all the bill payments. Although Dr. Arnold is not identified further, he is almost certainly Samuel Arnold, composer, organist, and music director, who had been awarded a doctor of music from Oxford in 1773 (Zöllner 2004).

The *Tomahawk* was entertaining and cheap. Articles on current theatre, opera, and musical events appeared regularly, no doubt with the help of Samuel Arnold. Some of these articles mentioned "Dr. Arnold." Poetry also appeared on the pages, often witty, political, and patriotic. There was much political commentary, with the Whig leaders Charles James Fox, Richard Sheridan, and Charles Grey, later Earl Grey, often skewered. Here are some titbits:

> Mr. Fox is looked upon, now again, as a man of some principle, by those who have no memory.

> Mr. Sheridan was the most improper person of all the Reformers, to present the petition from *Dunce*!
>
> … Mr. Grey's ridiculous and ill-timed *motion* for peace.

As to cheap, the *Tomahawk* sold for 2½ pence per issue, compared to, for example, the *Sun* at 4½ pence.

The *Tomahawk* shared elements of the scandal sheet. The nastiest attack to be found occurred late in 1795, but not against the Whig trio. During December, over six issues of the *Tomahawk*, the writers took swipes at William Augustus Miles (*Tomahawk! or, Censor General*, 7, 8, 10, 12, 18, 19 December 1795). Initially the object of their ire seems to have ignored these swipes. Miles was a political writer who had supported the government to the point that he served as a British spy during the first part of the French Revolution. By 1794 Miles had turned against Pitt and his conduct of the war with France. He was outspoken in print, publishing in 1795, for example, an almost libellous and highly critical pamphlet on the Prince of Wales. The final swipe against Miles appeared in the *Tomahawk* on 22 December. It descended into an ad hominem attack on Miles, accusing him, among other things, of having an affair with a woman in Frankfurt.[10] With that, Miles had had enough. He wrote to Windham on 1 January 1796 complaining that the *Tomahawk* contained "disgusting scurrillities and impudence" as well as "atrocious falsehoods."[11]

It is difficult to tease out exactly what Playfair wrote for the *Tomahawk*. There is only one reference to one article he wrote. Philippe, Duc d'Orléans, was guillotined in November 1793. An article entitled "Character of the Late Duke of Orleans" under Playfair's name in the *Tomahawk* on 3 February 1796 was scathing. With the addition of a concluding paragraph, it was lifted completely from Playfair's *History of Jacobinism*. A week later, there is an anonymous article on Jacobin clubs in the *Tomahawk*, much of which is a partial copying of, and expansion on, pages 172 and 173 of the *History of Jacobinism*. The next day in the *Tomahawk* there appeared an article titled "Republican Cruelties!" that also adapted from the *History of Jacobinism*, this time deep in the text on pages 802 and 803. A couple of articles not on the French Revolution might be attributed to Playfair. In two early issues of the *Tomahawk* (29 October and 3 November), there is a column with the headline, "Morning Conversation, at Stockdale's." Stockdale was Playfair's publisher for the last two or three years. Both conversations are about the *Tomahawk*. The real Stockdale must have salted his conversations with profanity. Every time Stockdale opens his mouth in the two articles, he starts with "G-d dang it," usually as he begins a sentence.

The *Tomahawk* ceased publication after 113 issues. In the farewell article on 7 March 1796, with its pages bordered in mourning black, the proprietors claimed that they were being "prosecuted by some persons belonging to the Stamp-Office." Playfair wrote that since the paper had no advertisements or news, it was not subject to the Stamp Duty.[12] Although he was correct in saying that the *Tomahawk* had no advertisements, it is a matter of interpretation whether or not some of the articles in the paper were news. Arnold's articles on the arts scene could be considered news. In any case, Playfair said that after making several attempts to resolve the situation, the proprietors folded the paper. The paper never achieved the circulation that Playfair hoped for or predicted. At the beginning, the *Tomahawk* had a circulation of 1,000 copies, declining to 300 when it folded (Warren 1896, 26).

Within ten days of the demise of the *Tomahawk*, Playfair had prepared a propaganda pamphlet. Printed by Charles Whittingham, printer for the *Tomahawk*, it was for sale in Stockdale's bookstore in early April 1796 (*Sun*, 4 April 1796). The pamphlet was titled *For the Use of the Enemies of England: A Real Statement of the Finances and Resources of Great Britain* (Playfair (1796a). It was catchy in the sense that the title seemingly implied that Playfair was exposing a British weakness that could benefit France. But upon perusing the pamphlet, readers would find that Playfair was showing exactly the opposite. The pamphlet intended to show that Britain had the resources to continue the war with France. It was written in the face of the allies' reverses on the battlefield in the previous two years, and of opposition to the war in Britain led in the House of Commons by Fox, Sheridan, and Grey, who had been agitating for peace with France since 1792.[13] France now occupied Holland and Belgium, as well as areas on the left and right banks of the Rhine. The latest from the anti-war trio were motions in the House by Sheridan and Grey, supported by Fox, and presented on 9 December 1795 and 15 February 1796.

In making his argument to continue the war, Playfair analysed the British financial situation. He was responding to the argument that British finances were in such dire straits that peace should be made on France's terms. He came down decisively on the side that Britain could afford to raise additional money through loans in order to continue the war. Reporting on a preview of the pamphlet, one newspaper put the sum that could be raised at £50 million. In the pamphlet, Playfair argued that the financial burden of the war on the public was less burdensome in 1795 than ten years before, basing his argument on analysis of interest paid on the national debt. In order to compare the debt in 1795 and 1785, it would be necessary to adjust the 1795 figures. This would

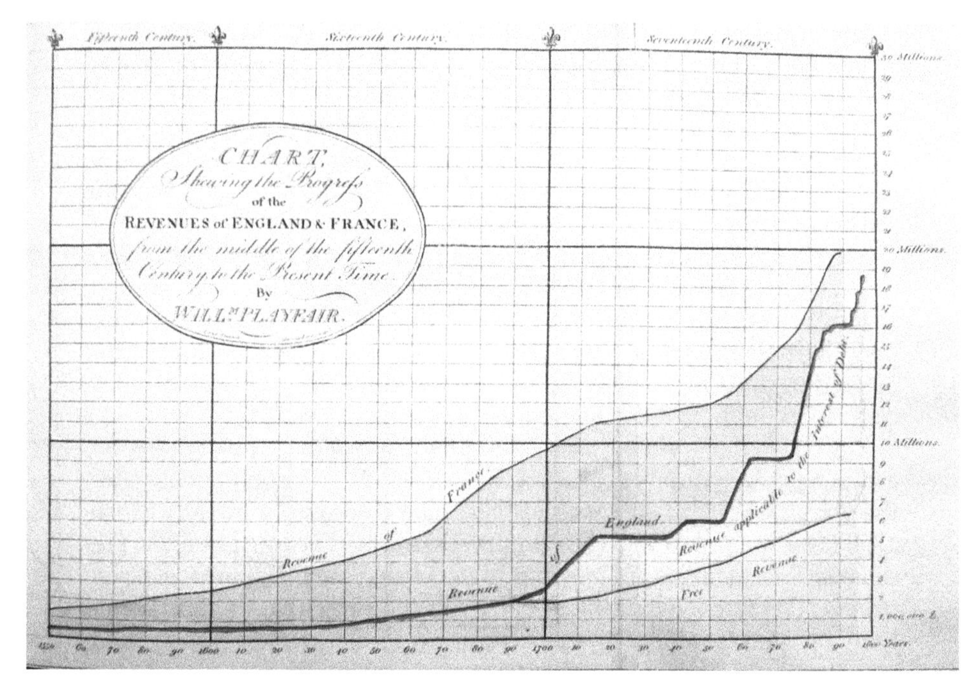

Figure 10.2. Revenues of England and France. Courtesy of Stephen M. Stigler, University of Chicago.

take into account inflation, and the increase in revenue from duties on exports and the taxes imposed on the purchase of goods by the public. For the latter, Playfair assumed that interest on the debt earned by the public would be spent on consumer goods that would be taxed a certain rate. After making these adjustments, Playfair showed that the burden of the debt actually decreased between 1785 and 1795, supporting his finding with two of his signature statistical graphs. One shows the yearly amounts of British exports and their increase over time. The other graph, given in figure 10.2, shows the revenues of England and France. The revenues are divided into two parts: the part that must be paid to service the debt, and the remainder that can be used for other purposes, including the continuing conduct of the war. Playfair sent Windham a copy of the pamphlet in its proof stage; he added that if Windham were interested he would send him a dozen copies in a few days.[14]

One of Playfair's bugbears about the French Revolution and its possible transfer to Britain was Thomas Paine and his *Rights of Man*.

Playfair saw Paine as a threat to Britain's internal peace and stability. It was a red flag when Playfair read the first part of a pamphlet Paine printed in the *Morning Post* on 30 April 1796, titled *Decline and Fall of the English System of Finance*. It was a popular pamphlet, going through at least fourteen editions in 1796. Paine argued that the British system of finance through the national debt would eventually collapse. He started his argument with examples of the devaluation and collapse of paper money in the United States and France, continental currency in the former and assignats in the latter. Since interest on the national debt was a form of paper money, he argued, the British system was also doomed. Unrestricted printing of paper money had led to disaster. Consequently, unrestricted growth in the debt, with the accompanying interest payments, would also lead to disaster. Paine made his main argument by extrapolating the national debt into the future. He noticed that in the six wars in which the British had engaged between 1697 and 1796, the debt increased by a factor of 1.5 with each war. In another six wars, the interest on the debt would be enormous and unsustainable.

Paine's article appeared on a Friday morning. By Monday afternoon, Playfair had a rebuttal in the *Oracle* (2 May 1796). He mocked Paine's abilities in finance and tried to connect his shortcomings in finance to excesses of the French Revolution. Playfair argued that with a sinking fund to pay the debt, which had been established earlier but now in abeyance because of the war, Britain was not on the verge of collapse and would not be in the future. Appearing as a seeming non sequitur to this article, the newspaper included Playfair's chart of yearly British exports, which had appeared in *For the Use of the Enemies of England*, as well as a lengthy excerpt from the pamphlet explaining the chart.

The second part of Paine's pamphlet appeared in the *Morning Post* on 2 May – the same day that Playfair's article appeared in the *Oracle*. Here Paine argued that paper money was bringing down the value of gold and silver. Both payments on the national debt and taxes were being made in paper money (Bank of England notes, in particular). Since the financial system was based on the equality of paper money and specie, and since more paper money was being produced through financial instruments such as exchequer bills, the whole system would collapse. Paine gave it twenty years. He ended the pamphlet by predicting that, as in the United States and France, the collapse of the financial system in Britain would result in a new form of government.

Playfair's next response to Paine came on 5 May, again in the *Oracle*. He had read the remainder of Paine's pamphlet in the *Morning Post*. He began, "Your attributing the depreciation of gold and silver to the

Funding system alone, must proceed from want of information on that subject, for the Funding System has not taken place in every country, yet the precious metals have lost of their value in all." He went on to illustrate Paine's lack of knowledge of how paper money worked and included assignats as an example of a growing gap between the value of paper money and specie. He concluded his response to Paine in a third article in the *Oracle* on 6 May. There he questioned Paine's motives. Paine had pointed out a problem with the British financial system but had not suggested any solution. Playfair accused Paine of trying to destroy the British government by prophesying only the collapse of the financial system. Paine would be better and more honourably employed if he worked to improve the French system rather than tearing down the British one.

The three articles in the *Oracle* were quickly combined and published as a separate pamphlet (Playfair 1796b). With one exception, this draws Playfair's work as political propagandist to a close. The exception is tied with the issue of Playfair possibly forging assignats. His role, or lack of it, in the schemes to forge assignats is closely tied to the claim that he was a British spy. This requires further analysis, which will appear in chapter 11 and the appendix.

Chapter Eleven

Playfair Gets Involved with Forged Assignats

As attested in chapters 9 and 10, it has been claimed in Bruce Berkowitz's *Playfair: The True Story of the British Secret Agent Who Changed How We See the World* (Berkowitz 2018) that William Playfair was a British spy. Central to the argument is the issue of forging the French Republic's paper money, or assignats, during the revolution. One tactic in the war against the French Republic was to flood France with forged assignats in order to destroy its economy. The British tried it, or aided in it. French émigrés tried it, as well as people in other countries. Berkowitz claimed that Playfair was the key player in the design and execution of the British operation to forge assignats. Dubbed Playfair's "Great Op," much of Playfair's career is tied to this operation, as a prelude to it or as a consequence of it. The main evidence is in three manuscripts by or about Playfair and assignats. Before examining them, it is useful to look at what was happening in England and elsewhere in the forging of assignats.

The French revolutionary government decided to issue assignats in December 1789. By early 1791, forgers were operating in several countries, including England, to take advantage of the ease with which they could produce false assignats and make a profit. In England, the efforts were small scale, uncoordinated, and scattered. The British government took police action against some of the small forgery rings. At about the same time, French royalist forces were gathering in Coblenz, organized by the Comte d'Artois to overthrow the revolution by force. Among the senior royalists in Coblenz, three decided to set up a forging operation: the Comte de St. Morys, his brother the Chevalier de Moligny, and Charles-Alexandre de Calonne, who had been comptroller-general of finances in France under Louis XVI until 1787. In Coblenz, St. Morys and Moligny began forging under Colonne's general direction, using the counterfeit assignats to defray expenses of their troops.

After the royalist troops were defeated at the Battle of Valmy on 20 September 1792, many royalists went to England, the trio of St. Morys, Moligny, and Calonne among them, as well as a British soldier, John Gordon Sinclair. In London, Calonne's brother, the Abbé de Calonne, joined them and their operation. The French émigré forging operation that was transferred to London gained in momentum in 1794, with the Comte d'Artois now taking part. Early in that year, the operation had British government support – moral support initially and then monetary later. The scheme gained steam with the arrival of Joseph-Geneviève de Puisaye, who had broken with the French revolutionists in 1793 and became a counter-revolutionary on the Continent before arriving in England in 1794. He quickly led in the forging operation and in November 1794 convinced the French royalist military council stationed in Brittany to bless the forging of assignats. In London, Puisaye had the ear of the prime minister, William Pitt, convincing him to mount a British-backed French émigré invasion of Quiberon, a bay on the south coast of Brittany. The émigrés were joined by French royalist troops on the Continent. The British recognized Puisaye as the leader of the invasion, whereas the French royalist troops on the Continent did not. St. Morys died in the invasion. Accompanying the invading troops in 1795 were many forged assignats, which were seized by the French government troops when the invasion failed, partly as the result of the divided leadership. The British minister most closely involved with the émigré forgers was William Windham, who had been appointed secretary at war in July 1794. The forging operation came to an end within a few months after the loss at Quiberon.

The émigrés employed several different papermakers and engravers. The most well-known source for paper used in the forgeries was at Haughton Mill, near Newcastle upon Tyne, while another was in the south of England, as well as engravers in Birmingham. Other than the centre at London, the whole operation appears to have been relatively diffuse.

What is the evidence that Playfair was a spy in this ring or at its centre? Most of it is found in three manuscripts that touch on Playfair's possible knowledge or vague awareness or even participation in operations to forge assignats. They are sometimes contradictory. The three manuscripts are (1) Playfair's proposal in 1793 to forge assignats;[1] (2) a manuscript by Sir John Swinburne in 1795, or later,[2] accusing the British government of running an assignat forging operation in 1793 and 1794 with Playfair's collusion;[3] and (3) a letter from Playfair to William Wyndham Grenville in 1811, and transcribed in the Fortescue papers, claiming that the British government was involved in forging assignats during 1794 and 1795.[4]

As a preliminary, Playfair had connections to Sinclair and to at least one in the trio of French émigrés in Coblenz. This association is revealed in the 1798 legal case that sent Sinclair to prison.[5] From Sinclair's perjury trial in 1798, it appears that Playfair was closer to Sinclair than Calonne, St. Morys, or Moligny. Prior to the trial, Calonne had asked Playfair to talk to Sinclair in an attempt to settle their differences in the dispute over Sinclair's claim that Calonne owed him money (Sinclair 1798). At the trial, Playfair was sympathetic to Sinclair's side and testified in his favour, claiming that Calonne had told him three different and contradictory stories about the debt. By the time of the trial, Playfair was a bankrupt and so his testimony may have been given little weight. Despite Playfair's testimony, Sinclair was found guilty.

Playfair's close connection to Sinclair is also evident from Playfair's financial situation in the mid-1790s. In about 1795 or before, he owed £397 to Lieutenant-Colonel Arthur Irwin (also spelled "Irvine" is some sources).[6] Irwin was in the York Hussars, a German unit raised on the Continent to fight alongside British troops. Prior to Irwin's embarkation, Sinclair paid off Playfair's debt and was now Playfair's new creditor.

In his letter to Grenville in 1811, Playfair wrote, "In 1794 and 95 a manufactory of assignats in imitation of those made in France was established at the back of Sloane Street, conducted by the Abbé de Calonne (brother to the Minister), Mr. de Puisay and St. Morrice, the same persons who were sent to Quiberon when France was invaded, where they had a principal command and took with them the assignats so manufactured, in great quantities and at a vast expense." Playfair seems to have been unaware of the forging prior to 1794, and the émigrés do not mention him in their activities in 1794–95. Nor does Playfair appear in any British sources except the Swinburne manuscript. But what Playfair wrote to Grenville shows that he was very knowledgeable about what was going on. The situation is similar to Playfair's reportage of the violent events in Paris on 10 August 1792. In his *Memoirs*, Playfair says that he stayed at home all day on 10 August. Yet his written reports of the events are very detailed, indicating that he talked to several first-hand witnesses.

The Puisaye Papers in the British Library indicate that Windham, and probably his undersecretary, Emperor John Alexander Woodford, were supportive, particularly financially, of the forging operation run by the Abbé de Calonne, St. Morys, and Puisaye. Naming only the abbé, Playfair corroborates this situation in his letter to Grenville. Other than the abbé, Playfair claimed not to have known any of the participants in the forging operation. But from Sinclair's lawsuit against Calonne in

1798 and Playfair's friendship with Sinclair, it seems highly unlikely that Playfair did not know about all the French players in the forging operation.

Playfair also states in his letter to Grenville, "When the expedition sailed to Quiberon [23 June 1795], the manufactory ceased, and the managers all went off except Calonne." Indeed, all did go off, except for Calonne. As noted already, St. Morys died on the expedition. Playfair's statement generally agrees with the Swinburne manuscript, which stated, "The following year 1795 this manufacture of assignat paper ceased at Haughton Mill." St. Morys's death may explain, at least partially, why the paper was no longer needed from the Haughton Mill.

In his letter to Grenville, Playfair claimed that he learned of the assignat forging and the government's involvement from the paper maker John Lightly. He also attested that he did not know the Abbé de Calonne. What he heard from Lightly was that when St. Morys and Puisaye sailed for Quiberon, they still owed Lightly £700 for watermarked paper left in the warehouse. Lightly threatened to sue the abbé for the debt. When Playfair learned of this and the embarrassment that it could cause the government, he went to Windham with what he had heard. Windham asked Playfair to settle the affair with Lightly, and he did.

Two letters from Playfair to Windham in 1795 support this version of events that Playfair expressed in 1811.[7] In one, dated 24 October 1795, Playfair wrote that the Abbé de Calonne lived at No. 112 Sloan Street near Chelsea. In the second, dated 30 October, Playfair described the situation in more detail: it was Lightly who was demanding payment from St. Morys and that the paper had never been sent to Sloan Street. St. Morys had intended to pay Lightly after he returned from the expedition to Quiberon, but never made it back. Playfair offered to resolve the problem. In return he wanted help, obviously financial but unstated, in getting his *Revolutionary Magazine* published, the magazine that became the *Tomahawk*.

Swinburne and Playfair agree on one thing about Haughton Mill – the name of the papermaker. Playfair claimed that after Quiberon, paper was left in a warehouse in Upper Thames Street and that the bill of about £700 to manufacture the paper had not been paid. Playfair agreed to be the go-between to settle the account. When the amount was agreed upon, Playfair claimed that "at 2 o'clock on that next day two Frenchmen called on Lightly, paid him the money and took a receipt in full." So Playfair identifies the papermaker as a Mr. Lightly. In a brief note at the end of his manuscript, Swinburne writes, "Mr Lightley clerk to the warehouse in London." This is at variance with some nineteenth-century historians, mentioned in the appendix of this volume, who identify the papermaker as Christopher Magnay on the basis of oral evidence presented by the descendants of the owner of Haughton Mill. The differences can be reconciled. Magnay had a paper warehouse at 72 Upper Thames Street in London. The partnership of the wholesale stationers Lightly and Atkinson did not get underway until 1799. Their business was at 178 Upper Thames Street (Maxsted n.d.). It appears reasonable to assume that Lightly was working for Magnay in 1795, perhaps even running the warehouse. Then Lightly started his own business as a stationer with Atkinson a few years later. The connection also ties Haughton Mill to the French émigrés.

Turning to assignats, Playfair's proposal to forge them[8] is dated 20 March 1793. This was a time when Playfair was going back and forth between continental Europe (to be near the allied army when they were campaigning) and England (when the armies in Europe were in their winter quarters) (Playfair 1808, 6). In March, the armies would be emerging from their winter quarters with some of the British officers wintering at home in England. It was also at the beginning of a series of anti-Jacobin publications that Playfair wrote. It was already speculated that Playfair was gathering information for his publications while he was near the army.

Playfair's proposal was not addressed to anyone in particular. Some candidates that Berkowitz puts forward are the chancellor of the exchequer, William Pitt, or the home secretary, Henry Dundas. Both Pitt and Dundas were unlikely recipients.[9] But the addressee can be surmised from marginalia in the proposal. The name "Col. Bullock" is written in the top left of the first page of the manuscript. This is the MP John Bullock who sat in Parliament from 1754 until his death in 1806 with the exception of the years 1774–80. He was a Whig politician who "seems to have been prepared to give ministers a minimal support when war with France broke out" and was also on good terms with William Windham (Thorne 1986, 300). His name appears on the page likely because the manuscript collector Alcide Donnadieu, who is the earliest known owner of the manuscript, obtained the plan from among Bullock's papers (Donnadieu 1851, 126). A reasonable interpretation is that the plan was sent to Bullock, as Playfair similarly had sent one to Sir John Sinclair. Rather than passing it on, as Sinclair had done, Bullock sat on the plan. It was never physically sent anywhere and remained among his papers until Alcide Donnadieu obtained it.

As with other Playfair proposals, such as those of 1788–9 to fix the machine de Marly at Versailles, it was short on detail and long on salesmanship. Playfair's assignat plan, as expressed in the manuscript, was to destroy the financial credit of France and thereby bring an end to the war that had begun in 1792. It consists of a mixture of altruism and self-interest. It contains several suggestions, both practical and impractical, for carrying out this financial destruction. As to altruism, using a fifth of the profits from this venture, Playfair proposed that a fund would be set up to support the French clergy living in England who had been disenfranchised by the confiscation of church property in France during the Revolution. Then a little self-interest crept in. After providing for the clergy, money would be set aside for the support of the British Army "after reserving for myself and those who assist me a sum *sufficient to be above want* [italics mine]." On the operational side, the forged assignats would be sent to all the armies fighting the Republic to use as spending money. That was the practical side. On the impractical side, Playfair somehow planned "that part of the sums received … also shall be destined to the discrediting of the assignats … I shall sell letters of Change upon Paris at an under price upon the Change of London and thereby ruin the assignats." It appears that Playfair was planning to have British bills of exchange issued at a discounted price in Paris compared to London so that it would be attractive to French investors. When a French purchaser sold the bills of exchange in London at the higher price, payment would be made in forged assignats so that there would be an imbalance in real payment in favour of London. How Playfair might achieve the difference in the purchase prices is hinted at in his 1795 *History of Jacobinism*. Claiming that the Paris Bourse was easily manipulated, Playfair wrote that the bourse "was no better than keeping a gaming table with false dice" (Playfair 1795a, 89). Playfair also planned to bribe members of the French National Convention. There is no mention in the proposal as to how Playfair was planning to forge the assignats.

Playfair's proposal bears some similarities to the scheme of Calonne, St. Morys, and Moligny that had started in Coblenz in 1791 or before. Playfair may have learned of the operation from Sinclair and expanded on it in his proposal. Commenting on the proposal, someone (possibly Alcide Donnadieu) has written on the manuscript in French "misérable coquin!!!!" followed by "A.^{de} D." The French phrase can be roughly translated as "wretched rascal." Since by the time Donnadieu obtained the manuscript Playfair was a known propagandist supporting the British government, the comment can easily be interpreted as a dismissal of the proposal as propaganda.

Consequently, this manuscript evidence may be interpreted either as a proposal that went nowhere or as a propaganda piece that was filed away by its recipient. The proposal is on par with Playfair's earlier suggestion that British entrepreneurs build warehouses along the Ohio

River to compete with American businessmen from the east coast of the United States.

We are left with the most difficult manuscript, the Swinburne manuscript that involves Playfair in forging operations. The manuscript is undated, but opens with "In the years 1793 & 94" and refers to events that happened in those years. All that is certain is that it was written after March 1794 when Richard Sheridan asked a question about forged assignats in the House of Commons. The manuscript was likely written several months after March 1794, since Swinburne says, "Sheridan once alluded to my Letter," giving the impression that it was not in the recent past. Consequently, the manuscript may be dated to early 1795 or later.

The manuscript is mostly in the hand of Sir John Swinburne, an English baronet (transcribed in Philipson and Isaac 1990). Swinburne briefly sat as a Whig MP for Launceton from 1788 to 1790. After that, he became involved in Northumberland politics, where he lived (less than twenty miles from Haughton Castle). Swinburne was closely associated with Charles Grey, MP for Northumberland from 1786 to 1807. As previously mentioned, Grey was part of the Whig coalition aligned against Pitt. He also strongly opposed the monetary support for French troops given by the British government. In the manuscript, Swinburne says that he wrote to Grey about the forging operation at Haughton. It was Sheridan, a close associate of Grey, who raised the matter in the Commons.

Swinburne's information about the forging operation was second hand, obtained from a man named Richmond who lived near Haughton Castle.

It is useful to provide some quotes from the Swinburne manuscript to try to make sense of what was claimed. After describing his investigation into papermaking at Haughton Castle in Northumberland through Richmond, Swinburne writes:

> The Manufacturer's name at the above place is Smith. He has a partner in [London] of the name of Watson; [insertion in other hands – "quere Brook Watson yes"] if these people were not employed by government, how came they connected with so notorious a hireling as Mr Playfair author of a still more notorious pamphlet on the partition of France by the Allies?
>
> ....
>
> This man [Richmond] likewise informed me, the paper had for some time, been stamped by Michelson the Exciseman at Hexham under the denomination of fancy paper, but he at last refused to stamp it any more, as the duty on that paper was very low, & he was sensible of the imposition,

> upon this Mr Smith went to Town, where he had several meetings with Playfair.
>
> ……..
>
> I wrote to Mr Grey about it, & I believe Mr Sheridan once alluded to my Letter in the House of Commons, since which it has passed unnoticed – about the same period I was informed by very good authority an application had been made to Bewick the engraver at Newcastle, by a considerable person at Birmingham, to engrave several plates, of which the broken pieces were sent to him, begging him they might be cut very sharp, as a great number of impressions would be wanted – Bewick on putting the plates together, found they were assignats of different values, upon which he very honourably declared he would have no hand in such business.

In notes at the end of the manuscript there are two additional items. The first consists of two lines in Swinburne's hand:

> Mr Watson, partner to Smith in Wapping
> Mr Lightley clerk to the warehouse in London

The second is in a hand other than Swinburne's:

> Mr Thelusson is the person who pays the Papermakers for the French paper & Mr Playfair, the Author of several political Pamphlets, is the Agent who manages the Business & who lately went over to Ostend with Mr Brook Watson.

In his comments on Playfair, Swinburne's Whig allegiance shows through. Swinburne talked of Playfair as the "author of a still more notorious pamphlet." The pamphlet, *A General View of the Actual Force and Resources of France* was popular enough that it went through two editions during 1793 (Playfair 1793b). In the pamphlet, Playfair was highly critical of the republican government: "Let us throw aside the veil which their republican gasconade has thrown upon the facts, and we shall soon see that France is only superior to other nations in vanity, injustice, cruelty, and misery; in these, we will allow, she is superior to the whole human race; but neither in freedom nor in force." Later in the pamphlet, Playfair suggested that a better strategy against France would have been "making war on the credit of France instead of combatting her troops." Such statements would have rankled Swinburne with his Whiggish sympathies to France.

Some elements in Swinburne's manuscript make sense and some take some time to make sense. Pitt and Dundas ignored Playfair, while

Windham took Playfair under his wing. Playfair's involvement in assignat forging would have occurred after July 1794, when Windham took office. Brook Watson's involvement, although Swinburne is the only one to mention it, makes some sense. He was commissary-general to the Duke of York's army campaigning on the Continent from 1793 to 1795. Watson could have stocked the army with forged assignats that he obtained from the French émigrés who were also joining the British troops. Swinburne says that Playfair "went over to Ostend with Mr Brook Watson." The problem is that the British abandoned Ostend on 29 June 1794 and the Austrians abandoned all their territories in the Netherlands (Ostend being a part) on 27 July 1794 (Sherer 1836, 8; and Smith 1825, 331). Since any other evidence of Playfair's involvement in assignat forging points to 1795 or later, we are left with a conundrum. The solution may be that Watson and Playfair sailed together to Ostend probably in the spring of 1794 when Playfair was following the army on its summer campaign, Playfair's purpose being to gather information for his anti-Jacobin pamphlets and book. Richmond may have been overzealous in his report to Swinburne about Watson and Playfair, exaggerating his report, knowing Swinburne's stance on the war against France.

To put the assignat forging operation into a little perspective, consider what happened at Ostend. When the British left, the French claimed to have found 2,270,000 livres in forged assignats. In the list of forged assignats in the Puisaye Papers, the average denomination of the assignat was about 80 livres. With a little arithmetic, this results in approximately 28,375 pieces of paper. This, in turn, translates into more than 140 200-page unbound books. However widely inaccurate these calculations may be, the forged assignats were not something that Playfair, or more likely Watson, carried with him in a suitcase on the way to Ostend. A similar calculation shows that French claims about the assignats seized at Quiberon in 1795 were a gross exaggeration. The largest denomination of assignat mentioned by Puisaye is 400 livres. Printing only this denomination, the total of 10 billion livres seized translates to 125,000 200-page unbound books.

A second name, mentioned only by Swinburne, is Thellusson. This may be a mistaken reference to Woodford, who was part of the wealthy Thellusson family. There is strong evidence to support Woodford's involvement in the French émigrés' forging operation. Or it may be Peter Thellusson, the patriarch of the family, who was behind a large long-term loan to the government in 1793 that supported the British war effort, which Swinburne opposed (Cassis and Cottrell 2015, 89). Another possibility is Peter Isaac Thellusson, son of the patriarch, who was elected an MP in January 1795 to Midhurst, a borough controlled by George Wyndham, Earl of Egremont. Thellusson probably paid a substantial sum in order to represent Midhurst. The new MP supported Pitt and "acquired a wholly deserved reputation as a peerage-hunter" (Thorne 1986, 362–4).

There is one final loose end to tie up. Part of Berkowitz's argument is that a surviving mould for producing paper from Haughton has a watermark for the Original Security Bank. This was also the name of Playfair's short-lived bank that opened and collapsed in 1797. Playfair became bankrupt as a result of the bank's collapse. Berkowitz says that Playfair must have used the same paper mill for his bank as for his forged assignats (Berkowitz 2018, 211). The source on which Berkowitz bases his statement is mainly from the article in *Archeaologia Aeliana*. The problem is that in the article, it is stated that the mould to make the paper was almost certainly made after 1809, well after the collapse of the bank. Given the date, there could be several explanations for the existence of this mould. Legal wrangling was still going on in 1809, and beyond, over the bankruptcies related to the bank's collapse. There was a least one attempt to make false financial claims on the bankrupts' estates. Perhaps the false claimant wanted to reproduce the bank's bills. Or perhaps Robert Kennett, a known forger and one-time Playfair partner, was seeking revenge on Playfair for not supporting him when Kennett and his son went to trial in 1805. Or perhaps someone was trying unsuccessfully to start a bank under the same name. There are many plausible possibilities.

At best, support for the claim that Playfair was a British spy is very weak. It is known that Playfair, in essence, worked for Windham in 1795. The earliest known surviving letter from Playfair to Windham is dated 24 October 1795.[10] *The Tomahawk! Or Censor General*, co-edited by Playfair and financed by Windham, first appears on 27 October 1795. Playfair may have worked for Windham as a courier, one possible interpretation of what Playfair wrote in his 1811 letter to Grenville. For assignats, Windham would have left the heavy work and financing to his trusted relatives. On the other hand, the evidence overwhelmingly supports Playfair's claim that the French émigrés were central to the assignat forging operation financed by the British government. Moreover, the operation predates Playfair's choice of 1794 as the year when the clandestine operation commenced.

The evidence points away from Playfair ever being a spy. He may have played a bit part as a courier in the forging operation late in the game, and he may have been a wannabe spy, but that is quite different from being a spy. He was definitely not central to the assignat forging operation that was going on at the time.

Chapter Twelve

Playfair Starts a Bank and Goes Bankrupt

Opportunity knocked and Playfair answered. The Original Security Bank, with Playfair as the day-to-day manager, opened for business on Wednesday, 4 January 1797, at 28 Norfolk Street, a street that ran off the Strand near Drury Lane.[1] No blue plaque that commemorates either the event or Playfair will ever be placed at that address by English Heritage; the street disappeared in a redevelopment of the area in the early twentieth century. The genesis of the bank was a month earlier. The initial meeting to establish the bank was held at the Crown and Anchor Tavern on Wednesday, 30 November 1796 (*Times* and *Oracle and Public Advertiser*, 2 December 1796). At the meeting, a simple business plan was put forward. Depositors could submit any government security and receive in return several notes in small denominations (£5, 5 guineas, £10, £15, £20, and £25), which totalled the value of the security submitted. For the depositors' convenience, they could lodge their securities with their own banks. It was intended that the Security Bank's notes would circulate as a medium of exchange; the holder of the note could sign it over to someone else, as in a bill of exchange. The notes would pay 2 per cent interest per annum and could be redeemed at the bank on the day the notes fell due. Government securities at the time paid about 5.5 per cent so that the differential would cover the expenses of the bank plus a sum to those who devised the plan. By the beginning of February, for those customers interested, the 2 per cent interest payment was dropped in exchange for dropping any deductions necessary to cover the expense of the transaction. Some altruism and general goodwill were also expressed – after all costs were covered, it was intended that any profit would go to charity.

The time seemed ripe for such a bank. Although there were several banks in England during the eighteenth century, only the Bank of England, by law, provided banknotes that circulated as a medium of exchange. Their notes could be redeemed on demand for gold bullion. The Original Security Bank was intended to provide an alternative, or a complementary service, to those banknotes issued by the Bank of England. By the 1790s the Bank of England had not achieved the status of a modern central bank with its control of the money supply. It did have the role of banker to the state, providing loans to wage the numerous wars that occurred throughout the eighteenth century, and acting as a clearinghouse for government paper. In the 1790s the Bank of England's role as the sole provider of banknotes was coming under question. As early as 1790, Sir John Sinclair suggested the establishment of a second bank to break the Bank of England's monopoly. Described as an "agricultural improver, politician, and codifier of 'useful knowledge,'" Sinclair is best known today, at least in statistical circles, as the author of *The Statistical Account of Scotland*, which probably popularized the use of the word "statistics" in the English language (Mitchison 2015). By the middle of the 1790s, a few plans had been put forward by others to set up a rival to the Bank of England (Cope 1946). None came to fruition. The issuing of notes by one of the proposed banks was to be secured on gold, silver, Bank of England notes, and bills of exchange, while the second proposed bank was to be secured on mortgages on real property. The Original Security Bank began operation only a few months after these other proposals fell through.

In theory, through his banking experience in Paris, as well as his earlier publications on the rate of interest and on paper money, Playfair had the background to carry out the project. He just did not have the capital to put behind it. He was already in debt and so could provide no financial backing; as mentioned in chapter 11, Playfair owed an associate, John Gordon Sinclair, £397. There were three partners in the bank: John (or Jan) Casper Hartsinck, The Rev. Julius Hutchinson, and Playfair. According to Playfair, Hartsinck provided the initial capital.[2] From the records and publications that are extant, Playfair acted as the day-to-day bank manager and wrote letters and articles to promote and defend the bank. Hutchinson was a latecomer to the partnership, so it appears that the Original Security Bank was the brainchild of William Playfair.

Playfair was not a fan of Sinclair, though he was a creditor. He thought the man troublesome and wrote to William Windham expressing that opinion. In March 1796, Playfair had learned that Sinclair was planning to cause trouble for one of Windham's colleagues. Playfair offered to keep an eye on the situation and keep Windham abreast of negative developments.[3]

Playfair, Hartsinck, and Hutchinson came from varied backgrounds. Playfair we have seen. Hartsinck was Dutch (Day 1911, item 436). His career in the 1790s was closely connected to William V of the Netherlands. William held the title of "Stadtholder" and was effectively the head of state for the Netherlands. Under William, Hartsinck held diplomatic positions with responsibility for William's interests in Hamburg and in Lower Saxony. Hartsinck also had business interests and a house in Hamburg. Early in his career Hartsinck had been a partner in the banking firm of Hope & Co., which had its head office in Amsterdam. As a result of the Batavian Revolution in the Netherlands, culminating with the proclamation of the Batavian Republic in 1795, William V fled to England. Hartsinck followed soon thereafter. In England, Hartsinck maintained his diplomatic titles under William V (Minister at Hamburg to their High Mightinesses the States General of the United Provinces and Plenipotentiary to the Prince of the Circle of Lower Saxony) and handled some of William's finances. Hutchinson, who attended New College, Oxford, beginning in 1768, was ordained a priest in the Church of England in 1774 (Foster 1888, 2:722). Although vicar for a small parish in Nottinghamshire, he seems to have been an absentee vicar and to have acted more as a lawyer. He also had landed property.

Hutchinson and Colonel Sinclair's paths crossed as early as 1785 when Hutchinson met Sinclair via a letter of introduction for Sinclair from Jean François Dutheil, who was serving as the London representative of the Comte d'Artois (Sinclair 1796, 70). There is also a possible early connection to Playfair. In an affidavit written to support Sinclair in a 1798 legal proceeding, Hutchinson commented on Sinclair's joining the royalist army in 1791 after military service elsewhere on the Continent: "Col. Sinclair was afterwards invited to Paris, and was much with M. De la Fayette, by whom, and by others of his party, I am well assured he was much tempted to enter into the service of the French Constitutionalists, but I, and some friends I then had at Paris, encouraged his disposition to join the King's Brothers, then assembling an Army at Coblentz, which he accordingly did" (26). The friends that Hutchinson "then had at Paris" may have included Playfair.

The proprietors were initially true to their word on the profits. The first notes issued by the bank were due on 31 January 1797, at which point the possessors of these notes were notified in the press to redeem them at the bank (*Observer*, 29 January 1797). About a week later it was reported that the Original Security Bank had donated £3 10s. to the Society for the Discharge and Relief of Persons Imprisoned for Small Debts. The society also reported that it had received £143 10s. in donations from which it had spent £103 15s. releasing thirty-eight debtors from prison (*Whitehall Evening Post*, 7 February 1797).

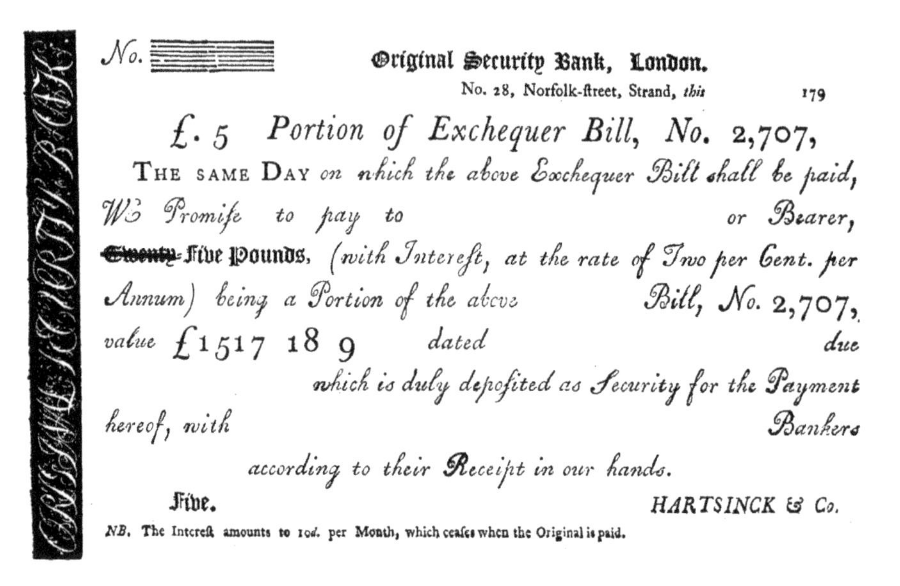
ORIGINAL SECURITY BANK.

No.

Original Security Bank, London.

No. 28, Norfolk-ftreet, Strand, *this* 179

£. 5 *Portion of Exchequer Bill, No.* 2,707,

THE SAME DAY *on which the above Exchequer Bill shall be paid,*
We Promife to pay to or Bearer,
~~Twenty~~ Five Pounds, *(with Intereft, at the rate of Two per Cent. per Annum) being a Portion of the above Bill, No.* 2,707,
value £1517 18 9 *dated due*
which is duly depofited as Security for the Payment hereof, with Bankers
according to their Receipt in our hands.

Five. HARTSINCK & Co.

NB. The Intereft amounts to 10*d.* per Month, which ceafes when the Original is paid.

Figure 12.1. Example note from the Original Security Bank. Courtesy of John Rylands Research Institute and Library.

Several government and company securities were acceptable to the bank. Among them are exchequer bills (government promissory notes that carried interest), Navy bills (negotiable instruments for fixed-term credit issued by the Navy), government debentures (payment orders made by the government), and India bonds (issued by the East India Company) (Original Security Bank 1796). Although these securities initially were issued to an individual or to individuals, they could be bought and sold on the open market. Their face values were too large to act as a medium of exchange and could not replace or complement banknotes.

An example of a note for £5 written against the deposits by the Original Security Bank is shown in figure 12.1. The example is taken from the Original Security Bank's prospectus, dated 28 December 1796. Here an exchequer bill has been deposited for £1517 18s. 9d. In the bank's prospectus, the bill has been divided into notes of all the denominations available, totalling £1517 15s. with 3s. 9d. in cash remaining.

The Original Security Bank had its detractors. As early as February 1797, a newspaper article about the bank (*Morning Chronicle*, 2 February 1797) mentioned Playfair specifically, the writer mockingly suggesting that he and the bank's proprietors invest their own money by buying a million pounds worth of exchequer bills to back their scheme. With exchequer bills paying 5.5 per cent and the bank's notes paying 2 per cent, the profits are obvious, the writer said. Later in August 1797, another writer commented, "Whatever be our opinion of the bank of England, and the wisdom and integrity of its directors, we certainly do not think Mr. P.'s bank in Cornhill is it's [*sic*] rival in consequence. We seriously question the propriety of any bank, similar to those in existence, for the assistance of commerce. There is no rational idea which attaches to the establishment of a bank but that of a loan office, or pawn broker's shop, where a man carries his pledges, and receives money" (*Analytical Review* 1797, 188). What the writer meant by "assistance of commerce" was the issuing of banknotes. Although he had a conservative view of the operation of banks, as was prevalent in the late eighteenth century, he was perhaps unknowingly correct that there should be only one central bank controlling the money supply.

The Bank of England had its problems. From the time that war began with France in 1793, gold had been flowing from England to the Continent to pay the troops and the other necessities of the war. By February 1797 rumours were circulating in England that France would invade England. This caused a further drain on the gold supply as country banks and other London banks converted their Bank of England notes into cold hard cash. On 27 February 1797 the Bank of England announced that their notes could no longer be converted into specie, i.e., gold and silver coins. This switch from a commodity-based standard, in this case gold, to a flexible exchange rate lasted until 1821 when the Bank of England returned to its policy of converting its notes for gold on demand. The 1797 decision resulted in a shortage of coins as well as low-denomination notes; the smallest denomination was a £5 note that began circulation in 1793 when the war began. The contemporary debates among economists about this issue is known is the Bullionist Controversy (Laidler (2000). For his entire career, Playfair was firmly committed to the position that a paper currency had to be backed by a commodity, preferably gold.

Playfair pounced on the opportunity that the events of 1797 opened. On 28 February 1797 he wrote to James Mackintosh, a fellow Scot, but also a barrister at Lincoln's Inn,[4] asking Macintosh for his opinion on some acts of Parliament regarding the negotiability of promissory notes. An act of 1775 forbade the issuance of promissory notes and bills of exchange valued at less than 20s. Two years later in 1777 another act was passed with further restrictions. Notes and bills issued for £1 or more, but less than £5, must have the name of the payee appear on the note or bill. These financial instruments were not transferable to another person. A third act from 1791 placed a 3d. stamp duty on notes in the range of £2 to £5. Playfair wanted to know from Mackintosh whether there was any kind of sunset clause on the first two acts and whether the duty imposed by the third act implied that notes of £2 could be transferable to someone other than the payee. Playfair may have heard rumours of changes to the first two acts or was following parliamentary proceedings very closely. On 10 March 1797, an act was passed to suspend the provision of the first two acts from 1 March 1797 to 1 May 1797. The suspension was later extended until 8 July 1797.[5] The reply from Mackintosh is not extant. However, on the same day that Playfair wrote to Mackintosh, he placed an advertisement in the *Oracle and Public Advertiser* announcing that the Security Bank would issue promissory notes in a £1 denomination (*Oracle and Public Advertiser*, 28 February 1797). In advertisements placed on 3 and 4 March, these promissory notes would be issued in exchange for Bank of England notes (3 and 4 March 1797). Current circumstances, the Security Bank claimed, made small notes necessary. It was noted in the advertisements that this was done as a service to the public; the Security Bank was not sure if any profit would be made from offering this service.

The choice of the *Oracle and Public Advertiser* to advertise may have been deliberate. Playfair later claimed in about 1806 that, as part of the Security Bank's assets, or perhaps as security that had been deposited in return for the bank's notes, the Security Bank held a one-quarter share in the newspaper.[6]

The Bank of England was not amused once the Original Security Bank had made this small change to their original business plan. Although the Bank of England began issuing £1 and £2 notes on 2 March, it saw the Original Security Bank's notes in competition to the banknotes issued by the Bank of England, which was the only organization legally charged with issuing such notes or any other similar notes. The issue came to a head on 7 March. Security Bank clerks were exchanging the Security Bank's notes for Bank of England notes in the Bank Coffeehouse, conveniently located in Threadneedle Street opposite the Bank of England. The coffeehouse was also a convenient place of business frequented by stockbrokers and bank clerks. Playfair was with his clerks ensuring that all was going smoothly. The Bank of England's solicitor and some employees confronted Playfair and his clerks, demanding under what authority the Security Bank's notes were issued. One of the notes was seized. Playfair was then asked, along with the two clerks, to accompany the solicitor to Mansion House, residence of the lord mayor of London and a very short walk away. Not seeing the mayor on arrival, they were requested to wait in a public room. After waiting about fifteen minutes, Playfair went to find the mayor and the solicitor. When he did find them, the solicitor, a magistrate, and the mayor had already concluded that the seized note was intended to imitate a Bank of England note, so Playfair and his clerks were arrested. The mayor left to consult colleagues on the amount of bail that should be set (Playfair 1797a).

Playfair quickly wrote a note to Hartsinck from Mansion House at 3:00 p.m. He reassured Hartsinck that the Bank of England was only trying to intimidate them in order to close the Original Security Bank. Playfair was convinced that the law was on their side and that the Bank of England would be embarrassed in the long run. With a little spite running through his veins, he also suggested that the Bank of England could be ruined if they were able to present half a million in notes to be redeemed in gold.

At 4:00 p.m. the mayor returned and refused to tell Playfair the amount of the bail. At 4:30 he changed his mind and gave Playfair an amount. Whatever Playfair offered the mayor (Playfair writes in his recollection of the events, "I asked if the money would do and he answered no") was not acceptable, and Playfair and perhaps his clerks were committed to Giltspur Street Compter. This was a prison built mainly to hold debtors; situated near Newgate, it was soon to be a familiar stamping ground for Playfair. They were in the prison for two hours before bail was arranged. Meanwhile Hartsinck presented officials of the Bank of England with £1,000 of their own notes for redemption in gold. When this was refused, a lawsuit was commenced against the Bank of England for non-payment. Playfair also claimed that when the time came for him to be put on trial for the alleged banknote imitation, he would show unequivocally that the Security Bank's notes bore no resemblance to those of the Bank of England.

On 9 March, an article appeared in the newspapers outlining the Original Security Bank's position after they faced "criminal prosecution aggravated by violence and insult" (*Morning Chronicle*, 9 March 1797). The Security Bank was providing a service to the public in a time when cash was in short supply. To allay suspicions that the Security Bank was imitating Bank of England notes, the Security Bank made changes to its own notes. Previously, its notes had stated that they could be traded for "Bank of England Notes." This was changed to read "Bank Notes" only, so that the phrase "Bank of England" appeared nowhere on the new promissory notes. It was also stated in the article that a "fair statement" of what had happened would appear the next day in print. The publication, *A Fair Statement of the Proceedings of the Bank of England against the Original Security Bank on Tuesday the Seventh of March, 1797*, written by

Playfair, gives a blow-by-blow account of the events of that day (Playfair 1797a).

The atmosphere in the Security Bank was quite tense over the next week or so. The crisis peaked on 17 March at a seemingly innocuous, but probably sumptuous, party. That evening, William V, the stadtholder in whose service Hartsinck worked as a minister, threw a birthday party for himself at Hampton Court. Among the invited guests were the Prince of Wales, the Duke of York, and Baron Anne Willem Carel van Nagell, the Dutch ambassador to Britain, as well as Hartsinck. There was a heated discussion between van Nagell and Hartsinck about the Bank of England and its recent actions, and early the next morning the two met for a duel in Hyde Park. In preparation for the duel, Hartsinck called his solicitor, Charles Carpenter, to prepare his will. Carpenter, greatly concerned, informed another gentleman in Hartsinck's house, who went to inform the magistrate at the court in Bow Street of what was happening. Thomas Carpmeal, one of the Bow Street runners, forerunners of the London bobbies, arrived in time to prevent the duel. The duellists were brought before the court. There, after an amicable settlement was reached, the magistrate recommended "oblivion to the past" (*Observer*, 19 March 1797; *True Briton*, 20 March 1797).

The Original Security Bank continued to operate in defiance of the Bank of England. On 25 March 1797, the bank advertised that, in response to the shortage of coins, they were offering notes of value 2.5, 3, 5, 6, 10, and 20 shillings in exchange for banknotes from the Bank of England.[7] The low-denomination notes were now legal by the recent act of Parliament. At the very end of March, the Security Bank and the Bank of England settled their differences. To thumb their noses at the Bank of England or because it was a better business location, within a week of the settlement the Security Bank set up offices at 35 Cornhill, very near the Bank of England and in the heart of London's banking district. Figure 12.2, dating from 1790, shows the approximate location of the bank and its proximity to the Bank of England. The Bank of England is on the left and the Royal Exchange in the middle. To the right, and on the opposite side of the street from the Royal Exchange, is 35 Cornhill, clothed in shadows. If one were able to zoom out from this picture, Mansion House would be in the foreground.

At this point, at the height of his banking career, Playfair became a freemason. On 1 May 1797, he was admitted to the Lodge of the Nine Muses (see *An Account of the Lodge of Nine Muses* 2012; Dawson 1970). The lodge was founded in 1777 by Bartholomew Ruspini and six other men. Ruspini was a prominent London dentist, serving as dentist to George, Prince of Wales. He was active in the lodge until his death in

Figure 12.2. The Bank of England, 1790. Courtesy of Bank of England.

1813. The members of the lodge tended to support the social ideals of Edmund Burke. Like Playfair, Burke was opposed to the French Revolution. Although many foreign émigrés joined the lodge prior to the French Revolution, there were no French members. The members also supported financial reform, and Playfair's work in the Original Security Bank fits the pattern well. Playfair's membership is of interest not only because of the financial and political interest of the lodge. During the summer of 1797, Ruspini had some dealings with the Original Security Bank.[8]

No sooner was the affair with the Bank of England settled than a new but less important problem arose. On 24 March 1797, Sir William Pulteney, baronet and MP, spoke in the House of Commons about the necessity of the Bank of England to open its doors again and to redeem its banknotes for cash (Great Britain 1797, 146–7). He said that he was ready with a remedy should the Bank of England keep its cash coffers closed for another month. Three days later it was reported that the "plan is to establish a Security Bank, issuing its own notes, hypothecated by the title deeds of estates" (*Oracle and Public Advertiser*, 27 March 1797). This sounds very similar to the Original Security Bank, the only difference being that the new bank's notes would be secured by property, while the three-month-old bank was already secured by government securities. The proposal is reminiscent of an earlier proposal in 1796 for the establishment of a similar bank. There was no response from the Bank of England and Pulteney rose in the House of Commons on 30 May to move to have a bill to create a new bank if the Bank of England did not pay its notes in specie by 24 June. The motion was defeated with only fifteen voting for the motion and fifty against (Great Britain 1797, 657–66).

One MP who spoke against Pulteney's plan on 30 May, or was at least lukewarm to it, was Richard Brinsley Sheridan. Sheridan may have been in a possible conflict of interest with this issue. He was, and if not, was soon to be, on the Original Security Bank's books for a loan resulting from Sheridan's financial needs in rebuilding the Drury Lane Theatre between 1791 and 1794. Rather than putting the wheels in motion now, Sheridan wanted to wait another month to see if the Bank of England would change its policy. He noted the gravity of the current situation; already Bank of England notes were being discounted for hard cash. He did concur with Pulteney in condemning the Bank of England for its actions.

Playfair tried by force of argument to stop Pulteney from going ahead with the new bank (see Playfair 1797b; *Morning Post and General Advertiser*, 3 June 1797; *Oracle and Public Advertiser*, 7 and 9 June 1797). He wrote to Pulteney about the proposed bank on 23 May, a week before Pulteney proposed it in the House. Although the motion failed, Playfair went public with his letter. He may have thought that the issue was not completely buried; or he may have wanted to advertise the services of the Original Security Bank. By 3 June, advertisements appeared in the newspapers announcing a new publication by Playfair, *A Letter to Sir W. Pulteney, Bart. on the Establishment of Another Public Bank*. The only other information given in the advertisement was that Playfair was the inventor of lineal arithmetic and that the plan for the Original Security Bank was also available free of charge. The cost of the new publication was set at one shilling. Other than the preface, the entire letter was published over two issues, 7 and 9 June, of the *Oracle and Public Advertiser*, the newspaper in which the Original Security Bank had a probable financial interest.

On the basis of his recent experience, Playfair was no friend of the Bank of England. In his letter to Pulteney he wrote that competition for the Bank of England would be a good thing, echoing Sir John Sinclair from 1790. With the loss of its monopoly, the bank could no longer operate despotically. Playfair wrote, "The commercial men of London are more completely under the control of the Bank than the Pachas [*sic*] in Turkey are under that of the Grand Signior." Despite the advantages of breaking up a monopoly, Playfair made two major points against the establishment of a bank that issued notes in competition to the Bank of England. Both were based on the securities upon which the banknotes would rest. The first, Playfair said, was that the security of banknotes in the current system rested on gold and confidence in the system. There was not enough additional gold to be had and confidence seemed to be lacking. The second spoke directly to the idea of securing banknotes on landed estates. The problem was the liquidity of land. Should it become necessary, land is not easy to sell quickly, and when sold quickly it was often at a reduced price.

From his experience with recent events at the Original Security Bank, Playfair had another suggestion for Pulteney. Notes could be issued on goods. He did not say so in his letter, but the Original Security Bank had already been doing this since the end of February. This was one more seemingly small change to the bank's original business plan. Later, Playfair did not follow his own advice to Pulteney. Beginning in early September 1797, the Original Security Bank advertised in the London newspapers that notes and agreements to accept bills of exchange would be issued on real property – another small change to the business plan (see, for example, *Kentish Gazette*, 5 September 1797). It was these two changes, along with some other bad business decisions, that proved fatal to the operation of the bank. Playfair later claimed that it was Hartsinck's idea to issue notes on goods and property. In view of what Playfair had written to Pulteney about security based on property, the claim rings true.

Although it was Hartsinck's idea to issue notes on goods, Playfair probably thought that that change in the plan was a good idea. This was how his bank in Paris, the Compagnie de commission, had operated. In his letter to Pulteney, Playfair supported issuing notes on goods by carrying out a risk analysis of this kind of banking business. He started well but ended poorly; he made a blunder in an elementary probability calculation and made a crucial assumption that was not true. Here is Playfair's scenario. Goods, wherever stored, could be represented by a bill of exchange. The person "depositing" the goods in the bank would deposit a bill of exchange co-signed by other solvent business people. Presumably, the bill of exchange would be a solid security since several solvent people backed it. Suppose, Playfair wrote, that there were six bills of exchange of £100 each. Each bill of exchange would be co-signed by three solvent people. Playfair then considered the possibility that in two months the £600 would be worth only £400, a loss of £200. For some obscure reason, Playfair claimed that nine (not six) of the eighteen solvent people must become insolvent. He then calculated the probability that nine co-signers would become insolvent. Again, he started well. There were about 50,000 people in trade in England and in any one week there were about 20 bankruptcies. Over eight weeks there would be 160 bankruptcies so that the ratio of people to bankruptcies over two months is 312.5 (50,000/160). Assuming the probability that an individual becomes bankrupt is 1 in 312, he claimed the probability of the £200 loss over the two months would be 1 in 79,872. This likelihood was so low that Playfair concluded that it was a "moral certainty" that insolvency would not happen, assuming that bankruptcies are independent of one another. Independent of the assumption of independence, Playfair wandered into erroneous probability calculations. He claimed that the probability that 2 people would go bankrupt in two months is 1 in 624 or 1 in 312×2 (the correct calculation is 1 in 312×312 under the independence assumption), that 3 people would go bankrupt in two months is 1 in 1248 or 1 in 312×2^2, and so on. Carrying this to 9 people, the probability that all nine would go bankrupt is 1 in 312×2^8 or 1 in 79,872. The blunder in the calculation was not the important issue since the probability is much smaller when calculated correctly. The critical assumption was one of independence. For example, when the Security Bank collapsed, it took all three partners and a number of its customers with it. Their bankruptcies were not independent of one another.

Thanks to Playfair, a brief balance sheet of the Security Bank's operation at the time of its demise is available. Playfair wrote it about eight years after the bank's collapse (circa 1806) as part of a statement he made, explaining his actions to an assignee in the bankruptcy case,[9] probably Samuel Wadeson, one of the lawyers involved for many years in settling the bankruptcy.

Notes had been issued, Playfair said, totalling £40,000 and matched against securities deposited. The securities were not in a balanced portfolio; 65 per cent were from only three different groups. Playfair did not specify what each group offered as security. Whatever they were, the securities had risks attached to them. One that was high risk was £8,000 for a group, whom Playfair dubbed "Irish adventurers," who had begun a bank in Bath. According to Playfair, the "Bath adventurers stopt payment without having done any business." Of lower but still significant risk was the £9,000, Playfair claimed, issued to Edward May; one account in the press concerning a lawsuit from 1804 puts the amount at £14,000, another at £8,000.[10] Whatever the sum was, May was given the amount in Original Security Bank notes. According to the lawsuit, May put up a leasehold estate as his security. Over a term of seven years, May was to pay cash for the bills coming in against his account. Generally, May was not regarded as a respectable individual; but what raised May's risk level was his family connection. In 1795, May had bailed out George Chichester, Earl of Belfast (a courtesy title, on the death of his father he became Marquess of Donegall) from debtors' prison (Maguire 2002, 19–21). The cost of this transaction was that Chichester marry May's illegitimate daughter, Anna. Lord Belfast continued to have financial woes throughout the 1790s. The third and final group comprised Richard Sheridan, John Grubb, and Joseph Richardson. The latter two were Sheridan's friends and fellow proprietors of the Drury Lane Theatre, who were on the Security Bank's books for £9,000. Playfair did not state what they used as their security, but later evidence indicates that it was related to the theatre itself. The £9,000 may have come in piecemeal to the theatre over the course of the Security Bank's brief operation. A letter from Sheridan to Grubb states, in part, "I think Peake could [get] £500 at six months from Playfair." In the published edition of Sheridan's correspondence, the date of the letter is given as "1798?," but it must be 1797, prior to the bankruptcy; the editors also picked the wrong Playfair as the potential lender.

Indications that the wheels were beginning to fall off the new bank appear in early October 1797. A person known only by his initials "A.B." advertised in the *Times* that he was willing to buy the Security Bank's notes at a discount (*Times*, 2 October 1797). Rumours began to circulate that some of the bank's notes were issued contrary to the bank's prospectus. On 25 October, the Security Bank advertised in the *Oracle and Public Advertiser* for people to come to the bank to inspect the securities on which their notes had been issued, but there seems to have been no general panic. In the same issue of the *Oracle and Public Advertiser*, an individual advertised that he would take the Security Bank's notes for goods sold at a fair price at his warehouse in Fetter Lane (*Oracle and Public Advertiser*, 25 October 1797). The impending collapse was in sight when the Security Bank advertised a request for their note holders to meet in the London Coffeehouse on 30 October at 7:00 p.m. "in order to settle the best mode of selling the Securities [i.e. what backed the notes], that they may be immediately paid, and a new Firm go on."[11]

Things went from bad to worse. Beginning early in November, advertisements appear in the newspapers asking those holding the Security Bank's notes to tell the bank the amount of the note, the due date, and where the securities for these notes were held. As the firm was sinking, the office appeared to be in chaos; this was information that should have been on the bank's books. This was probably Playfair's fault. His poor bookkeeping dates back to his Boulton and Watt days. Bankruptcy of the bank was announced on 14 November 1797 in the *London Gazette*.

There was a process involved in the declaration of bankruptcy. Creditors wanting their money would go before the commissioners of bankrupts to prove before a tribunal what the creditors were owed (Welbourne 1932). Several tribunals met simultaneously in a crowded and noisy room in Guildhall. Not all creditors would take this route since there was a one-guinea fee each time the commissioners met to consider the case. And even for a creditor, there was a stigma attached to being connected to a bankruptcy case. Once the debts were proved, the commissioners could then declare the debtor a bankrupt. The decision would be published in the *London Gazette* so that other creditors would be aware that there was a method to recover some of their debts. On declaration of bankruptcy, the creditors appointed a small group from among themselves to track down and keep abreast of the debtor's assets. These were called the assignees for the bankruptcy case. Periodically the assignees would meet to disperse among the creditors any recovered funds from the sale of the debtor's assets. There was also a game to be played among the creditors. If a debtor was declared a certified bankrupt and received a certificate of discharge, then all assets were seized and given to the creditors. This certificate was issued only on the permission of the creditors. Once the assets were dispersed, the creditors had no call on any future earnings of the debtor. However, if the creditors decided not to pursue certification, the debtor remained an uncertified bankrupt. In that case, the creditors did have a call on future earnings. It was a decision between taking some of the money now, or spending some money in the future in hope of recovering a greater fraction of the debt. As well as being labelled an insolvent debtor, Playfair was an uncertified bankrupt from the 1790s on.

By examining more closely the Irish adventurers, one of the bank's three major clients, some insight might be gained into the reasons for the Original Security Bank's demise. As stated already, two of the reasons come down to the liquidity of its capital by allowing real property as security and the poor risks that it took on when providing notes in exchange for these securities. The transactions with the Irish adventurers seem complex and the sources are a little vague on some points.

Playfair's creditor, Colonel Sinclair, along with an obscure Frenchman named Pierre (or Peter) Vertaul, was closely involved with the bank in Bath. The Sinclair-Vertaul connection includes Julius Hutchinson, so it was probably Hutchinson who was the Security Bank partner who brought in, or worked with, the Irish adventurers. In Sinclair's legal proceeding, mentioned earlier, in which Hutchinson had written an affidavit in support of Sinclair, Vertaul pops up as a go-between trying to resolve the dispute between the plaintiff and the defendant, Charles-Alexandre de Calonne and Sinclair respectively. Playfair also acted as a go-between. Vertaul was a highly questionable character sometimes operating under an alias, Harrel La Vertu (or La Virtee). By the time Vertaul went bankrupt in February 1798 he was in the Fleet for debt. On 17 October 1799 he was committed to Newgate prison by the secretary of state after receiving a government order of expulsion in the previous April.[12] Sinclair also wound up in Newgate in 1799. Sinclair claimed that Calonne owed him £725 for his military service and sued Calonne for it in court. Sinclair first claimed in court that he had been paid in forged French assignats and then changed his story to not being paid at all. When Calonne produced a receipt for the payment, Sinclair was charged with perjury, found guilty, and sent off to Newgate for a year. The shadiness of Vertaul is alluded to in a letter of 18 September 1805 to Samuel Wadeson in which Playfair defended his actions describing the situation as "the imprudent & unfortunate in the affair of Vertaul which involved me in very confused & irregular transactions."[13]

The problems between the two banks are laid out in an exchange of letters between Playfair and Julius Hutchinson, with Playfair writing to Hutchinson on 23 August 1797 and Hutchinson responding soon thereafter.[14] The Security Bank had originally advanced £8,000 to the bank in Bath, to be repaid in instalments. Notes from the Bath bank came to Hartsinck. Vertaul was given £2,050 of the notes to discount, and he kept back at least £460. For some reason he was able only to obtain £850 on discounting the balance of what he had. This was unknown to Playfair, who had been against the whole transaction in the first place, but Hutchinson had approved it. The transaction came to Playfair's attention when Sinclair brought in the £460 as restitution for the shortfall in funds. Playfair had to use some creative accounting to put it all through the Security Bank's books, and the £2,050 may not be accurate. At one point, in an undated letter, Playfair wrote in French to Vertaul, "Mr Playfair begs Mons Vertaul to take only the least possible pounds to finish the affair of 2850 pounds."[15] Carrying out the business in this way could not be good for the bottom line.

The bank in Bath was most likely the short-lived Union Bank Bath. In the letters between Playfair and Hutchinson, there is mention of "Giles & Co," "Crow & Co," and "ye Union Bank," all as clients of the Original Security Bank.[16] On banknotes in the Somerset Archives, the partners in the Union Bank Bath, at least in August 1797, were Crowe, Giles,

Lyttleton, and Holt.[17] This bank had its troubles. Giles left the partnership by mutual consent in early December 1797 and within a couple of weeks went bankrupt (*London Gazette*, 2 January 1798). According to the *Bath Chronicle*, the Bank was no longer at its premises in February 1799 (*Bath Chronicle and Weekly Gazette*, 28 February 1799).[18] Further, all surviving dated banknotes of the Union Bank that could be found on the internet and in the Somerset Archives, are from July or August 1797. Hence the assumption that the bank was short-lived and that this was the Bath bank of the Irish adventurers.

Of the other two major clients of the Original Security Bank, one was able to ride out the crisis and the other was not. The one who fell was Edward May, and he probably brought another with him. May went into bankruptcy on 2 January 1798, very shortly after the collapse of the Security Bank (*London Gazette*, 2 January 1798). The possible collateral damage was a merchant named Isaac Bernal. In addition to the leasehold estate that May put up for security to obtain notes from the bank, Bernal granted a warrant of attorney on May's deal. When the assignees for the bankruptcy of the Security Bank did some accounting, it was found that May owed about £2,000 to the bank. With May a bankrupt, Bernal was now liable for the debt. Bernal fought the matter in court and lost his suit in November 1804 (*Morning Chronicle*, 22 November 1804; *London Courier and Evening Gazette*, 23 November 1804). He took May and his son-in-law, now Marquess of Donegall, to court in 1805 (Maguire 2002, 12–13). Bernal went bankrupt on 2 August 1806 as result of this legal dispute or perhaps others in which he was involved at the time (*London Gazette*, 2 August 1806).

Richard Sheridan and his associates rode out the crisis. Early in 1798, it appeared that Sheridan's account would be quickly settled. It became the responsibility of Playfair, as bank manager for the Security Bank, to track down some of the bank's unresolved debits and credits. On 3 February 1798, Playfair wrote to Samuel Wadeson, "Mr Sheridan has promised if he can get the whole account of what is owing to Hartsinck & Co he will immediately set apart a sum for paying it. To this it is necessary I fancy that the operation of separating what is a legal claim from what is only a claim on his honour should be performed as you seemed to think when I was with you at your office. I shall wait upon you when you please to assist in doing this. I am in possession of Mr Sheridans letters (to myself) requesting renewals of his bill & the whole matter in my mind is easily to be settled."[19] It was not that easy. In 1802, part of Sheridan's debt was paid by selling a share in the theatre in Drury Lane (*Morning Post*, 8 September 1802). The purchaser would receive £1 for every performance taking place in the theatre and up to four free admission tickets to performances. The share would be valid from 16 September 1795 to 16 September 1894. But this was not enough. In the statement circa 1806 that Playfair gave, probably to Samuel Wadeson, Playfair had this to say about Sheridan's, as well as Edward May's, debts: "It is scarcely to be doubted that by my plan of settling the debts every thing would have been arranged except about 4000£. Our debts were of an uncommon nature & ought to have been liquidated in a peculiar way. As it [is] the great business remains unsettled. Neither Mr May nor Mr Sheridan's property has been brought to market tho' meetings upon meetings have been held and great expenses incurred."[20] And it continued for several more years. As late as 1813, notices appeared in the *London Gazette* that money was still owed from Sheridan's account (*London Gazette*, 5 January 1813).

Cleaning up the mess after the bank collapsed was not a straightforward task. As the former manager of the Security Bank, Playfair was fairly closely involved in the clean-up under the supervision of the assignees. The original assignees, as elected by the creditors, included William Curtis, William Hutchins, and the law firm of Wadeson, Hardy, and Barlow. Curtis and Hutchins were short-lived as assignees, for reasons we shall soon see. Playfair dealt mostly with Samuel Wadeson. The letters from Playfair to Wadeson begin on 12 January 1798 and continue throughout the year with a small cluster in 1799 and another in 1805. They are typically formal and polite, with attempts to be friendly and helpful.

The first order of business for the assignees was to track down what was owing to the bank and what the bank owed. Notices in the newspapers appeared as early as 14 December (*Morning Chronicle*, 14 December 1797). Those who had originally made deposits to the Security Bank, and those who now held the bills, were to come to two separate meetings "to be apprized of the real state of their affairs." Meetings of this kind continued for a few months. Subsequently, some creditors' claims were rejected by the assignees. Indeed, vultures appear to have been circling over the Security Bank's carcass, and some of the claims were outrageous. In letter to Wadeson on 29 June 1798, Playfair complained that an attorney named Isaacs had claimed that Playfair personally had accepted the liability of paying bills of exchange to the tune of £17,000. Playfair flatly denied this and nothing more seems to have come of it.[21]

In his first letter to Wadeson (12 January 1798), Playfair tried to set a friendly tone. Someone was interested in the silver plate held by the Security Bank and wanted to take an assay to determine its value before bidding on it. Playfair referred the person to Wadeson. He continued three days later with suggestions about how to reduce the labour in collecting and putting all the necessary financial information in order, and promised soon to have a list of the holders of all the outstanding notes issued by the bank. He wrote a similar letter to William Hutchins the same day.

At first, things seemed to be going relatively smoothly for Playfair. During February and March, Playfair and his former partners met with the commissioners appointed for the bankruptcy in order to be examined about the finances of their bank. These meetings continued into April and May, at which time things took a turn for the worse.

The first problem surfaced when the *Times* reported on one meeting between the partners and the assignees: "When the assignees came to enquire of Mr. Playfair about certain concerns of the partnership, which are of a most extraordinary nature, no answer could be returned until the books were made up" (*Times*, 3 April 1798). Playfair replied four days later in the *Oracle and Public Advertiser* (the Security Bank was probably still a part owner in the paper at that time). He considered the statement made by the *Times* unfair. He was not the bookkeeper for the Security Bank; and when he asked the bookkeepers present at the meeting, they said it would take five weeks to complete the required work.

A different and more serious problem arose in the following month. Playfair was to meet with Wadeson and Hardy at their offices on 8 May. That morning he wrote to them that he would be late, because "a friend who promised me a lend of some money of which I am very much in want goes out of town this day & necessity obliges me to see him."[22] For some unknown reason, he completely missed his next meeting with the commissioners for the bankruptcy, scheduled to be held at Guildhall on 22 May. For that offence he was arrested and held without bail in Newgate prison in order to answer the commissioners' questions.[23] The commissioners took their time. Their meeting with Playfair was held at Newgate on 21 July 1798 (*London Gazette*, 14 July 1798).

It is difficult to say whether Playfair spent the whole time in Newgate between 22 May and 21 July. Several letters that he wrote during this time give no address. Only one specifies that it was sent from Newgate; and it is dated 23 May 1798, the day after his arrest. He probably did remain in Newgate the whole time; in a few of the letters he expresses his desire that the commissioners would meet soon.

Most letters written at this time contain polite requests to the assignees for information, or for copies of material related to the Security Bank's operation, so that he could prepare for his meeting with the commissioners. These are typically written to Wadeson. When he wrote to Curtis and Hutchins, he was more testy. To Hutchins, Playfair wrote that he wanted those who had made false accusations against him to be examined by the commissioners and also confronted by Playfair himself. One accusation was that Playfair would escape the whole mess and go to France. What he wrote in a postscript to a letter to Hutchins on 17 June 1798 underlines his political beliefs that he expressed in his post-revolution writings:

> The acrimony of two or three of the small creditors must have been observed at Guildhall. One of them whom I only know by name went so far as to say that I was a Democrat & would go to France. Unfortunately for him my political principles are too widely known for that to obtain belief. I have written many things in favour of this Government & have received scores of letters of thanks on the subject. Some of these have been translated on the Continent and reprinted in America as well as circulated in this country at the expense of government.[24]

He concluded that "calumny may be carried too far" and that he would find the originators of the slander against him. Despite his testiness, he appealed to Hutchins on compassionate grounds, saying, "My family are in a really disagreeable situation." Later, Playfair believed that Hutchins was the cause of the problem.

After some of the dust had settled on the bankruptcy case, at least two of the partners, Hartsinck and Playfair, began playing the blame game. Once Hartsinck thought he had settled his accounts and was free to continue with his life, he wrote to his stadtholder, William V, on 14 November 1798:

> I should consider myself lacking in the gratitude which I owe Your Serene Highness for the kindness which you have shown in sympathizing with me in my misfortune, if I did not seize the first opportunity of informing Your Serene Highness that, *notwithstanding the exertions of my former partners and their infamous friends*, I have at last had the good fortune to justify my conduct in such a way that nearly all my creditors, the assignees appointed by them, the commissioners who, by the laws of this country, are employed in such cases, and the Lord Chancellor have acquitted me of all blame. (Day 1911, letter 244; italics mine)

Playfair had a different view, as seen in a letter that he wrote to Samuel Wadeson: "When you came to be solicitor to the bankruptcy all was confusion, prejudice was at its height, misrepresentation had gone forth, falsities had been invented, & in that Hartsinck had made a party to save his reputation at any expense."[25] In his 1806 statement, again probably given to Wadeson, Playfair was more explicit.[26] He placed much of the blame for the failure of the Security Bank on Hartsinck. Initially, he said, Hartsinck brought about £16,600 to secure the bank by depositing these funds in another London bank. Soon, however, Hartsinck withdrew his money and committed only about £5,000 to the Security Bank. Then he began speculating in the financial market and lost most of the money committed to the Security Bank. This happened about two months before the bankruptcy. Playfair succinctly described what was done next: "Notes were then issued for the purpose of raising money & it was bursed at a great loss & by irregular means." If this claim is correct, it also provides another major and perhaps overriding reason for the demise of the Security Bank.

At the bankruptcy proceedings, Hartsinck hid about £5,000 of his property from the bankruptcy commissioners, so Playfair claimed. Playfair had further accusations. Through his relatives and powerful friends, Hartsinck manipulated the choice of the assignees, arranging the appointment of William Curtis and William Hutchins. Curtis was a London alderman who, according to Playfair, "pretended to be a creditor for a few panes of plate glass he had left in the shop." Hutchins was a former bankrupt who, again according to Playfair, "lived by selling peoples accounts."[27] The assignees quickly settled Hartsinck's account. As a result, Hartsinck was given a certificate of bankruptcy so that no further financial claims could be made upon him. He would be free to start again. The timing of Hartsinck's certificate ties in with a change in assignees. The change was made during September 1798; Hartsinck's friends, Curtis and Hutchins, stepped down as assignees on about 15 September 1798, perhaps as soon as they had obtained Hartsinck's certificate of bankruptcy for him.

The surviving evidence points to the truth being mostly on Playfair's side. Over 1801 and 1802 there were three lawsuits against the three partners that exposed some of Hartsinck's underhanded activities (Espinasse 1804, 44–6 and 102–3). Part of the evidence is given in a report of one of the cases, with a judgment rendered on 27 April 1802: "One of the Defendants, *Playfair*, let judgment go by default; the two others pleaded bankruptcy, having obtained their certificates. Their certificates were impeached, on the grounds, of the two Defendants having been guilty of concealment: having lost money by gaming in the stocks: and that money had been given to induce creditors to sign their certificates" (102). Playfair did not attend the trial and hence the judgment went against him. It did not really matter; he had no money to pay, whichever way the judgement went. Both Hutchison and Hartsinck had obtained their certificates of bankruptcy; hence their defence. The judge found in favour of Hutchinson, but against Hartsinck. Hartsinck's certificate of bankruptcy was revoked and it was now open season on Hartsinck for any of the Security Bank's creditors who had not previously been paid. Hartsinck soon fled to the Continent to avoid his creditors and to protect what was left of his property. By mid-May 1802 he was in Paris with his family (Day 1911, letter 270).

As for Hutchinson, Playfair claimed that he made every effort to avoid the bankruptcy. Hutchinson's main asset was an estate in Essex called Shelhaven, a 400-acre farm by the sea. After the bankruptcy, it remained in Hutchinson's name and he may have drawn some income from it. Most rents from the farm went to paying the creditors in the bankruptcy. The assignees in the bankruptcy became lax in handling Hutchinson's case. Since the property was still in his name, in 1806 Hutchinson tried to put a mortgage on it to raise money for himself. His application went to court, which ruled that Hutchinson could not have any interest in the property until the creditors were all paid twenty shillings on the pound.[28] Hutchinson died in 1811 and left the farm to his family, once the lien on the farm had been removed. But the family never inherited. The farm was eventually sold in 1825 to pay off the remaining debt.[29]

After Hartsinck fled, he lost what he had left behind. It took until 1805 for his property to be seized and sold. The furniture from his house went up for auction on 25 May 1805, to be followed three days later by an auction of his wine cellar, which included bottles of Chateau Margaux and Chateau Lafite, as well as a then popular hock wine, all with vintages dating back 50 years. Perhaps the oldest wine in Hartsinck's cellar was a 150-year-old Tent, a sweet wine from Moselle. The last of the property to go was Hartsinck's extensive library, which went up for auction on 20 September. The library included works in English, French, and Dutch, covering history, politics, divinity, law, biography, and science (*Morning Post*, 22, 25 May, 19 September 1805). Despite these losses, when he died in 1835, Hartsinck left an estate of £15,000 in 3 per cent consols to family members living in England.[30] Playfair was correct in his assessment of Hartsinck being able to hide some of his assets.

Both Hutchinson and Hartsinck lost considerably in the whole affair, but they remained far from destitute. It was Playfair who was left penniless. Documents that survive from his 1801 imprisonment in the Fleet for his debt show that he was living nearly hand to mouth at the time.[31] When he applied to be discharged from the Fleet under a recently enacted Act for the Relief of Certain Insolvent Debtors, he submitted a list of his assets to the court. There were only two. The first was a fee of £25 due to him from a shady character named Robert Kennett for writing

Playfair had done for Kennett. The second was half the profits to be earned from the edition of the *Commercial and Political Atlas* published in 1801. He was not to get the entire half; there was also an unspecified lien on his profits, which had been lodged with his lawyer. In the same document, he described his situation when he arrived at the prison:

> At the time of my first imprisonment I was possessed of several articles of furniture being in the apartments rented by me in Great Brook Street aforesaid, the particulars whereof I cannot fully ascertain but believe the value of the whole to be under twenty pounds; and since my imprisonment the greater part has been sold and disposed of, and the money arising thereby spended in the support of myself and family of a wife and six children. And a small part of the furniture has been removed to No. 8 Northam's Place, Kentish Town, where my family now resides.

Financially as a bankrupt, Playfair had hit bottom.

This was Playfair's last venture into banking. His career in private enterprise was in tatters. Although he tried to initiate some business ventures, what was left to Playfair was writing. As he told his kinsman Peter Roger in 1802, a career in letters did not pay well. It was an insightful comment; Playfair spent the rest of his life trying to stay out of debtors' prison using tactics he had learned in England early in his career and had later honed in France.

Chapter Thirteen

Playfair Ekes Out a Living as a Bankrupt

Although he was bankrupt, none of Playfair's creditors made an immediate move to put him in debtors' prison. He had no money that they could lay their hands on. For them, it was better to leave him free in order for him to make some money to pay off his debts. Hartsinck and Hutchinson were better bets, and neither of them was put in debtors' prison. The easiest financial path open to Playfair was working as a writer; bankruptcy would have closed many other paths. Up to the time that Playfair started the Original Security Bank, he had been publishing opinion pieces and pamphlets related to the French Revolution. After his bankruptcy, Playfair went back to working as a writer. He also tried to restart his career in manufacturing.

When Playfair returned to his literary pursuits, he first tried two that he considered to be safe paths. Very quickly he decided to update and reissue his most successful piece of propaganda, the *History of Jacobinism*. With the previous success of the *Commercial and Political Atlas* in mind, he tried a new publication to highlight his graphical work, to which he gave the title *Lineal Arithmetic*.

With the fall of Robespierre on 28 July 1794, there were political changes in France. There was a new constitution. Similar to Britain and the United States, there were now two legislative bodies and an executive body in France. The executive, called the Directory, was a five-member council. The war between France and Britain, along with Britain's allies, continued. On the battlefields of Italy, Napoleon Bonaparte was on the rise. By late 1797, Bonaparte was advocating an invasion of Britain.

Playfair responded with a proposed update to his *History of Jacobinism* to be published by John Stockdale (*True Briton*, 6 February 1798). What Playfair had previously published covered the period up to the fall of Robespierre. The proposed update, first advertised in the newspapers on 6 February 1798, would add the period spanning the death of Robespierre to the return of British diplomats from Lille in September 1797 after failed peace negotiations with France. The intended publication date was 1 May 1798, about three months from the initial advertisement. Like the first edition, it would be sold by subscription. This time the subscription cost was set at 7s. 6d. a copy, a shilling less than the first edition. With an appeal to patriotism, Playfair's and Stockdale's profits would be given to the Bank of England to be applied as aid to the government against the threatened invasion of England.

When the new edition came to print, it was not new, but it did have a new printer, John Wright. Having experienced the first edition, perhaps Stockdale thought it too great a risk. Wright appears to have been an appropriate choice. With premises at 169 Piccadilly, his premises were not far from Stockdale's. More importantly, Wright was the printer for the *Anti-Jacobin, or, Weekly Examiner* (Life 2021). Like the *Tomahawk*, it attacked the opposition Whigs and any sympathizers of the French Revolution. It also had government backers, possibly Pitt himself, as well as government contributors. And like the *Tomahawk*, its publication life was short, ending in early July 1798. As one anti-Jacobin project came to an end, perhaps Wright saw merit, and possible government backing, in publishing Playfair's *History of Jacobinism*.

The new edition of the *History of Jacobinism*, which appeared in two volumes, cannot have been a great success. No advertisements for it could be found. Likewise, there were no reviews of it in the literary journals. It did not live up to its hype. None of the additional promised material appears in it. In fact, it appears to be a reprint of the old edition to the point that the type appears to be exactly the same, with identical spacing and placing of words on the page. Only the two title pages are different. Perhaps the new edition was made up of unsold copies of the old one, with the title page replaced, and perhaps some others, or the type for the old edition was kept intact and just transferred from Stockdale to Wright. It is unlikely that Stockdale would keep the type intact for three years, and more likely that Stockdale sold his remainders from the first edition to Wright.

Playfair's second attempt at rejuvenating his literary career was in the publication of *Lineal Arithmetic* in 1798 (Playfair 1798b). Printed for the author by Alexander Paris, it was a repackaging of the *Commercial and Political Atlas*. In *Lineal Arithmetic*, Playfair covered the same topics and had graphs very similar to those in the *Commercial and Political Atlas*. In some cases, Playfair rewrote the text to go with each graph, whereas in others he mostly copied what he had written in 1786. Mainly shading and stippling were used in the graphs of *Lineal Arithmetic*, whereas in others colour was added. Reflecting the cost of producing the graphs, the 72-page book, including thirty-three graphs, sold for 10s. 6d. (half a guinea) compared to 7s. 6d. for 814 pages of the *History of Jacobinism* (Playfair 1798a). Similar to the retread of the *History of Jacobinism*, no advertisements were to be found for the *Lineal Arithmetic* in the London newspapers. Nor were reviews of it found in the literary journals. Hawking the book on his own, *Lineal Arithmetic* did not sell well. This may explain why Christie's auction house described the book as "very rare" in their sale of it as part of Edward Tufte's book collection in 2010.[1]

The next year, Playfair made one more attempt at his literary renaissance with *Strictures on the Asiatic Establishments of Great Britain* (Playfair 1799). Written late in 1799, it was first advertised for sale in late January 1800 (*Morning Chronicle*, 25 January 1800). Later in 1801, he followed with another edition of the *Commercial and Political Atlas* and a new work, the *Statistical Breviary*, the first published earlier in the year and the second at the beginning of October (Playfair (1801b). The latter two together are beloved of historians of statistics, and several others for their innovative graphics.[2] As they had done with Playfair's recent work, the critics ignored the new edition of the *Commercial and Political Atlas*. Both *Strictures on the Asiatic Establishments of Great Britain* and the *Statistical Breviary* came under heavy criticism in some of the literary journals, the first for its length, verbosity, and political stance, and the second for inaccuracies in the data portrayed in it.[3]

Henry Dundas was deeply involved in the affairs of India and its relation to the East India Company. In 1799 he was president of the Board of Control, the British government body that supervised the East India Company's affairs. Prior to the publication of *Strictures on the Asiatic Establishments of Great Britain*, Playfair tried a little scam on Dundas in order to make some extra money. It was the old technique, perfected in the 1780s and before by newspaper editors: "I have some damaging information on you; pay me not to publish it." Playfair wrote to Dundas on 3 August 1799 enclosing a short manuscript (not extant) that he thought he might include in the book he was writing. He was concerned that what he wrote in the manuscript about the East India Company might be harmful to the company and wanted Dundas's opinion on it. Dundas wrote back on 18 August probably advising (the letter is not extant) Playfair not to publish what he had been sent. The attempted scam becomes obvious from Playfair's next letter to Dundas on 7 September 7.[4] He told Dundas that he had gone to the East India Company offices asking to be paid (in his words "indemnified") for not publishing the manuscript. He talked to Hugh Inglis, the deputy chairman. Inglis and one of the directors, John Manship, read the manuscript and "saw not any danger from its publicity & could do no steps about it." Playfair was writing to Dundas hoping that he would put pressure on the East India Company: "I believe the interests of the country in some degree involved & whatever indemnity I might expect from a trading company, from my country I have no wish to obtain any."[5]

Prior to devoting 120 pages to commentary and 100 pages to appendices in *Strictures on the Asiatic Establishments of Great Britain*, the source of Playfair's concern about India can be seen in the graph that appears at the beginning of the book (figure 13.1, which is an update of figure 5.1). For 100 years the British had experienced a trade deficit with India and over the last two decades it had grown substantially after the suppression of the smuggling of tea into Britain by lowering the tariffs on tea through the Commutation Act of 1784. The act also gave the East India Company a virtual monopoly on the importation of tea into Britain (Mui and Mui 1963).

Playfair's suggestions, which drew the ire of the reviewers, were manifold. Most importantly, and positively according to modern views, Playfair wanted to end the East India Company's monopoly on trade. He also wanted a change in the way in which the British possessions in India were governed; the East India Company had changed its original focus from commerce to revenue generation gained by administering territories acquired in India. Curiously, Playfair wanted the East India Company to maintain its administrative control over the British territory in India, but subject the company to more regulation and control, with all surplus revenue being sent to the British government. Some of his suggestions were even more questionable. In order to redress the trade imbalance, he wanted India to ship raw goods to Britain for processing there, thus putting established manufacturers in India out of business. Further, Playfair thought that exportation of British woollen goods, small-calibre guns, muskets, ammunition, hardware, and objects such as anchors would help redress the balance of trade.

Likely getting back into business was always on Playfair's mind. Some time during 1800 or before, Playfair found a new business partner – Nicholas le Favre, who came to London from Dublin. While in Dublin, le Favre petitioned the Irish Parliament in 1789 to build wet and dry docks in Dublin on the River Liffey (Ireland, Parliament, House of Commons 1790, 390). During the 1790s, he ran a lottery office in Dublin (*Dublin Evening Post*, 29 October 1796). It was his experience, or at least interest, in shipbuilding that le Favre brought to the partnership. Playfair and le Favre obtained a patent in 1800 for a moveable wave baffle attached to the bow of the ship (Great Britain Patent 1800/2455). It was claimed that the invention would increase the ship's speed in rough

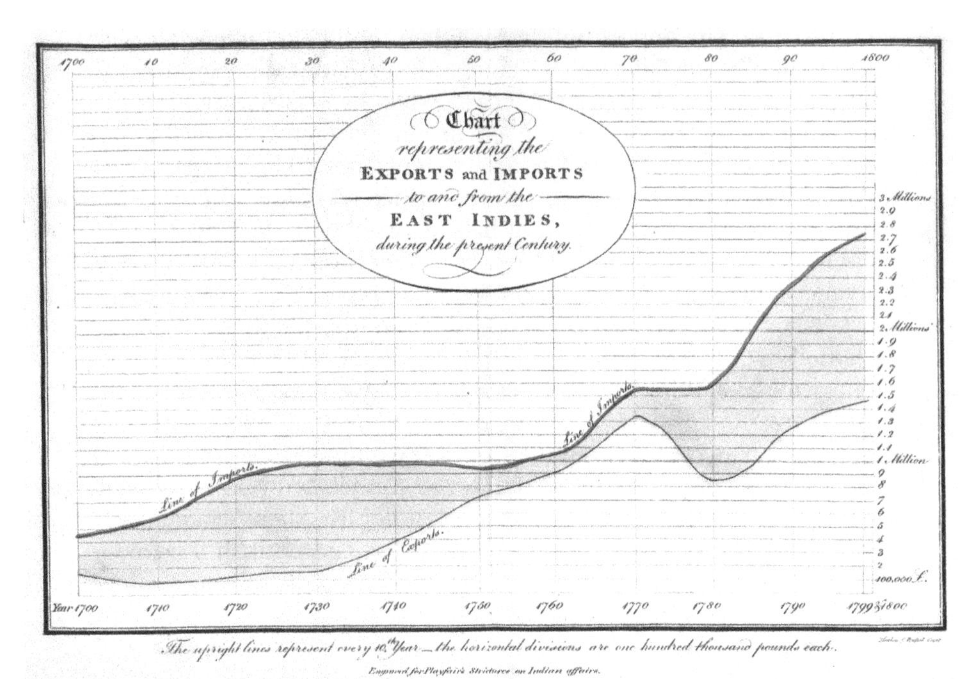

Figure 13.1. Playfair's graph on trade with India. Courtesy of London School of Economics Library.

waves. There is no evidence that Playfair and le Favre took their invention beyond the drawing board. Further, the evidence is slight, but le Favre may have had financial problems similar to Playfair's, which make sense in light of Playfair's bankruptcy and the stigma attached to it. Le Favre's problems pop up in reports of lawsuits in Dublin in 1802 and 1804, when, as a result, some of le Favre's property was seized and sold (*Saunders's News-Letter*, 2 July 1802 and 4 June 1804).

Whatever his business activities in 1800, Playfair found time to translate Jakob Bötticher's *Statistische Uebersichts-Tabellen aller Europäischen Staaten* (Bötticher 1789) from German into English (Boetticher 1800). As a book of statistical tables with a plethora of information, it was a topic that made full use of his talents. It was not a word-for-word translation. Playfair rearranged material, added some of his own in places, and changed all the units of measurement to English units.

William Windham, who had financed Playfair's publication of the *Tomahawk*, continued to be on Playfair's mind. Perhaps Windham might help Playfair again if Playfair could establish some kind of quid pro quo. The opportunity came when John Gordon Sinclair once again walked across Playfair's stage. Sinclair had been committed to Newgate on 9 May 1798 for one year.[6] When he was released in 1799, he continued to pursue debts owed to him by French royalists. This time it was the royalty itself. Sinclair went after Charles, the Comte d'Artois, for money he had lent to the prince while the prince was in England and for expenses Sinclair had incurred while raising the regiment in Coblenz. He also sued Charles's elder brother, Louis Comte de Provence (later Louis XVIII), who was not living in England at the time. Sinclair had d'Artois arrested for debt. Playfair wrote to Windham on 21 January 1801, after he learned that Windham was helping d'Artois monetarily in settling the lawsuit. He advised Windham that if the British government paid d'Artois's debt to Sinclair, it would look very suspicious.[7] The case went to the Court of Common Pleas in February 1801 (*Morning Post*, 7 February 1801). The comte's lawyer argued, and won, in court that Charles was protected under the Alien Act and could not be held in custody for debts contracted outside British territories. Playfair's letter seems to have had little impact and no evidence can be found that he received a quid pro quo.

Playfair continued to be plagued by financial difficulties. The merchant Isaac Bernal had Playfair committed to the Fleet Prison for a debt of £42 2s. 6d., a minor amount compared to some of Playfair's other debts. Five years later, Bernal himself went bankrupt, probably a victim of Playfair's venture into banking that began late in 1796. Whatever Playfair had or would earn from the new edition of the *Commercial and Political Atlas*, it would be lost to his creditors.

A little more than a month after Playfair was sent to the Fleet, Parliament passed An Act for the Relief of Certain Insolvent Debtors (Great Britain 1801), which allowed for the release, under certain conditions, of prisoners whose debts were less than £1,500. The debtor was required to submit a list of his real and personal assets to the prison keeper. Then he was to advertise in the *London Gazette* his desire to seek a discharge. The advertisement was to appear three times over a three-week period. The three weeks were to allow creditors who had sued the debtor for recompense, such as Bernal, as well as any other creditor who had not made suit, to view the list held by the prison keeper, and to lay claim to whatever assets were on the list. Once this had been done, the prisoner was free to go. At this point, unless the list had been fraudulently compiled, the debtor was left with a clean financial slate. Creditors had no further claim on any future earnings or assets of the debtor. The debtor could not call in any loan that he had made, as it was now either void or the property of one of the creditors (Welbourne 1932). And yet the freed prisoner carried the stigma of being labelled an insolvent debtor for the rest of his life.

Playfair was quick off the mark to get out of debtors' prison. His first notice appeared in the *Gazette* on 4 July 1801. His was one item in a list of several prisoners seeking discharge. Then there was a glitch. The next notice appeared a week later, but it was incorrectly labelled "Third Notice" rather than "Second Notice." It did not fulfil the legal requirements, so it was void. The corrected "Second Notice" appeared on 18 August with the proper "Third Notice" on 22 August. The *Gazette* appeared twice a week. In the 11 July issue, the printer had stated that any error in the notice lists found by an imprisoned debtor would be corrected free of charge in the following issue. If Playfair had noticed his flawed notice earlier, it could have been corrected as early as 18 July. Playfair was discharged from the Fleet on 3 October 1801. Despite his release from prison and clean slate, Playfair remained an insolvent debtor in the eyes of the law.

On Playfair's release, the *Statistical Breviary* came out in print. With the new act for insolvent debtors and the successful handling of his release from prison, any remuneration received from the sale of the book was his and not his creditors'. Playfair used his translation of Bötticher's work as the basis for the *Statistical Breviary*. He saw the need to simplify Bötticher's information and to make it more accessible to the reader. He also went beyond Bötticher by including information on India that he thought was of interest to the British reading public, given their stake in the area. For each country, Playfair gave a one-page description and commentary. Then followed a page of statistics for each country showing items such as the size, population, revenues, number of land forces, number of seamen, extent of the seacoast, and more. The simplicity of presentation made it easier for the reader to digest. Playfair went a step further by drawing two highly innovative graphs showing four and five variables at one view. One of his multivariate graphs is shown in figure 13.2. The circles show the areas of each country in square miles. The colour on the circle shows the nature of the military power of the country: green for a maritime power and red for a land-based power. The red band around the green circle for France shows the landlocked territories gained after the French Revolution. The red line to the left of each circle indicates the population size in millions of people and the yellow line to the right indicates the revenue in millions of pounds sterling. Although one reviewer was generally critical of the *Statistical Breviary*, claiming a lack of accuracy in some of the data, he did find that in the graphs the data were shown "in a very clear manner" (*Critical Review* 1802, 76). Although the presentation was revolutionary in its

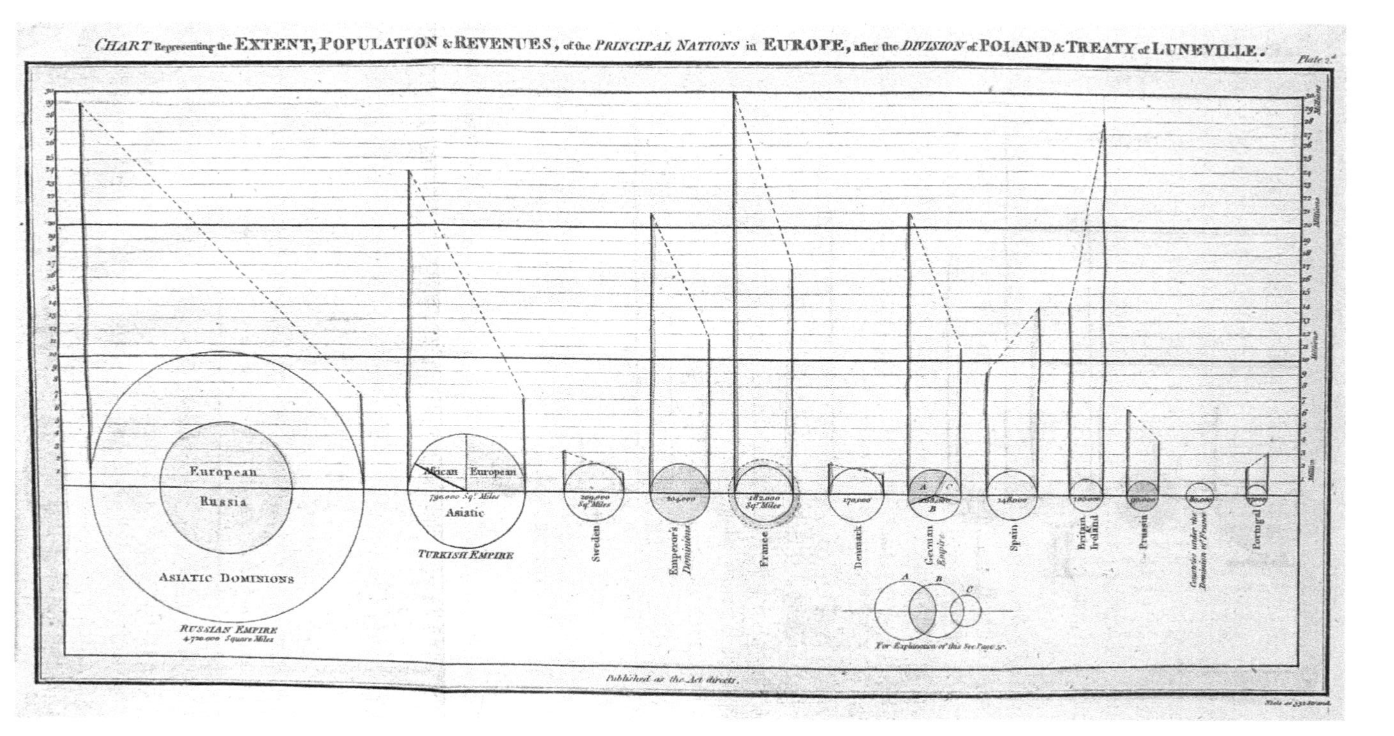

Figure 13.2. Playfair's depiction of four variables simultaneously. Courtesy of Stephen M. Stigler, University of Chicago.

day, William Cleveland has commented in a modern look at Playfair's graph that the sizes of the nations in figure 13.2 might be plotted more effectively in a dotplot, which is very similar to a bar graph (Cleveland 1993, 142–3).[8]

As for questions of accuracy, the problems that the reviewer points out are sometimes due to Playfair's uncritical use of his source, Bötticher's *Statistische Uebersichts-Tabellen*. Referring obliquely to the fall of Constantinople in 1453, the reviewer corrects Playfair's statement that "the finest portion of the world is in the possession of the Turks since the year 1000." What the reviewer ignored was that the Seljuk Turks had taken much of Asia Minor during the eleventh century, with the Byzantine Empire still holding territory hugging the edges of the Mediterranean and Black Seas. In the original, Bötticher was a little vague about the claim, but did mention the year 1000, and Playfair overly interpreted his statement. Where Playfair took his information directly from Bötticher was a list of the produce of Germany. Included on Playfair's list was olive oil; Bötticher uses the word *baumöl*, which means "inferior" or "second press olive oil." The reviewer raised his eyebrows at this item in English on Playfair's list; Bötticher validly included the item since at the time northern Italy was part of Habsburg lands, and therefore part of Germany. The confusion comes because both Playfair and Bötticher have two separate entries: the German Empire and the Emperor of Germany's Dominions. Italian possessions are listed under the second entry. While the produce of every other dominion is given in the second entry, there is no mention of the produce of Italy. One quibble in which both Playfair and the reviewer were both wrong was in the population of London. Playfair pegged it at 1,100,000 without mentioning his source. Quoting what he thought was a reliable source, the reviewer thought the number was below 700,000, while his source actually gives 650,000 for 1798 (Middleton 1798, 452). Without a census, population estimation at the time was notoriously inaccurate. The 1801 Census of England and Wales places the number at 864,845, nearly halfway between the source and Playfair (Great Britain 1802). Whatever the shortcomings of the data, Playfair's use of graphs was brilliant.

The *Statistical Breviary* made its way to Paris, where it was seen and translated into French by the statistician Denis-François Donnant in 1802. Little is known of Donnant other than that he translated several works from English to French and German to French. He was a member of several learned societies in France, including the original Société de statistique (Hildebrandt 2003, 465–7). Donnant had left France in about 1790 and spent a few years in the United States. To his translation of the *Statistical Breviary*, Donnant added a large amount of material on the statistics of the United States and titled the whole work *Élémens de statistique*. He also commissioned his own engraver in Paris to replicate Playfair's original graphs. Late in 1802, Donnant sent a copy of *Élémens de statistique* to a bookstore in London with a covering letter to Playfair. It was about six months later when Playfair actually had the book in hand (Playfair 1805, iii–iv).

The revenue from the *Statistical Breviary* must have been meagre or spent very quickly.

In late February 1802, Sir Richard Phillips applied to the Literary Fund for financial help on Playfair's behalf. Phillips was familiar with Playfair's work. An author and bookseller himself, Phillips had established and run *Monthly Magazine*, a literary journal, in 1796. As a result of Phillips's intercession, on 23 March 1802 Playfair received £10 from the Literary Fund.[9] To put this amount into perspective, according to the historian Lucy Worsley, an annual income of £100 would have kept Jane Austen, her sister, and her mother, who were all living together in 1808, in a "limited but genteel and independent lifestyle" (Worsley 2017, 285).

In his letter, Phillips mentions that Playfair has a wife and six children to support. Playfair also mentions this in his application to be released from prison under the Act for the Relief of Certain Insolvent Debtors.[10] Three have been mentioned so far: sons John, born in about 1780, and Andrew William, born in 1790, and a daughter Zenobia, born in 1795.[11] In addition, Elizabeth was born about 1782 and nothing is known about Louisa.[12] Neither Louisa nor the sixth child appears in any records, it can be assumed that they died before reaching adulthood, as was common for this time period.

Dim though it was, there was some light at the end of the financial tunnel for Playfair. And it was all due to Napoleon Bonaparte. On 9 November 1799, the Directory was overthrown in a bloodless coup and quickly replaced by the Consulate with Napoleon as first consul. In the *Statistical Breviary*, Playfair saw this as a positive event, stating that after the excesses of the French Revolution and successive tyrannical governments, "at last a general of uncommon personal merit and abilities, has ventured to ameliorate the state of the people, and to govern with mildness and moderation" (Playfair 1801, 32). Like Beethoven, Playfair admired Napoleon until war, which had ended briefly in 1802, recommenced about a year afterward, and Napoleon proclaimed himself emperor in 1804.

To consolidate his power in France and to give him time to institute reforms, Napoleon agreed to make peace with the British. For their part, the British, straining financially under the prolonged war with

France, wanted to rebuild their economy and to restore their trade with Europe. The war ended, temporarily, with the Treaty of Amiens on 25 March 1802. Soon after the signing, a wide variety of Britons flocked to France, including Playfair. In addition to artists, clergymen, scientists, and soldiers, many bankers and merchants "were eager to recover confiscated property and old debts, or to revive business relations" (Alger 1899, 740). Although he described himself in his current profession as a journalist, Playfair more likely came again to France for another reason. There was the matter of the 3,200 French livres (£140 approximately, a substantial amount for the time) that Playfair had posted as a bond for the Société Gerentet et Playfair. It was still there, at least on paper, as part of the funds seized from James Carrey, the English banker in Paris. It would be another fifteen years before Playfair could put his hands on his money.

The East India Company reappears on Playfair's radar in 1804. He took up a new issue concerning the company after reading a French newspaper or a book that he saw probably in Stockdale's bookstore. The background to this issue was a ship in the East India's fleet that had been captured by a French privateer – the Peace of Amiens was over and the war was back on. There were passengers on the ship, the Admiral Aplin, and the privateer seized all the British correspondence that could be found, nearly ninety letters in all. The letters were taken to Paris, translated into French, and published in the 16 September 1804 issue of the *Moniteur*, which had become the official journal of the French government. The ship was taken as a war prize. Someone in London read the letters in the *Moniteur*, translated them back into English, and published them through a London printer under the title *Intercepted Letters. Letters Intercepted on Board the Admiral Aplin*. This was where Playfair came onto the scene. The translation was available at Stockdale's bookshop, as well as two other places. On reading the letters in the *Moniteur* and the translation, probably at Stockdale's, Playfair decided that the translation was incorrect in several key places. On 30 September 1804, he wrote to at least two newspapers about the issue. His letter appeared on 2 October, the same day that the translation was advertised for sale in the same newspapers (*Morning Chronicle* and *Morning Post*, 2 October 1804). His interpretation of the British publication of the letters was that "they were introduced by some rebel British subjects now in Paris, for the purpose of disturbing our peace, and destroying our confidence." Before the end of October, Playfair had published his analysis in *Proofs Relative to the Falsification of the Intercepted Correspondence in the Admiral Aplin, Indiaman* (Playfair 1804). Playfair was probably off base in his analysis. In his letter to the newspapers, he wrote, "The expression, for example, of 'stubborn fools,' applied to the Court of Directors, should have been 'very decisive Gentlemen,' which, though not a term of panegyric, is not either so strong, so grating to the ear, or so ill applied as the others." The phrase that actually appeared in the English translation was "obstinate fools," which, along with "stubborn fools" is an accurate translation of the phrase *"entêtés imbéciles,"* which appears in the *Moniteur* (*Gazette nationale ou le Moniteur universel*, 16 September 1804).

During 1801, Playfair met Robert Kennett, a bankrupt whose money problems dated back to 1785. In June 1794, he was the subject of a small ad campaign in the newspapers placed by the Society of Guardians for the Protection of Trade against Swindlers and Sharpers. According to the society, he was "guilty of transactions grossly usurious" (*Oracle*, 7 June 1794). The accusations were based on reports of a 1794 trial against Kennett in the Court of King's Bench with Lord Kenyon, the lord chief justice, on the bench. In a report of the trial, "Lord Kenyon observed, he wished it be known to the Public, that Mr. Kennett was a man who did business in so an iniquitous manner, that the person who did so, ought to be burned in the hand, and held up to public scorn" (*Morning Post*, 9 June 1794). Even an artist's rendition of Kennett prior to his execution for forgery (not involving Playfair) in 1813 depicts him as bad character (figure 13.3).[13] Why would Playfair become involved with such a man? It probably had much to do with the general attitude towards debtors in the late eighteenth and early nineteenth centuries. There was a stigma attached to debts that could not be paid; debtors were generally viewed as criminals. There was no distinction made between those who went into debt as the result of misfortune or poor decision, and those who went into debt because of a fraudulent action. Playfair and Kennett probably became business associates because of their mutual difficulties in finding solvent partners. Although Playfair brought legal and financial baggage to the partnership, Kennett brought even more. They were a matched pair.

Despite Kennett's bad reputation stemming from his 1794 law case, in the summer of 1803 Playfair and Kennett went into partnership as gun carriage makers for ships in the fleet of the East India Company. It must have appeared to be an easy business to enter. As seen in figure 13.4, a naval gun carriage was a fairly simple device. Both Playfair and Kennett had manufacturing experience, Kennett as a one-time cabinetmaker. The problem was that the manufacture of gun carriages was slowly being outsourced. By the end of the Napoleonic Wars, nearly all gun carriages for the East India Company were made in India, not in England (Great Britain 1814a, 32–3). The teak from India was more durable and less likely to rot than other timber sources available through English timber merchants. Playfair and Kennett had entered a dying

Figure 13.3. Robert Kennett. Courtesy of British Museum.

Figure 13.4. Naval cannon in its carriage. From https://en.wikipedia.org/wiki/Gun_carriage. Image available via Creative Commons.

business, and they wound up their partnership in 1804. The termination of their partnership may be related to Playfair being committed again to the Fleet for debt on 13 February 1804; he was discharged on 17 September.[14]

Just prior to forming their brief partnership, Kennett involved Playfair in a questionable affair related to Kennett's finances. Since the middle of the 1790s Kennett had been involved in an intrigue to keep his property out of the hands of his creditors by hiding his assets. His scheme involved manipulating the legal process of bankruptcy cases in the selection of assignees – the group of creditors appointed to keep abreast of the bankrupt's assets. Kennett's response to his mounting debts was to replace his former assignees with ones he thought were more malleable to his interests. He arranged for his son Henry to obtain power of attorney for a senile relative, John Parlby, living in Portsmouth. Henry declared that Kennett owed money to Parlby. While Playfair was in the Fleet, or just prior to his incarceration, Kennett had hired Playfair to write something for him. The fee of £25 was unpaid so that Playfair was also one of Kennett's creditors. Parlby, through Henry Kennett with his power of attorney, and Playfair applied as creditors to the commissioners of bankrupts to be made assignees to administer Kennett's debt and were successful in their request. The problem for Playfair was that as soon as he was discharged from the Fleet in 1801 under the Insolvent Debtors Act, he then had no claim on Kennett's debt to him. The debt had become one of Playfair's assets to be claimed by Playfair's creditors.

Kennett's scheme to hide his assets from his creditors soon went awry.[15] The dubious scheme fell apart when Robert Kennett and Playfair decided to wind up their partnership, then fought over division of the assets in the gun carriage business. Playfair now decided to take his role as assignee seriously, so Kennett tried to have him removed. Soon Daniel Duff, who ran a school attended by Kennett's four children, appeared. Kennett owed Duff about £100 for school fees and there were other new creditors as well. Duff and the other creditors discovered the questionable change in the assignees, and it became apparent to them that Playfair was a bankrupt discharged from the Fleet and that Parlby was senile. Since these creditors had received no funds and since Duff was also convinced that Kennett had property that he was hiding, Duff and several creditors took their case to the High Court of Chancery to change the assignees so that they could have greater control in administering Kennett's debts and assets. The decision came down on 12 May 1804; both Playfair and Parlby were removed as assignees.

This was not the end but the beginning of more problems for Playfair.

Chapter Fourteen

Playfair Has a Good Year during 1805 with Hints of Ending Badly

The year 1805 is known in British history for the Battle of Trafalgar and the death of Nelson. For Playfair, the year marked his most productive literary year, with three or four published books, two of which contained his innovative graphs. It also ended with an impending dark side, a foreshadowing of his incarceration in Newgate, described in chapter 1.

One of Playfair's publications from that year, or perhaps a non-publication, may quickly be dispensed with. In his obituary of Playfair, Thomas Byerley lists forty-one of Playfair's publications. The last two entries are ([Byerley] 1823, 172):

> 40. Montefiore on the Bankrupt Laws
> 41. European Commerce, by Jephson Ody [*sic*], Esq.
>
> These two works, though published under the names of the gentlemen last mentioned, were written by Mr. Playfair.

European Commerce, Shewing New and Secure Channels of Trade with the Continent of Europe, published in 1805, will be dealt with first. Authorship on the title page is attributed to J. Jepson Oddy. Oddy was a London merchant who had gone bankrupt in 1802 and had spent time in 1804 in the Fleet Prison for debt (*London Gazette*, 4 August 1804, 26 February 1805). By 1809 he had recovered sufficiently to attempt to obtain a seat in Parliament for the borough of Stamford (Sweet 1998).

In the preface Oddy states that he had begun work on his book in 1804 and, when he found he needed more information, he travelled to Europe to obtain it (Oddy 1805, vii). The book itself contains several tables of statistics but no graphs, which Playfair likely would have included had he written the book. Once the book was published, on 14 August 1805, Oddy had an audience with the king where he presented the king with a copy, and they talked at length about the topics covered in the book (*Morning Chronicle*, 16 August 1805). The next day, the king sent Oddy a letter thanking him for his book, which he considered an important work. He also thanked Oddy for all the effort that he had put into writing it (*Morning Chronicle*, 17 August 1805). Five years later, Oddy wrote a closely related book with the title *A Sketch for the Improvement of the Political, Commercial and Local Interests of Britain* (Oddy 1810). Finally, Oddy's brief obituary mentions only two things about him: he wrote *European Commerce* and he had been a candidate for the borough of Stamford (*Leeds Mercury*, 10 September 1814). At most, it could be believed that Oddy talked to Playfair about his work for *European Commerce* and perhaps Playfair obtained some background information for it. Byerley's claim that Playfair wrote the book is not credible ([Byerley] 1823, 172).

It is also doubtful that Playfair wrote Joshua Montefiore's 1806 *The Creditor and Bankrupt's Assistant*. Montefiore was trained as a solicitor and so was better qualified than Playfair to write it. Advertisements for it mention Montefiore's legal background (*Morning Post*, 21 January 1806). Prior to 1806, he wrote four other books on law and business. The contents of *The Creditor and Bankrupt's Assistant* also fit well with Montefiore's background. The writer of *The Creditor and Bankrupt's Assistant* knew the law well and his recommendations on how to reform the whole system of bankruptcy went well beyond what Playfair later wrote on the subject in *Political Portraits of a New Æra* (see Playfair 1806). Montefiore also dedicated, with permission, *The Creditor and Bankrupt's Assistant* to Lord Eldon, who was the lord high chancellor in the Court of Chancery. This meant that Montefiore asked permission to make the dedication and Eldon agreed, but that would be unlikely to happen for a ghostwritten book. It is possible Playfair knew Montefiore. A William Playfair living at Lisson Green (the only instance of this address that has been ascribed to Playfair that can be found) in London subscribed to Montefiore's 1803 *Commercial Dictionary* (Montefiore 1803, xxxvii). Both Playfair and Montefiore were in the Fleet Prison for debt in 1801 (*London Gazette*, 11 July 1802, 832). The two may have discussed bankruptcy laws, and some of Playfair's ideas and perspectives may have wound up in *The Creditor and Bankrupt's Assistant*. It is hard to believe that Playfair ghostwrote the book for Montefiore.

By 1805, Playfair had established his reputation as a writer on economics and politics, and it was not high. One reviewer of his work wrote, "Mr. Playfair is by no means a luminous writer; he is one of those who boast of the advantages of connection and arrangement only in the index and table of contents. At the same time, however, we must confess, that we have occasionally found a portion of original matter and valuable research, though conveyed in a method extremely intricate and perplexed, and incumbered with useless and irrelevant speculations" (*Critical Review*, ser. 3, vol. 8 [1806]: 2). As had happened in the past, Playfair continued to have mixed reviews of his work even up until the time of his death.

The first literary adventure in 1805 that can definitely be ascribed to Playfair is his *Statistical Account of the United States of America*. Although advertisements for the book do not appear until August, the book had been published by April.[1] It was a translation of, and an addition to, parts of Denis-François Donnant's *Élémens de statistique* published in France in 1802 (Playfair and Donnant 1802). The first part of Donnant's 1802 book is a translation of Playfair's 1801 *Statistical Breviary*. This was not an unusual activity for Donnant. From 1800 to 1805 he was very active in translating books, usually from English to French (Quérard 1870, 821). Following his translation of Playfair's work in *Élémens de statistique*, Donnant moved on to his own work on the statistics of the United States. This was the part of Donnant's book on which Playfair focused. Prior to writing *Élémens de statistique*, Donnant had spent some time in America (Martin and Behar 2003, 465). Once he finished his book, Donnant sent Playfair a copy. He had no address for Playfair in London and so he sent it to "a Bookseller's in Paternoster Row," according to Playfair (Playfair 1805, iii). The bookseller was James Wallis. He appears first in the list of booksellers on the title page of the *Statistical Breviary* and is the only one with a street number attached to his address, 46 Paternoster Row.

Playfair's major innovation, inserted into his translation of Donnant's work, was the now ubiquitous pie chart. He claimed that he invented it specifically for this publication. The chart, shown in figure 14.1, depicts the relative sizes of the states and territories of the United States. The pie wedges show the fraction of the whole area of the country taken by each member of the Union. Written in most of the wedges is the area of the state or territory. At first glance, the chart appears to be incorrect in its depiction of the relative sizes. For example, New York is smaller than Vermont. A comparison of the pie pieces to an 1800 map of the United States also brings into question the accuracy of the graph (Adams 1909, 211). The solution to this oddity is that the numbers in the graph correspond approximately to the square mileages given in a table in Playfair's *Statistical Account* (Donnant and Playfair 1805, 20). Upon closer inspection, Playfair's inaccurate numbers for areas are derived from a table in Donnant's *Élémens de statistique* (Playfair and Donnant 1802, 84). There, Donnant has given the length and breadth in miles of

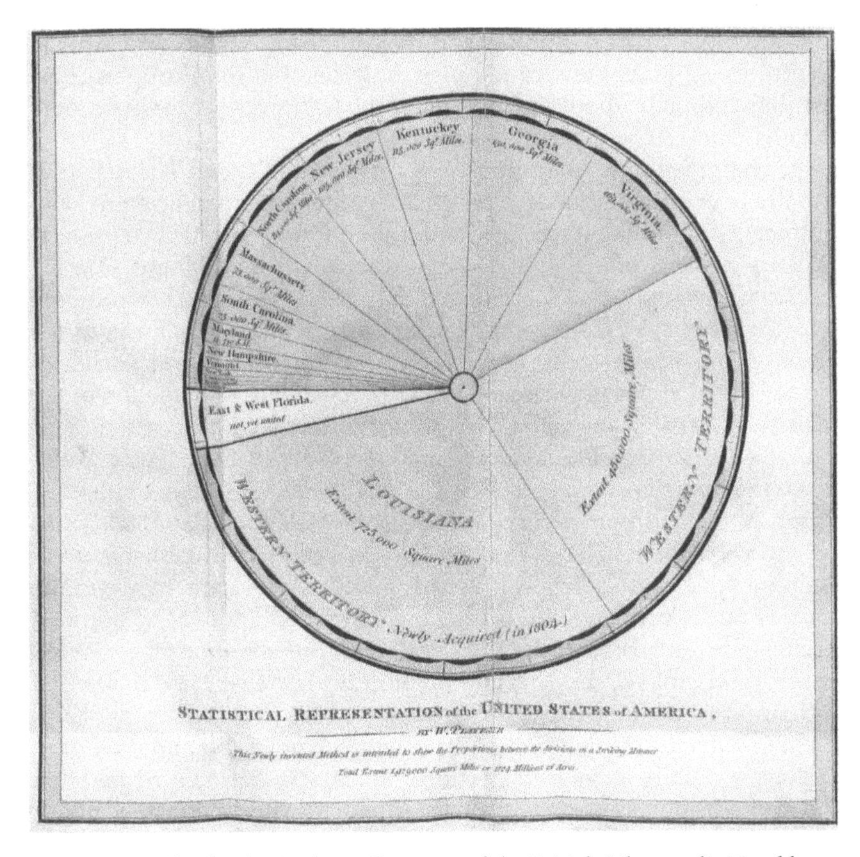

Figure 14.1. Playfair's pie chart. Courtesy of the British Library, digitized by the Google Books project.

each state or territory. Playfair has merely multiplied the two together to obtain his area in square miles, assuming that the states are all rectangular. Besides a few multiplication errors, Playfair mistakenly switched the values for New York and New Jersey. Further, his calculation for Louisiana in his figure (725,000 square miles) does not correspond to the value obtained by multiplying the length and breadth of Louisiana in Donnant's table (130,760 square miles). Playfair did not fully accept Donnant's calculations; Playfair's computation for the size of Louisiana is closer to the accepted size of the Louisiana Purchase, about 827,000 square miles.

Playfair anticipated his own pie chart in the *Statistical Breviary* published four years earlier. In figure 13.2, the Turkish Empire is divided into three pie shapes, representing its European, Asiatic, and African territories. A quick view of the pies indicates that the European possessions comprise about one-quarter of the territories, the Asiatic more than half, and African territories the remainder.

Pie charts are now ubiquitous. So is criticism of them. The debate is usually over whether to use a pie chart or a bar chart (or the equivalent dotplot) to show the data. The simplest and most succinct discussion appears in an introductory statistics textbook by David Moore – *The Basic Practice of Statistics*. Moore says that bar charts are used to compare sizes of groups, while pie charts emphasize a group's relation to the whole. Playfair's pie chart of the sizes of various states and territories of the United States does indeed give a quick and easy depiction of the relative sizes of the areas that comprise this new country.

One element that Playfair added to Donnant's treatment of America in the French original *Élémens de statistique* is an update of his 1793 proposal appearing in *Better Prospects to the Merchants and Manufacturers of Great Britain*. In 1793, Playfair had proposed the establishment of warehouses along the Ohio River to provide Americans living in the interior of the country with British manufactured goods. The updated proposal in 1805 is basically the same, with the warehouses extended to the Mississippi River. This is another indication that Playfair wanted to return to a business career. Playfair based his proposal on two false premises: that the United States, especially the newly acquired western regions, would be primarily a country devoted to agriculture; and that the immense market opening up in America would end the jealousy between European nations. On the other hand, Playfair did predict correctly that Florida and the mouth of the Mississippi would become part of the United States (Donnant and Playfair 1805, 46–8, 70–2).

Resurrecting the warehouse proposal had been on Playfair's mind for at least six months before publication of *Statistical Account of the United States of America*. On 9 October 1804, he wrote to Henry Dundas, now Viscount Melville, about trade with America.[2] Ten years earlier, he had written to Dundas as home secretary about a propaganda campaign against the French. Now he was writing to him as first lord of the Admiralty. Playfair wanted to quell the rising ambitions of America, especially in trade with the West Indies, fearing that American success in trade might lead to U.S. possession of some of these islands. To cut into American trade and to protect British interests, Playfair suggested to Dundas that America's newly acquired western territories would soon have a greater population than the states on the Atlantic coast, and their main activity would be agricultural. The navigable rivers in the western territories did not flow into the Atlantic Ocean, so American products would be expensive to send to the interior. Therefore, he proposed to Dundas that Britain should promote its own products along these rivers in order to stem American trade. It would have the further benefit of keeping the economic activities of these territories tied to agriculture for years to come. If the Americans treated the British unfairly, Britain could blockade their one seaport. Playfair did not seem to be current on American seaports. Nevertheless, nothing seems to have resulted from his proposal. And none of the tone of this proposal saw its way into his warehouse proposal in *Statistical Account of the United States of America.*

Once he published *Statistical Account of the United States of America,* Playfair sent twenty-five copies of the book to Thomas Jefferson. In the covering letter, Playfair told Jefferson that he was dedicating the book to him.[3] He also reminded Jefferson of their encounters in France, referring only to Playfair's 1789 book in French, *Tableaux d'arithmetique lineaire* (Playfair mentions only "You were then pleased to bestow on my application of lines to matters of finance"). Nothing was mentioned of the Scioto fiasco. The letter is published with the book. At the time, Jefferson was serving as president of the United States. Although it was nowhere stated in the book, perhaps Playfair had in mind that Jefferson would somehow be part of Playfair's proposal to build a string of warehouses in the newly acquired western lands and Louisiana. He was probably trying to play both the Brits and the Americans in order to get back into business.

Next in line during Playfair's banner year of publications is his edition of Adam Smith's *Wealth of Nations.* In 1804, the copyright ran out on Smith's book and, with Smith having died in 1790, the publishers were looking for a way to have some ownership in the book's high status. They decided to bring out a new posthumous but updated and edited edition. This would be a way to extend the copyright (Tribe and Mizuta 2002, 31). The task fell to Playfair.

Playfair's edition of the *Wealth of Nations* was available to the public as early as the beginning of August 1805 (*Morning Post,* 2 August 1805). Five years later, he described the nature of his involvement with this classic work.

> About five years ago I wrote Notes, and a Supplement to the celebrated Works of Doctor Adam Smith, on the Wealth of Nations, which I sold to Messrs. Strahan and Cadel, the original proprietors, for a considerable sum, after they had the manuscript examined by a proper judge. I differed in opinion with that great writer (whom I had the honour to know personally and to esteem highly), on the subject of monopoly of provisions, of which he denied the possibility, which I maintained it was not only

> capable, but real. This brought a number of writers upon me, as being a man guilty of something approaching to sacrilege in disturbing the system of so celebrated a writer. It is true, no one tried to refute what I maintained, but it was found easier and more convenient to affect to treat me, together with Lord Kenyon and all those who believed in monopoly, like ignorant Dolands. In short, Sir, we were all consigned to contempt with a single dash of the pen. (*Morning Post*, 6 December 1810)

Several points in this quote are worth noting. First, by 1802, the publisher was Thomas Cadell Junior and William Davies. Cadell senior died in 1802, and quickly thereafter Andrew Strahan left the business. Playfair's slip is one of several indications of the general lack of care he took in many of his publications. Second, the "proper judge" was probably the political economist Thomas Robert Malthus. In a letter to the publishers of 16 December 1804, Malthus comments on a pre-published manuscript he was sent that was a subset of the full Playfair version: "Some of the notes appear to have merit, others not, and some of the most important points are not discussed at all" (quoted in Tribe and Mizuta 2002, 30). Later in the letter Malthus writes, "The style is not very correct, and is written without punctuation." Several literary critics from Playfair's day support Malthus's first observation. As for the second, a full reading of Playfair's manuscripts and manuscript correspondence leads one to sympathize with Malthus. Despite his criticisms, Malthus did not oppose the publication of Playfair's edition. The third and final point to note from Playfair's description of his involvement with the publication is that he acknowledged severe criticisms of what he had done. Indeed, the critic for the *Edinburgh Review* was scathing in the opening to his review: "In the whole course of our literary inquisition, we have not met with an instance so discreditable to the English press, as this edition of the Wealth of Nations" (*Edinburgh Review, or Critical Journal* 7 [1806]: 470). For Playfair, the review went downhill from there.

What brought "a number of writers" on Playfair was the first supplementary chapter that Playfair added to *Wealth of Nations*. Titled "On the Commerce of Grain, Monopolies, and Forestalling," it was a response to Smith's position on monopolies with respect to the market in grain.[4] What Smith meant by "monopoly" is not how we understand the word today. Smith's discussion uses the words "competition" and "monopoly." "Competition" meant that there were no restrictions on the freedom of trade, whereas "monopoly" meant that there were artificial restrictions to this freedom of trade. The restrictions often resulted from government intervention such as special privileges given to trading companies, the East India Company being one example. From these concepts Smith argued that there could be no monopoly in the sale of grain. Grain growers sell their crops to the local community and to grain dealers ("corn factors" in British terminology). This group is too large to join to elevate the price of grain artificially, even during bad harvests when lack of supply raises the price. Smith argued that though a bad harvest is a type of restriction to the freedom of trade, it is not an artificial one. Therefore, he concluded that monopolies that restrict supply of grain seldom, if ever, occur. It was this conclusion that Playfair criticized and consequently had the ire of several reviewers rain down upon him.

Playfair argued that monopolies did indeed occur on the sale of grain. His argument was based on the rising popularity of newspapers. In former times, without newspapers, a grain grower and a grain dealer sold locally or fairly close by, so the price was determined by local or regional supply and demand. Prices could vary across the country. With newspapers, information became available that allowed sellers to withhold grain or to buy stocks in advance in order to withhold it (known as forestalling and regrating) in order to force up the price of grain. Playfair gave two examples in Britain. In 1798 there was a shortfall in the harvest. When some parliamentarians raised the alarm in Parliament that there would be severe shortages in grain, sellers hoarded the grain and drove up the price, which tripled, although many granaries were full. In 1804, Parliament enacted a Corn Law that imposed a duty on imported grain to protect British farmers. The passing of the act late in 1804 and the reporting of it in the press was a signal, Playfair claimed, for growers and dealers to repeat their actions of 1798. Despite Playfair's protestations, the 1804 act had little effect, since the price of wheat was relatively stable, until 1813 when there was a bumper crop (Barnes 1930, 89).

This supplementary chapter was one of six that Playfair added to *Wealth of Nations.* What Smith originally wrote was preserved in its entirety in the three-volume 1805 edition. In addition to the six supplementary chapters, Playfair's contribution was a biography of Smith and many footnotes that comment on or clarify what Smith had written. Several of the footnotes and some of the supplementary chapters mention errors that Smith allegedly made, creating another source of discontent among the reviewers. In the words of one modern writer, "He [Playfair] wrote supplementary chapters to discuss the issues which he thought Adam Smith would have wanted to discuss, had he lived so long, and shared the same historic experience" (William Rees-Mogg, introduction to Smith and Playfair 1995, 11). Four of the six supplementary chapters deal with topics Playfair had written about or was writing about: the national debt and paper money, education, commercial treaties, and causes of the French Revolution. Two of these four topics, education and the national debt, would appear in *Decline and Fall of Powerful and Wealthy Nations*. Of the chapter on monopolies in the grain market, Playfair would return to this topic in newspaper articles published in 1810 and a pamphlet backed up by seven appendices published in 1821. The final supplementary chapter was written in support of Smith on the concept of productive and unproductive labour. Playfair was keeping up with the economic literature. James Maitland, 8th Earl of Lauderdale, criticized Smith in his *Inquiry into the Nature and Origin of Public Wealth*, which came out in May 1804 (*Morning Post*, 15 May 1804). In his updating of Smith, Playfair missed the changes in technology brought about by the Industrial Revolution and the rise of the factory system. Playfair wrote no supplementary chapters on those topics.

In 1787, at a meeting of the General Chamber of Manufacturers of Great Britain, Playfair had spoken against the trade treaty (known as the Eden Treaty in Britain) between France and Britain. On the basis of the longstanding rivalry between France and Britain, and French support for the Americans during their revolution, Playfair claimed that France could not be trusted. But in his supplementary chapter on treaties in *Wealth of Nations*, Playfair seemingly changed his tune. He saw the treaty, which had lasted only four years, as initially fair but eventually disadvantageous to France. On the basis of the credit offered by English manufacturers for their goods, their French counterparts could not compete, thus causing French manufacturing to go into decline. Not being mutually beneficial to both countries, the Eden Treaty was therefore not a good one. It was a deft dance. He had justified his opposition to the treaty on both occasions with contradictory reasons separated by eighteen years. At the end of his discussion of treaties, Playfair reiterated the same mistake he made earlier in the year in *Statistical Account of the United States of America*. He claimed that the opening of the new western territories in America would reduce trade tensions in Europe. They would never industrialize and would remain devoted to agriculture so that American agricultural products would stimulate demand for European manufactured products.

The title of one supplementary chapter, "On the Increase of National Debt, the Bank of England, the Sinking Fund, Including Such Occurrences in Finance as Have Taken Place since This Book Was written," is a little deceptive. It is really about paper money: money issued by the Bank of England when specie was scarce in the early 1790s, American paper money issued during the American Revolution, and French assignats issued during the French Revolution. Playfair's résumé entries for paper money are extensive. He was closely involved in two banks – one in Paris and one in London – that dealt in paper money. His Scioto venture in Paris depended on buying American bonds at a discount. When assignats were introduced in France, Playfair wrote a pamphlet about the nature of paper money. In two later publications he commented unfavourably on assignats. After describing the British, American, and French experiences with paper money, Playfair's main comment was concerned with how paper money was issued and its effect. Using the American Continental currency (used during the American Revolution to pay military expenses) as an example, Playfair notes that currency, as it depreciates, is like a tax. Devaluation of the currency has an inflationary effect. It generally favours the borrower over the lender. For those running the country, for example, government debt is reduced in value even when no payments are made to reduce it. Playfair goes on to say that if paper money were issued without any backing such as gold, then consumable property would pass from individuals to those in power. To back this claim, Playfair uses French assignats as an example, after they were no longer backed by land that the government had expropriated from the church.

In his treatment of banking, Playfair completely ignored the 1802 publication *Paper Credit* by Henry Thornton, which was the most important contribution to monetary economics between about 1750 and 1900 (Laidler 2008). Compared to what Thornton wrote, Playfair's treatment of paper money is superficial. Further, Smith's *Wealth of Nations*, originally written in 1776, had become dated by 1805 with respect to banking (Thornton and Hayek 1939, 37–8). The number of country banks had increased exponentially in response to the increase in manufacturing outside of London (Ashton 1955, 179, 183). In addition to its original purpose that Smith outlined in *Wealth of Nations*, the Bank of England became the bank for bankers, the lender of last resort. There was also a rapid growth in the use of cheques for which a central clearinghouse was established in about 1775 among London bankers. It seems odd that Playfair mentions none of this, considering the fact that Playfair had previously run two banks.

The supplementary chapter, which follows the chapter on paper money, is about French economists. Earlier in his *History of Jacobinism*, Playfair had seen one root cause of the French Revolution as philosophy that was based on theory alone and not on practical experience. In this supplementary chapter, he exchanges French economists for philosophers and makes them one of the root causes for the same reason: theory without practice. Of the four economists Playfair mentions, only two were active in the Revolution: Comte de Mirabeau and Marquis de Condorcet.

Initially, Playfair was taught at home by his older brother, John. At some point in his youth, probably around age fourteen, he was apprenticed to the engineer Andrew Meikle.[5] With this as his background, it is no surprise that Playfair would support apprenticeships as a form of education. Besides his own training, he did have another reason for this support, which he succinctly (for a change) laid out: "Dr. Smith has so fully, and so well discussed the subject of education with regard to the higher orders of society, and those destined for the learned professions, that it is not necessary to add to what he has said; but he has omitted the education of the middling ranks intended for business, or of the lower orders destined to manual labour, on which greatly depend the happiness and prosperity of a state" (Smith and Playfair 1805, 239). Smith criticized apprenticeships and had advocated for universal education and public funding of schools for all. It was on this that Playfair disagreed. He saw a difference between education and training. It was training, not education, that would result in commercial prosperity. For that reason, Playfair wanted the apprenticeship system expanded to other areas, including agriculture, business, and manufacturing.[6] Playfair returned to this topic within a few months in his *Decline and Fall of Powerful and Wealthy Nations*.

Among the footnotes in *Wealth of Nations*, there could have been added a lengthy discussion of Britain's usury laws. Playfair had written *Regulations for the Interest of Money* in 1785 advocating changes to these laws; throughout several editions of *Wealth of Nations*, Smith defended fixing a legal rate of interest. In Britain the legal rate was 5 per cent, which Smith had defended on the grounds that money lent to the government was at 3 per cent and money lent to those with good security was at 4 and 4.5 per cent. In 1785, Playfair had argued that the interest rate should be determined by the amount of risk associated with the loan; not doing so stifled business. Curiously, Playfair's footnote in 1805 only hints at Smith being in error and makes no reference to *Regulations for the Interest of Money*. Playfair merely argues that when government borrowing had risen above 5 per cent interest, the legal rate did not change, making it difficult for individuals to borrow money.

During 1805, there was a lingering residue from Playfair's bankruptcy. A small cluster of letters from Playfair to the solicitor Samuel Wadeson in 1799 is concerned mostly with small business items and information about the bankruptcy, with Wadeson holding the upper hand as the assignee in the bankruptcy. Another cluster appears between the two in 1805, when the two appear to be on more friendly terms.

Wadeson apparently contacted Playfair in September about Edward May, possibly about the lawsuit between Isaac Bernal and May, along with his son-in-law, the Marquess of Donegall. In a lengthy letter dated 18 September, Playfair begins by saying that he is enclosing a copy of his *Statistical Breviary*, published in 1801, which he thought would help Wadeson's daughters in their study of geography or modern history. After this introduction, Playfair says that he would be happy to help. Then he spends the remainder of the letter outlining how he had been badly treated by Hartsinck and Hutchins and how, thankfully, Wadeson had come to realize what was really going on. He concludes by saying that he expected some remuneration for his work; but for work related to Edward May's and Richard Sheridan's accounts he would work gratis. In a postscript, he mentions that he was hard at work; Wadeson could direct any letters to his publisher, where he would soon be in town to read proofs of his new book, *An Inquiry into the Permanent Causes of the Decline and Fall of Powerful and Wealthy Nations*, which was in print by mid-October.[7] Wadeson must have agreed to Playfair's request because, over the next month, Playfair was writing to him about Sheridan's account, as well as another related to the *Oracle and Public Advertiser*.

As acknowledged by at least one reviewer, Playfair's *Decline and Fall of Powerful and Wealthy Nations* was the first generalist treatment of ways that nations decline in wealth and lose their power: "In an age in which the objects of speculation have been so various and important, and the range of regular and philosophic inquiry so widely extended, it appears a singular fact, that the principles which influence the decline and fall of nations, have never yet been made the subject of systematic investigation. The only lights which have been thrown upon this grave and interesting topic of consideration, are dispersed over the pages of authors who have treated of collateral subjects, or of historians who have directed their attention to the detail of particular examples" (*Critical Review*, ser. 3, vol. 8 [1806]: 1).

The most famous historian who directed his attention to a particular example is Edward Gibbon in *A History of the Decline and Fall of the Roman Empire*. In his own book, Playfair mentions Gibbon by name only once with reference to the frontispiece of his book (figure 14.2) when he is discussing how long the Roman Empire lasted.

Playfair used the Scottish Enlightenment approach to carry out his analysis of the decline and fall of nations. By looking at several historical examples over three millennia, he formulated a general model for a nation's life cycle,[8] which falls in line with the human life cycle: birth, infancy, maturity, decline, and death, illustrated empirically in a graph he devised. Figure 14.2 shows the rise and fall of twenty-one

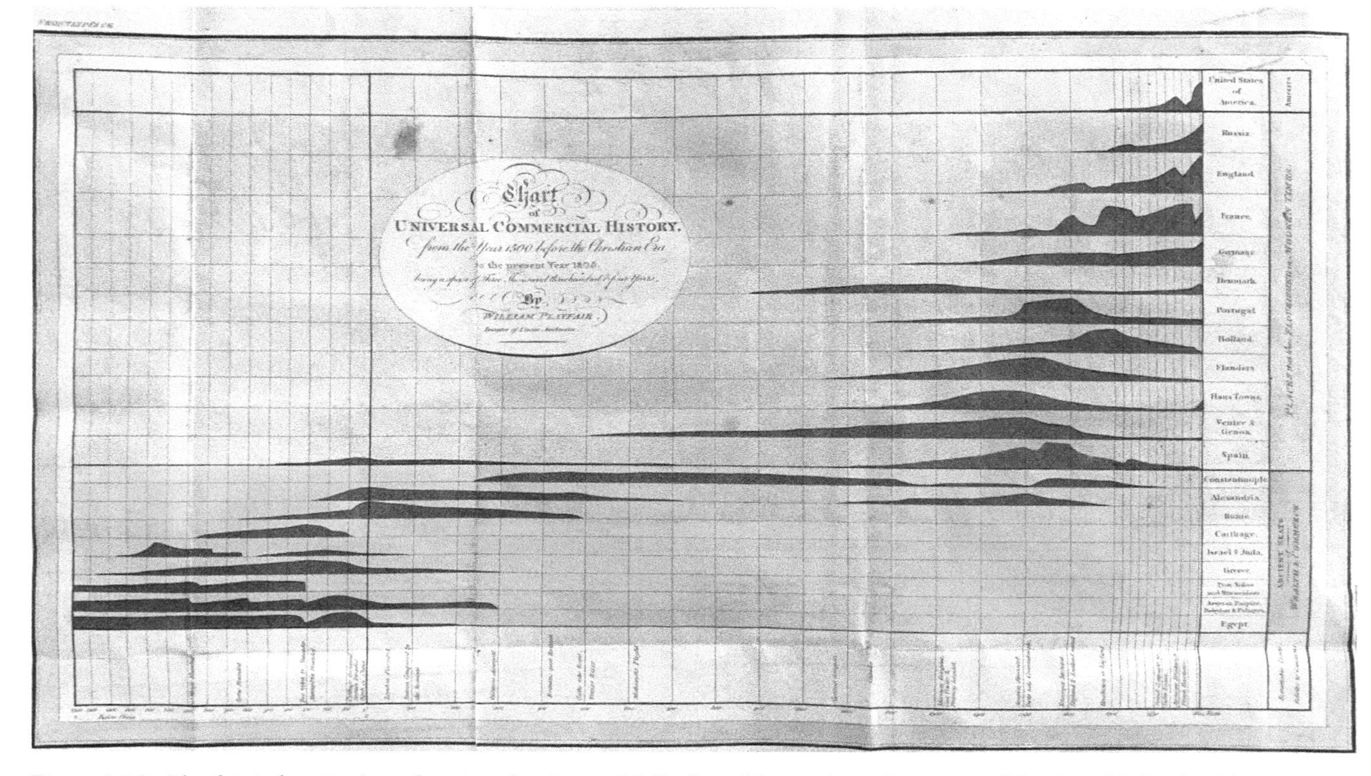

Figure 14.2. Playfair's frontispiece showing the rise and fall of wealthy nations. Courtesy of Stephen M. Stigler, University of Chicago.

countries or city states. Each is illustrated by what looks like a mountain range. The horizontal extension of each range tallies with the time a particular state began to achieve greatness to the time it fell. These lengths, Playfair claims, are exact. The height of a range indicates the amount of power and wealth of each state at a particular time. Playfair admitted that the heights were not exact, but estimated from available documents.

Playfair devotes the first part of *Decline and Fall of Powerful and Wealthy Nations* to descriptions of nations that have come and gone, with comments on why they have disappeared. He sees three different periods in which they flourished and fell: up to the fall of the Roman Empire in the West, from the fall of the Roman Empire to the discovery of America and the passage of the Cape of Good Hope, and from these European discoveries to the present. In the earliest period, the rise of a country was connected to force of arms. In the second, it was associated with the development of commerce. In the final period, it was related to military power and commerce. According to Playfair, gunpowder, printing, and the magnetic compass were invented late in the second period. But they had all been invented in China centuries before. Playfair was also not the first to recognize the importance of their introduction into Europe; several European authors mention them in the mid-sixteenth century.[9] Overall, Playfair's analyses of countries' rise and fall range from the reasonable to the ludicrous. For example, he claims that envy drove the Turks to attack the Byzantine Empire. Playfair's claim that Turkish envy stemmed from Byzantine riches is reasonable. But he also makes the ludicrous claim that another source of Turkish envy was that Byzantine women were beautiful and Turkish women were ugly (Playfair 1805, 50).

As usual with Playfair, his ideas are scattered throughout the book, so it is often difficult to see the forest for the trees. Henryk Grossman has distilled Playfair's analysis it into a succinct form. During birth to maturity in the life cycle of a nation, capital tends to concentrate in a small group of people within the country, the workers tend to become poorer, and the middle class tends to shrink in size. When sufficient capital has accumulated, decline begins, since surplus capital tends to flow outside the country in investments elsewhere (Grossman 1948).

Focusing on the third period in which nations flourish and fall, Playfair treats the causes for their decline in the second part of *Decline and Fall of Powerful and Wealthy Nations*. He enumerates eleven, discussing one per chapter. Causes are divided into internal and external, with most falling into the internal category. Among these causes, internal and external, most are related to the increasing wealth of a nation on the rise. In his discussion, he spends most of the time on economic factors and completely omits any religious component to the rise and fall of nations. Edward Gibbon, for example, attributes one causes of the fall of the Roman Empire to the rise of Christianity. Playfair instead focuses mainly on economic and military reasons. On the rise of Islam, for example, he completely ignores the religious fervour behind the early Arab conquests and makes no mention of early Islamic empires.

When dealing with rising wealth as a cause for decline, Playfair make several insightful remarks. His main premise is that entrepreneurial spirit builds a country, while at the same time creating power and wealth. Entrepreneurial spirit is driven by necessity, and wealth flowing from the drive is built on hard work and labour. As noted in Grossman's description of Playfair's model, decline begins as people grow richer. The entrepreneurial spirit declines and is not passed on to the next generation, who merely live off their inherited wealth. These people look for safer investments. They buy land and live off the rents, or they buy government annuities and live off the interest. They lose sight of the origin of their wealth – hard work and labour – so they cease to be productive members of society. Even though wealth may be rising in a country, there is an unequal distribution of it. As a result, the middle class – which Playfair considers to have the entrepreneurial drive – shrinks. Further, the bottom of the economic scale becomes disgruntled, and the result can be unrest and violence. Such stratification of society leads to more problems. The children of the rich are not brought up to know necessity, which drives the entrepreneurial spirit; the children of the poor are brought up in an atmosphere only of want, so they become unproductive members of society. According to Playfair, the ongoing health and stability of a country is directly related to the vibrancy of its commerce. This is a general view he had held since the 1780s, particularly in his correspondence with Patrick Colquhoun and especially in his *Regulations for the Interest of Money* and *An Essay on the National Debt*. In *Decline and Fall of Powerful and Wealthy Nations*, his view is more fully fleshed out.

Playfair also made insightful comments about taxes. He observes that taxes are necessary to pay for the armed defence of a nation and for diplomacy between nations. Without sufficient public revenue, a nation becomes vulnerable to external threats, which arise because a wealthy nation is envied by less wealthy ones. If the wealthy nation has good defences, other countries will only try to imitate it. However, if the nation has insufficient protection, other countries will try to conquer it. Beyond the necessity of taxes, over-taxing the people contributes to decline. Playfair contends that taxes tend to increase with the wealth of a nation. Taxes that increase too greatly cause discontent among the people, which could lead to rebellion. Further, excessive taxation causes the flight of people and capital from the nation, thus weakening it.

Playfair's comments on trade and foreign investment align with Grossman's description of the decline in a nation's life cycle. Playfair argues that surplus capital from a wealth country tends to be invested in less wealthy countries. It follows, he claims, that the capitalist's income is derived from investments only, decreasing entrepreneurial spirit. He also argues that when trading with a less wealthy country, giving long-term credit has negative consequences. The borrowed capital obtained from the availability of credit is invested in the manufacturing sector of the less wealthy country. The less wealthy nation then builds up its manufacturing sector to the point that it rivals the wealthier nation.

Less insightful is Playfair's comment on the diets of wealthy nations. His comment appears to have been inspired by the work of Thomas Robert Malthus on population growth and the food supply. Playfair claims rightly that as a nation increases in wealth, more people can afford to eat meat. He notes that it takes more land to produce meat than to produce grains and vegetables. Also correct. He concludes that when there is insufficient land to support such a diet, prices rise and negatively affect a large number of people. What Playfair missed was that during his lifetime Britain was seeing an increase in the productivity of the land. Also he neglected the possibility that a country could import grain. The food supply was not entirely constrained by the amount of available land.

Since Playfair reasons that all nations go through this cycle of birth to death, death in inevitable. Playfair's strategy is to forestall the advent of the Grim Reaper as long as possible. This, Playfair claims, is the role of government. Playfair's view of the form of the government is questionable. He sees government as a constitutional monarchy with checks on the monarchy (the executive branch) by the House of Commons to limit arbitrary power. Upsetting this system causes decline, Playfair claims. The upset can come from the encroachment of other "public bodies," of which the most dangerous are lawyers. Playfair says of lawyers, "There can be nothing more natural than that they should contrive to render the business, which they alone understand, of as much importance and profit as possible" (Playfair 1805, 119). In later publications, Playfair strongly defends the unreformed status quo of the House of Commons, an institution that needed more than a little upsetting. When Playfair wrote about it, the system had MPs returned from many so-called rotten and pocket boroughs and very little representation from the emerging middle class.

Also questionable was Playfair's view of education, which he had expressed in his edition of Smith's *Wealth of Nations*. Playfair contends that decline comes when the "useful parts" of education are abandoned and too much emphasis is placed on educating people for the "learned professions" and "fine arts." By "useful" he meant apprenticeships and training for careers in manufacturing, agriculture, and commerce.

As soon as *Decline and Fall of Powerful and Wealthy Nations* hit the bookstores, Playfair returned to another obsession – the power and encroachments of France. This time it was the reverse of the situation of the French émigrés, aided by the British, forging assignats to bring down France in the 1790s. Now Playfair thought he had uncovered a French plot to forge British banknotes.[10] On 28 October 1805, he wrote to John King, undersecretary of state in the Home Office. He claimed that the French government had employed an agent to obtain several denominations of Bank of England notes, as well as the notes of several private banks. The operation would cost about 300,000 florins and would be run out of Holland. The notes that the agent obtained were to be forged and then distributed by agents in Britain in order to instil mistrust in British paper money. Passing the forged money would occur in about two months, Playfair warned. Whatever his motivation – scam or public service – is unknown. No evidence could be found that France under Napoleon was engaged in counterfeiting operations as part of its war effort.

Hanging over all this throughout the year was the tail end of the court case involving Robert Kennett, Playfair's former partner. Playfair had been removed as an assignee in Kennett's bankruptcy case in 1804. There was more to come at the end of 1805 and into 1806.

Chapter Fifteen

Playfair Has Serious Legal and Other Problems

While he may have been basking in the success of his literary endeavours during 1805, Playfair's legal, financial, and now medical woes escalated.

The 1805 case to remove Playfair as an assignee in Robert Kennett's bankruptcy case was heard in the Court of Chancery, which considered cases concerning property and did not handle criminal questions. At the end of his summation in the Chancery case, the lord chancellor stated:

> Upon the whole, I should not do my duty, where there is so much reason to imagine fraud and latent combination, if I did not most solemnly declare it as my opinion, that in any manner to prevent the operation and effect a commission of bankruptcy is a most serious crime and misdemeanour, and one, which, I trust, will soon meet with the most exemplary punishment. In order to be satisfied whether there has been such a fraud and conspiracy in this case, I desire that the papers be laid before His Majesty's Attorney General, and that he inform this Court against whom a prosecution should be instituted at the expense of the public.[1]

The "papers" were sent to the then attorney general, Spencer Perceval. Perceval turned the case over to the solicitor general who presented it before a judge and jury in the Court of King's Bench on 21 December 1805.

Playfair was unprepared for the trial in the Court of King's Bench, or he was trying to avoid going to court. He had been served with a notice of proceedings against him early in 1805. His lawyer, a barrister by the name of Parker from Gray's Inn, had responded to the notice. According to Playfair, nothing happened for the rest of the year. Playfair claimed that he found out about the trial only by accident the day before it was to take place. In a written affidavit leading up to his eventual trial in 1809, Playfair commented on the 1805 proceedings: "As I had not been party to any conspiracy I had not previously thought of any defence and had not then time to prepare one. I had no Council nor time to Subpoena any witnesses. But not fearing punishment where I felt no Guilt I did not even make an affidavit to put off the Trial which I might have done & certainly should have done had I Imagined there was the least danger."[2] Perhaps he had not paid his barrister and was out of the legal loop. Or perhaps he thought everything would go away if he were to lie low and ignore the situation, as he had done before. Whatever the case, the jury decided that Playfair, Kennett, and his son were guilty of two conspiracies: one to deprive the creditors of their claims and the other to keep Kennett's assets out of the hands of the commissioners of bankrupts.

Robert and Henry Kennett went again before the court in June 1806. Henry received a sentence of one year in prison, while Robert was sentenced to one month in prison and to be set in the pillory in Oxford Street. The pillory, seen as a major punishment, was made from hinged wooden boards with holes for the head and hands that could hold two prisoners. The whole apparatus rotated around a post, typically one prisoner on each side. The prisoner was placed in the frame and made to walk in circles, usually for an hour. Prisoners in the pillory were often pelted with refuse and offal from the slaughterhouse. If the crowd chose more lethal missiles, the prisoner could die in the pillory. It was a shaming device usually placed in a very public place to destroy the offender's reputation.

After the Kennett sentencing, Playfair went into hiding. The court responded by issuing a writ to have him appear. A court document describes the lengths that were made to find him: "Altho' he [the officer serving the writ] made very diligent search of him and had reason to believe that he was living at or in the Neighbourhood of Brompton yet he took such measures to conceal himself that the Officer could not apprehend him."[3] Playfair's methods harken back to how he left France in 1792. It is hard to say why he was not tracked down during 1806.

Following the banner year of publishing in 1805, Playfair's literary output dried up for medical reasons. In June 1806, he became afflicted with rheumatic gout, with likely symptoms of pain and swelling in the joints of his feet and legs.[4] He was unable to write and had to walk with crutches. There continued to be mouths to feed – four children still at home and his wife. There were further complications: his daughter Elizabeth was blind. Playfair applied to the Literary Fund for help on 11 April 1807 and within a month received ten guineas (£10 10s.).[5]

Windham attempted a financial rescue for Playfair. With his patronage, Playfair began a new periodical, *Anticipation, in Politics, Commerce, and Finance, during the Present Crisis.*[6] With a first issue appearing on 12 March 1808, it came out weekly until 25 June 1808. Most of the articles were probably written by Playfair, and perhaps even some of the letters. There was also substantial self-promotion by the proprietor. With no advertisements in the periodical, there was an announcement and promotion of *British Family Antiquity*, on which Playfair was already working. There was also an article on Jepson Oddy, praising his book on European commerce that was written in 1805, possibly by Playfair, or with his help.

One feature in *Anticipation* was a column titled "Letters from Old Politicians in the Shades, to Young Politicians in the Sun-Shine." In the second issue of *Anticipation*, twenty-two letters were listed that would subsequently appear. As he had done in *Joseph and Benjamin*, Playfair's political views were expressed through the mouths of the characters appearing in the letters. The first letter was in print on 26 March, in the third issue. The first letter was "written" by William Pitt, who had died in 1806, to William Grenville, who had succeeded Pitt to form a national unity government called the Ministry of All the Talents. Windham, Playfair's patron, was part of this ministry. Grenville and his ministry resigned when George III rejected a move to relieve Catholics from legal strictures. From the words that Playfair put in Pitt's mouth, Playfair was not in favour of Catholic emancipation. Although twenty-two letters were promised, and although it was claimed that twenty-two were written, only twenty-one appeared in print. Of these, three were changed to other writers and recipients. Among the dropped letters, two were intended for the Prince of Wales (one from Henry V and one from Hamlet) and one from Alfred the Great to the people of England.

One letter, from one military man, deceased, to another, living, was expanded into a pamphlet (Playfair 1808). The two soldiers were the Duke of Marlborough, of the Battle of Blenheim fame, and George III's son, the Duke of York, who had commanded the British troops in the Flanders campaigns of 1793 and 1794 and was commander-in-chief of the British forces when Playfair was writing his pamphlet. Now in 1808, when Napoleon's troops had entered Spain and Portugal, a debate raged in the British press about whether the Duke of York should personally lead the British troops fighting on the Iberian Peninsula. According to Playfair, public opinion was against York's appointment. Over the course of the few pages of his pamphlet, Playfair sets out to change public opinion. He argues that the reverses experienced by the British in the Flanders campaign were not the fault of York, but rather of Prince Josias of Saxe-Coburg, the commander-in-chief of the allied forces, as well as the Prussian and Austrian governments. He urges readers to work to change public opinion and to give the Duke of York his rightful command. But it was Arthur Wellesley, later Duke of Wellington, who was put in command, then recalled after some early military reverses, and soon reinstated.

In a series of anonymously written articles in *Anticipation*, Playfair expresses his own appreciation to the Literary Fund, but in a roundabout way (*Anticipation* 1808, 113–15, 132–4, 159–60, and 177–9). He describes the good work done by people of the Literary Fund, how "men of letters" were important to society, how the fund was sustained by volunteer contributions, and how the fund amounting to about £1,000 annually was managed "gratuitously." Given that foreign governments, such as Prussia and the ancien régime in France, in some way financially supported their literati, Playfair suggests that the government contribute £500 yearly and encourage others to donate.

Another bee in Playfair's *Anticipation* bonnet was the establishment of an institution that he called the "Asylum for Female Servants, for the Preservation of Virtue." As part of the preamble to a call for financial support of this proposed institution, Playfair writes anonymously: "Female servants, a very numerous, and by no means unimportant, class of society, are, in London, exposed to unexampled hardships and dangers, and almost to certain ruin, by the circumstance, when they quit a place of service, they have no means of obtaining lodgings at a reasonable rate, or amongst respectable people. To associate with the vicious, and to run in debt, are almost inevitable; so that pressed on the one hand by want, encouraged by bad example, and stimulated by bad counsel, the unfortunate female yields to necessity, and is undone" (*Anticipation* 1808, 30).

Playfair's altruism is laudable; in this case there was likely also a personal motivating factor behind it. In 1808, Playfair's daughter Elizabeth, blind and living with her parents, was now about fifteen or sixteen years old. With little money in the family, one of her options was to go into service. Her younger sister Zenobia eventually became a governess.[7] That this general route was the probable family intention for Elizabeth is seen in events occurring seven years later. In 1815, Elizabeth was accepted as a boarder by the School for the Indigent Blind to attend their school in St. George's Field in Southwark (*Times*, 15 July 1815). There she was taught simple industrial work such as making the cords used for hanging sash windows.[8] Zenobia also received a reasonable education that allowed her to become a governess later. A bill for her schooling from December 1808 to March 1809 survives.[9] It shows her receiving instruction in writing, arithmetic, grammar, elocution, geography, history, and music.

Despite a search, only the first sixteen issues of *Anticipation* could be found, with the last published on 25 June 1808. In a reprint of an article on the Duke of York that appeared in *Anticipation*, Playfair stated that later numbers of the publication were to include a history of the military expeditions from 1793 onward (Playfair 1808, 4). That ambition was dated 27 July 1808 (22) and so the ambition died before it reached the compositor's table or even before Playfair put pen to paper.

Legal authorities eventually tracked down Playfair in 1808, probably as the result of Playfair's blunders. Circumstantial evidence is strongly in favour of this scenario. Between August and November 1808 Playfair was writing to Spencer Perceval, through Perceval's private secretary, John Herries. It was the usual scam – pay not to publish damaging information. By this time Perceval's position in the government had changed from attorney general to chancellor of the exchequer. Playfair was trying to alert Perceval to the possible publication of a pamphlet that Playfair had been given to proofread, titled *Curious & Interesting History of the Supposed Narrative of Evidence Given on an Inquiry Made into the Conduct of her Royal Highness the Princess of Wales.* The pamphlet concerned the marriage breakdown of George, Prince of Wales, and Caroline, Princess of Wales, accusations of her adultery, and the possible questionable actions of ministers in the government, including Perceval, concerning what came to be called the "Delicate Investigation" dating from 1806. Playfair offered to act as the intermediary in the purchase of the pamphlet so that it could be suppressed. Some toing and froing went on until November, when Playfair was invited to visit Herries in his office. Playfair was not satisfied with the negotiations and made his first blunder. As he had done in other scams, he threatened to publish a pamphlet of his own outlining his role in trying to help the government and at the same time putting Herries, as well as Perceval, in a bad light. Herries was angry and wrote back to Playfair on 23 November 1808: "The accusations against myself, though uncivil in manner, and in substance untrue, I would have answered explicitly, if you had presented them to me unmixed with other matters; but seeing that these alleged grounds of complaint against me are merely subservient to the introduction of other topics, with which I am not personally connected or acquainted, I must decline making any further observation on them."

Playfair's second error was to give his current address on his letters to Herries – 22 St. James Street, Covent Garden. This was nearly four miles from Brompton, his approximate whereabouts, according to the authorities in 1806. Perhaps Perceval responded to all this by having Playfair arrested after recalling Playfair's legal problems from three or four years before when he was attorney general.

When the law did catch up with Playfair, he was taken into custody in April 1809. At that point, desperation began to set in. He wrote to Perceval on 10 April requesting help. Through Herries, Perceval told Playfair that he could not interfere and that Playfair should make his request to the attorney general, Sir Vicary Gibbs. On 15 April Playfair wrote to Gibbs asking to be set free on his own recognizance with the promise that he would appear in court when called on. He alluded to correspondence with Spencer Perceval in which he had provided services that were of considerable importance to the government. He then enclosed an affidavit protesting his innocence of any wrongdoing, that unknown to him the conspiracy to deceive the creditors was between the two Kennetts. He gave his own twist to the story of what had happened. He concluded, "It is from the above favourable circumstances that I presume to hope that I shall be discharged on my own recognizance to appear when called particularly as I have a large family depending on my Exertions as I am not in good health & cannot, like those who really did conspire, live in prison on the Fruits of what had been taken by Fraud from the Estate of the Bankrupt. At the time sentence was passed on the 2 Kennetts they had upward 1400£ unfairly received but as before stated I never had but one Farthing which in addition to other Circumstances alters the Case Intirely."[10] Another draft of the affidavit has survived, but was not read in court. It went over the same territory as the draft read in court, but concluded with greater bravado:

> Your Deponent further saith that in addition to his being quite innocent in the affair from the Beginning to the End that he has a family depending on his Literary exertions that he is very Ill in health Requiring Great Care and that Imprisonment will probably be Fatal to him & in addition he begs leave to State upon Oath that for these number of years he has constantly by his writing defended and supported the Government of this Country against its enemies without fee or reward that lately namely in September Last he did Government a Real Service thereby Shewing himself a good subject and a deserving man & Furthermore your Deponent Declareth his Readiness to suffer any punishment if it can be proved if in any one act sale purchase transfer or concealment of property he acted with any of the said parties and as he has never been concealed during 4 years he is at a Loss to know why he is now arrested as it can be of no benefit to any party whatsoever.[11]

He was set free to return to court on 27 April 1809, at which time he received his sentence of three months in Newgate. Once in Newgate, Playfair wrote to Perceval, appealing to his sympathy, complaining that he was unable to go out in the courtyard of the prison because of a bad leg.[12]

Throughout this whole sordid episode, Playfair's literary work continued. That year Playfair published the first four volumes of his nine-volume work, *British Family Antiquity*, a compilation of histories of aristocratic and other landed families. Despite these successes, he remained financially embarrassed.

Chapter Sixteen

Playfair Dabbles Deeply in Family History and Political Biography

In his obituary of Playfair, Thomas Byerley, Playfair's friend and fellow journalist, briefly describes the most ambitious project of Playfair's literary career:

> The "British Family Antiquity," was an immense work, and published at a very extravagant price; it forms nine bulky volumes, and was published in royal 4to. [quarto] at 45 guineas, and the imperial 4to. at 75 guineas. The proprietors, who were not booksellers, became bankrupts in the undertaking, and it is only within the last few days that the remaining part of the copies of the work, have been sold by Mr. Saunders in Fleet Street. Of all others, it was a work for which Mr. Playfair was least suited. Industry and research were wanted, while Mr. Playfair supplied genius only, and was too volatile a disposition, even in literary pursuits, to confine himself to the dull labour of tracing out a family pedigree. ([Byerley] 1823, 171–2)

There is a lot to unpack from this brief description.

First, to touch on the more mundane, royal and imperial refer to paper sizes, and perhaps also to paper thicknesses. Imperial is larger than royal. Two printings of *British Family Antiquity* are held in the British Library. The copies of each volume from the library of George III, presumably in imperial, are slightly larger and at least twice as thick as the other volumes held in the British Library.

British Family Antiquity was sold by subscription. The final volume of the work contains a list of 510 subscribers. Taking the lower-priced book at forty-five guineas, subscriptions should have raised in excess of £24,000. The fact that there were copies left over means either that additional copies were made beyond the subscription numbers, or that some subscribers merely supported the project and were not interested in receiving their subscribed copy. One example of a subscriber not taking his books is Playfair's brother John. John Playfair is on the subscription list, but *British Family Antiquity* does not appear among the books in his library sold by auction after his death in 1819. It was unlikely that his heirs kept some "family" books back; some of John Playfair's own books were part of the auction (Playfair 1820).

Producing a nine-volume work was not Playfair's original intention. In his initial advertisement for the book, central to the publication was a set of chronological charts showing "the origin and progress of rank and honours of the British nobility" (*Anticipation*, 28 May 1808, 230). This was to be accompanied only by a large one-volume printed book (230–1). The whole enterprise was to be printed by one of England's most prestigious printers, Thomas Bensley. Playfair's original intention makes sense. He was good at producing informative graphs. As Byerley wrote, he was not good at genealogy. The research and other costs in the project were beyond Playfair's means. He began by going into partnership with Thomas Reynolds on 15 March 1808.[1] Reynolds got a half-share in the project on the condition that he put up the money for the project's expenses. Playfair extracted another condition. He was to receive personally all money obtained from any subscriptions he was able to generate from the Royal Family. Soon the whole project increased exponentially in complexity and length. The one volume soon turned into two, but only on the peerage of England. At that point, money must have been running low and Playfair sold his half of the project to Harvey Grace with the condition on Royal Family subscriptions remaining in effect. It was under Reynolds and Grace that the remaining seven volumes were produced, with Playfair being reimbursed for his research and other costs. For example, Playfair received a cash advance from Reynolds of £881 in 1808. From February to June in 1811 he was reimbursed by the partnership for £222 of his expenses.[2]

Playfair's approach to doing the research for *British Family Antiquity* was seemingly straightforward. He collected manuscripts; he wrote to families requesting genealogical, biographical, and other information; and then he sent what he wrote to each family for correction (Sharpe 1888, 386; Berry 1865, 386). What seems straightforward, especially to an amateur in the field, almost always turns out to be more complicated. Playfair spent about £5,000 on his research "expended in corresponding transcribing & consulting different sources of information."[3]

Part of the £5,000 would have been in postage costs. Until the postal reforms of the late 1830s and the penny post beginning in 1840, recipients of letters paid the postage, not the sender. Until the time of postal reform, postage could be costly and may have prevented a response from some of the families Playfair was researching. Volume 8 of *British*

Family Antiquity contains entries for about 120 baronets of Scotland. A list of Scottish baronets assembled by a genealogist in another publication has over 220 entries (Betham 1805, 41–5). In one place, Playfair described some of his respondents as "generally negligent" and in another commented on having "so little aid from families in general."[4] He lamented that without the assistance of these people or other knowledgeable people who could help, it would be impossible to provide correct family histories and genealogies (Sharpe 1888, 386). Some or perhaps even much of the Scottish baronets' failure to respond might be attributed to Playfair's lack of knowledge. In volumes 4 and 5, he missed four of the Irish peers – not an easy thing to do for anyone knowledgeable about the peerage (Malcomson 2000, 295).

In writing to so many families, Playfair had a lengthy one-page form letter typeset and printed, which he signed in his own hand. He also enclosed the preface and an advertisement for the book. In the form letter, Playfair outlines the proper role of the nobility in government by using two examples. He first alludes to the Spanish nobility and praises their resistance to Napoleon during the Peninsular War. He held this up as an example, which showed that "an ancient nobility when properly acting and properly esteemed are the surest Guardian of the Liberation of the People, as well as the Prerogatives of the Throne and of their own Honour and Independence."[5] On the other side, he criticized the old French nobility for their "levity and their indifference to the prosperity of the people, and disregard to public opinion." Playfair concluded with a request to the recipient to send him the names of the recipient's relatives who might be interested in subscribing to *British Family Antiquity*. Copies of the volumes could be purchased only through subscription, he said.

The preface and advertisement that Playfair included with the form letter was probably the same in substance as what he advertised in his *Anticipation* on 28 May 1808, in which his political philosophy is clear. He wanted the political status quo to be retained as it existed in 1808 in Britain: a sovereign at the pinnacle with checks and balances from Parliament constituted at the time from landed interests. He wanted to "maintain the present happy order of things, and the incomparable government of this country, which are so intimately connected with the preservation of the different ranks of society, and so dependent on the respect and esteem justly marked for the higher orders" (*Anticipation* 1808, 230). He was using *British Family Antiquity* to prove the truth of this philosophy. He used an argument relying on correlation and cause. There was a connection, Playfair claimed, between the great actions of men, as well as their good and virtuous conduct, and the titles given to them by the monarch.

The preface that appears in the first volume of *British Family Antiquity* covers the same arguments as, but is not identical to, the preface printed in *Anticipation*. The final version of the preface runs to eighteen large typeset pages, while the preface in *Anticipation* is truncated at five pages with a promise "To be continued," which never came to pass (Playfair 1809–11, 1:1–18; *Anticipation* 1808, 231–5). Much ink was spilt over nine volumes of *British Family Antiquity* laying out the great actions and exemplary conduct of the British nobility.

This philosophy and Playfair's justification of it are espoused in greater length in two of Playfair's lengthy pamphlets published in 1809 and 1810 (Playfair 1809b, 1810). His arguments bear all the hallmarks of wrong conclusions taken from an observational study. After looking at the family histories of titled people, he concluded that the titles given were based on merit and not favours given out willy-nilly by the Crown. The descendants of these original titleholders had more than their fair proportion of merit and talent (1809b, 44–51). To try to prove his point and to be fair, Playfair restricted the population of those without titles to the affluent in society. With very little data to go on, he reckoned that the ratio of the generally affluent to the titled was about 200 to 1. When he looked at politicians, military men, scientists, literary men, and men of "inventive genius," there were many more than the expected number among the titled population. The flip side was to look at France, where titles had been abolished. France went through a period of chaos and now had a strongman in charge. Philosophers took the first step into this abyss by ridiculing the idea of titles. Therefore, he concluded, following the philosophers in their ideas of reform was a false path; those who had titles were in a better position to sit in Parliament than those without.

In view of his agenda to justify the political status quo, Playfair tried to make *British Family Antiquity* different from other genealogical works, in part by "modernising the style and enlarging the biography."[6] This can be seen in Playfair's treatment of Sir John Sinclair, known to statisticians as the one who popularized the word "statistics" through his work *The Statistical Account of Scotland* published in twenty-one volumes, beginning in 1791. In the standard style of genealogical works, the entry in Arthur Collins's *Baronetage* for Sir John Sinclair states when he was created a baronet, when he married and to whom (Collins 1806, 494). Then follows a brief list of his ancestors: paternal grandfather, his spouse, their children and their spouses, followed by Sir John's father, his spouse, their children and spouses. The only additional biographical information is that Sir John Sinclair was a "member in the present parliament for Bute and Caithness in Scotland." For his part, Playfair includes the "begats" as Collins did, but does it in a prose style with biographical information on several of Sir John's ancestors, especially their contributions to politics. Regarding Sir John himself, Playfair spends eight pages describing Sinclair's political and other accomplishments, thus justifying his baronetcy (Playfair 1809–11, 7:322–32). The progression to Sir John's baronetcy was natural; several of his ancestors were public spirited and made worthy contributions to politics and the public weal.

Playfair complained to Charles Kirkpatrick Sharpe, an Edinburgh poet and antiquarian, about the lack of response and lack of help he had received from the families that would appear in *British Family Antiquity*. At the same time, two irritants for Playfair were families embellishing their backgrounds, and families that got information to him too late to be included in the publication (Sharpe 1888, 372–4, 386). Some families did respond to his enquiries. The Dowager Marchioness of Downshire wrote to Playfair advising him to add in additional titles held by her son, the marquess.[7] The family of the Marquess of Wellesley, elder brother of the future Duke of Wellington, sent in carefully and thoroughly done corrections.[8]

A significant portion of Playfair's budget must have been in purchasing manuscripts. One of Playfair's sources in this area was the clergyman and antiquarian William Betham of Dublin, who had published the five-volume *Baronetage of England* in 1805. From Betham, Playfair purchased Betham's manuscript on Irish baronets and wove it into the contents of volume 9 of *British Family Antiquity* (Sharpe 1888, 373). He also received help and advice from Sharpe.

To encourage responses and subscriptions, Playfair sent his newly devised charts to several of his respondents. The one shown in figure 16.1 was sent to the Dowager Marchioness of Downshire.[9] It shows English viscounts and barons, as well as peeresses in their own right and bishops who sat in the House of Lords. The lengths of the horizontal lines in the chart correspond to the times from the creation of the associated titles. Each line is colour-coded to indicate the nature of the service rendered that resulted in the creation of the title. Green, for example, indicates that the original peer was awarded the title on the basis of loyalty or attachment to the throne. Consequently, all episcopal peers are colour-coded green. The other colours are orange for service in the navy, red for service in the army, yellow for parliamentary service, and blue for service to the law.

The expense of the coloured charts may have been only a small contributing factor to the bankruptcy of Reynolds and Grace. Another was the cost of the rest of the printing. If the entry for Richard Colley Wellesley, 1st Marquess Wellesley, is anything to go by, the printing costs for

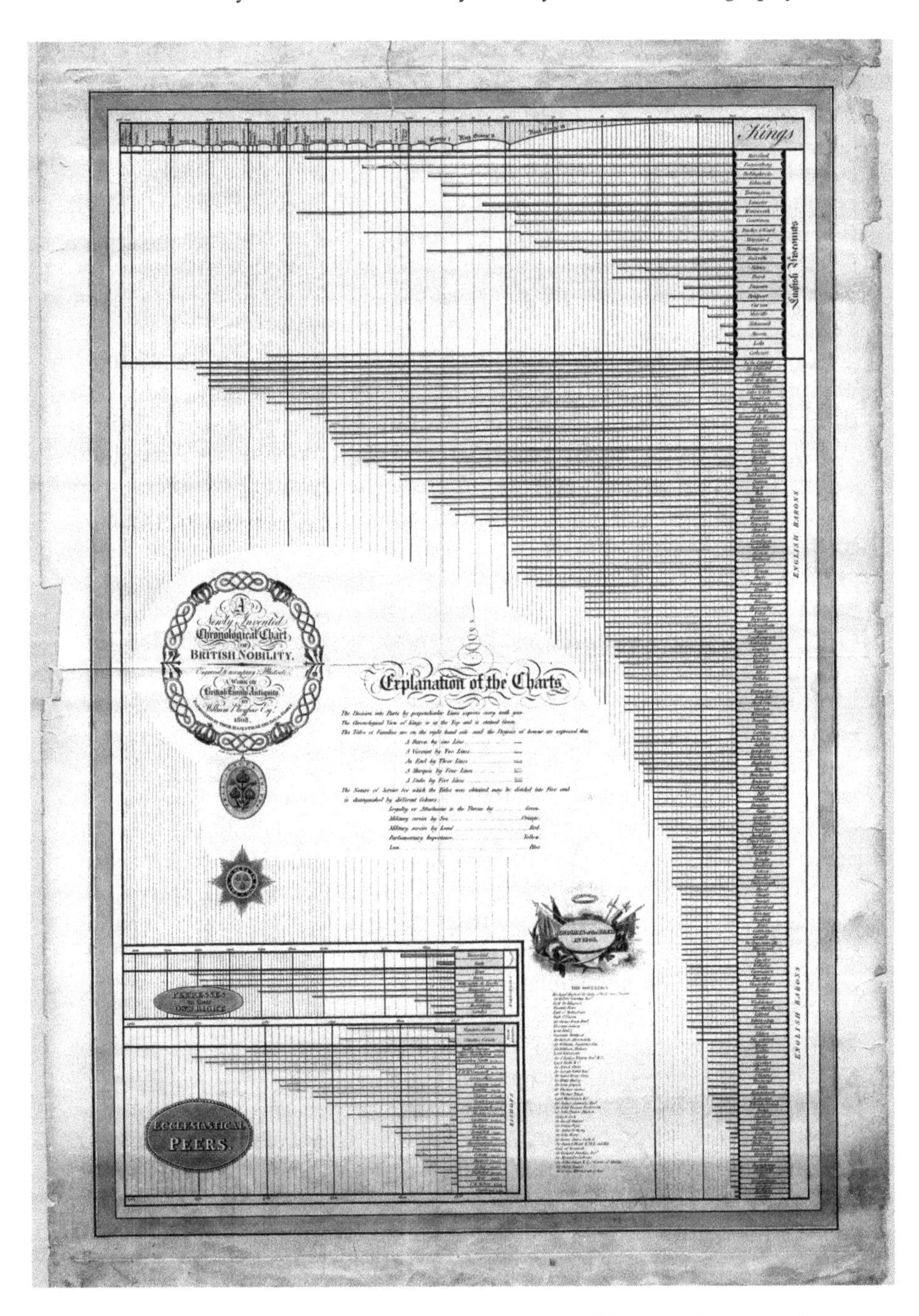

Figure 16.1. Playfair's chart of English viscounts and barons. Courtesy of Public Record Office of Northern Ireland.

British Family Antiquity could have been twice the normal cost of printing the volumes. The Wellesleys received several typeset pages for their entry in *British Family Antiquity*.[10] Someone in the family inserted corrections and changes on these pages. When what the Wellesleys received and what appears in *British Family Antiquity* are compared, it is obvious that the type in the final version had been completely reset. For example, the first paragraph in both versions contains the same words using the same font, but the layout is completely different. The early version of this paragraph is printed in twelve lines and the later version in eight.

Playfair's belief that there was a natural ruling class, so that no democratic reforms should be made, carried over to a certain extent into the next generation. Playfair's son, Andrew William, joined the British Army as an ensign on 25 April 1810. The Napoleonic Wars were good to him. Lacking money, he was promoted to lieutenant without purchasing a commission and transferred from the 32nd Regiment of Foot to the 104th Regiment of Foot (Hart 1840, 370; *London Gazette*, 9 November 1811). The promotion without purchase was probably due to attrition among the officers without the landed classes coming forward with replacements by purchase. With the 104th the young Playfair saw military action in Canada during the War of 1812. At the end of the war he took the option of taking a land grant in Canada rather than returning to England. In the colonies, he prospered. He operated several businesses, including a lucrative sawmill. He also served a term as a member of the Legislative Assembly in pre-Confederation Canada. His obituary states, "Loving, consistent, clinging loyalty to the crown and constitution of Great Britain was one leading feature in the character of Col. Playfair" (*Journal of Education of Ontario* 21 [1868]: 134). During the Rebellion of 1837, Playfair supported the Family Compact, the ruling group that stood against democratic reform in Upper Canada, what the reformers called "responsible government." In 1843, Andrew William Playfair's brother, John, and his sister, Elizabeth, and their children joined him in immigrating to Canada, all settling close by in an area of Ontario near Ottawa, now known as Playfairville.[11]

In support of Byerley's comments on Playfair's genealogical abilities, James Maidment levelled the first salvo against Playfair in that regard, about thirty years after publication of *British Family Antiquity*. Maidment was studying a manuscript that came from Charles Sharpe, the same man with whom Playfair corresponded about manuscripts and other genealogical enquiries. Maidment's was an early sixteenth-century manuscript, which contained the constitution, written in Latin, of the Convent of Saint Catharine of Sienna. Located in Sciennes, a district of Edinburgh, this was the only Dominican order of nuns in Scotland. Maidment transcribed the constitution and in a preface gave an extensive description of the life of Catharine of Sienna, as well as a history of the convent. Playfair enters into Maidment's work during a discussion of the subsequent owners of the property after the Reformation in Scotland. Maidment describes Playfair as "a wholesale manufacturer of pedigrees" (Maidment 1841, xxxvi). He is highly critical of Playfair's genealogical treatment of the forebears of one of these owners, writing of "the absurd attempt of Mr Playfair to magnify a respectable tradesman into a descendant of a Laird of that Ilk" (lxiii).

While carrying out extensive genealogical work on the Moncrieff family, George Seton appears to have done the most thorough dissection of Playfair's genealogical work (Seton 1890). Seton had access to three pedigrees of the Moncrieff family, Playfair's being one of them and discovered several discrepancies. He described Playfair's treatment of the Moncrieff family as containing many errors and some of Playfair's statements as ludicrous and irrelevant. He characterizes Playfair's approach to genealogical work as "free-and-easy 'magnificence' as a genealogist" (142).

What separates Playfair's work for *British Family Antiquity* from that of the professional genealogist is that Playfair did not consult any original source material such as parish registers or manuscripts in archives and libraries. For Scotland, Playfair could have, and should have, visited the Faculty of Advocates library, which is now regarded as the finest working law library in the British Isles. It would have been a treasure trove for Playfair's Scottish genealogical work. Instead, he often relied on information, possibly biased and inaccurate, offered by the families he was researching.

Beyond the professionals, at least one of those who appeared in *British Family Antiquity* was not impressed by it. George Stewart, 8th Earl of Galloway bought the first five volumes of *British Family Antiquity* as a result of mistaken identity. Normally, he did not subscribe to books. But in this case, he was under the impression that *British Family Antiquity* was written by John Playfair, William's older and more eminent brother. The mistake in the identity of the author probably came from a friend, Lord Webb Seymour, fourth son of the 10th Duke of Somerset who had made the same mistake. Seymour was a friend of John Playfair and, in the past, had worked with the elder Playfair on some geological projects. When he received his first five volumes (costing ten guineas each) of *British Family Antiquity*, Galloway was "grievously disapointed by the trifling stuff and fulsome flattery with which the production … abounded" (Berry 1865, 386–7). Galloway's younger brother questioned whether John Playfair was the real author. The two brothers appealed to Seymour for his opinion. Without knowing whether his friend was the real author, Seymour denied that John Playfair was the author for the sole reason that the "book, it seemed, was a bad one" (387). Seymour, it seems, had not read any of *British Family Antiquity*.

Just as it had reached completion, the whole family history project fell apart. The publishers of *British Family Antiquity*, Thomas Reynolds and Harvey Grace, were in financial trouble and although they dissolved their partnership on 1 November 1811, that action was of no help to them; they were declared bankrupt on 21 January 1812 (*London Gazette*, 2 November 1811, 21 January 1812). Although Playfair was financially at arm's length from the publishers, he had a minor brush with the bankruptcy and a more major brush with his own.[12] Part of Playfair's deal with Reynolds and Grace had been that any subscription money brought in from members of the Royal Family would go directly to him and not to the two partners. Needing money, Playfair wrote to the prince regent's secretary on 21 July 1812 asking for payment for the regent's volumes of *British Family Antiquity*. The bill was for seventy guineas as well as the cost of binding which totalled £104 14s. (about £8,500 in today's money). Unable to pay the rent on the house in which he was living, Playfair was sent to the Fleet Prison in October 1812. The prince regent's secretary visited him in prison and gave him £50 as partial payment of the account, and the bill was eventually paid in full by December. About a week after Playfair wrote to the prince regent's secretary, the trustees in bankruptcy for Reynolds and Grace also wrote to the secretary. From the partnership records, the trustees in bankruptcy had an unpaid bill dated 11 May 1809 that they wanted to collect. It was a bill made out to the prince regent for £107 2s., payable to Reynolds and Grace for the regent's copy of *British Family Antiquity*. Whatever confusion arose from this double billing, the second bill was also paid, but not until 26 April 1813.

Over the years 1798 to 1810, there appeared an annual publication under the title *Public Characters*, running to ten volumes. A careful researcher and biographer, Alexander Stephens wrote the first nine volumes (*Annual Biography and Obituary* 6 [1822]: 412–22). *Public Characters* contains over 330 biographies of living people who had made some impact on British society. Playfair's *Political Portraits* (Playfair 1813–14) seems roughly modelled on *Public Characters*. Published in two volumes over 1813 and 1814, about two years after the last volume of *British Family Antiquity* came off the press, *Political Portraits* contains biography with a twist. The emphasis is on political and economic commentary rather than complete biographical information. Playfair's *Political*

Portraits contains nearly 150 biographies, placed in alphabetical order, of which about one-third also appear in *Public Characters*.

The difference between *Political Portraits* and *Public Characters* may be seen easily by comparing the treatments of Richard Brinsley Sheridan, who appears in both publications (*Public Characters* 1799, 19–49; Playfair 1814, 346–51). Sheridan was well known to Playfair. Through the Original Security Bank, he had financial dealings with Sheridan relating to the Drury Lane Theatre. Through the *Tomahawk*, he had mocked Sheridan politically and anonymously. Appearing in the 1799/1800 volume of *Public Characters*, Stephens's biography of Sheridan covers Sheridan's birth and early life, his two marriages and children, his work as an author and actor, and his entry and work in Parliament, all treated very sympathetically. Playfair makes no mention of Sheridan's family life or his work as a playwright; instead, he skims through Sheridan's political career, focusing only on a few topics. By the time that Playfair wrote his article on Sheridan, Sheridan no longer had a seat in Parliament and was now fair game for his creditors. MPs had parliamentary privilege and could not be taken to court in civil actions or arrested for debt (May 1844, chap. 5). Early in 1813, the creditors in the Original Security Bank bankruptcy came to a settlement regarding how to deal with Sheridan's debts to the bank (*London Gazette*, 5 January 1813). One of the first things that Playfair wrote about Sheridan was, "Mr. Sheridan has had the misfortune of being ushered into the world under pecuniary difficulties, from which he had not the prudence, nor, perhaps the means of extricating himself, and from which his political associates had not the generosity or honour to liberate him" (Playfair 1814, 346–7). Playfair may have held Sheridan partially to blame for the failure of the bank. Throughout the article, he acknowledges Sheridan's talents, energy, and political integrity, but then mixes his praise with comments on Sheridan's political imprudence.

Similar to *British Family Antiquity*, Playfair wrote to some of his subjects who appeared in *Political Portraits*. One was John Wilson Croker, who was serving as secretary to the Admiralty. After Playfair had sent Croker his political portrait, Croker wrote back to say that no corrections needed to be made.[13] It may be assumed that Playfair did not write to those of whom he was highly critical.

Playfair uses *Political Portraits* as one more venue to flex his political opinions. He outlined his main points at the beginning of volume 1:

> The steps that seem to be most necessary to preserve tranquillity in this country, in either case are nearly the same. First. Economy in public expenditure. Second. Steady perseverance in Mr. Pitt's plan for paying off

> the national debt. Third. The abandonment of Mr. Pitt's plan about the price of provisions. Fourth. An effort on the part of the wealthy, to assist in reducing the debt. And last, though not least, our joining with a firm resolution to resist all theoretical reforms; together with a disposition to make practical and safe ameliorations, both in Britain and Ireland.
>
> On these important subjects it is that I have endeavoured to give my views in the portraits, and the notes, which, with the assistance of the index will easily be found.

Contrary to Playfair's claim, what is easily found in the index are only his thoughts on the national debt and the sinking fund that was established by Pitt to pay it off. On this subject he offers two seemingly contradictory opinions, one under Stephen Lushington, who was secretary of the treasury and the other under Nicholas Vansittart who was chancellor of the exchequer. Playfair had strongly supported Pitt's sinking fund and was concerned about the changes that Vansittart was making to financing the debt. In Lushington's portrait, Playfair expressed concern about what might happen when the debt was paid off. He feared that the capital freed up by paying off the debt would flee the country for other investment opportunities. He wanted it to be reinvested in Britain. To put it all together, he wanted a debt-free country but worried about what would happen when he had one.

Playfair's thoughts on Ireland are not easily found in the index and are best discovered by a keyword search in digitized copies of the two volumes of *Political Portraits*. What emerges is Playfair's fairly low opinion of the Irish and Catholicism, especially when put together. His view seems typical of the oppressor on the oppressed: "The Irish are generous, good-hearted, and affectionate, but they are hasty and inconsiderate; and, as they are peculiarly susceptible to the influence of eloquence, they have been terribly misled; and this has been the more easily effected, that the common people are kept in a wretched state of poverty and ignorance. The poverty and ignorance are not, however, owing to England, but partly to the proprietors of the soil, partly to the religion of the majority, and partly to the indolence of the people living under such circumstances" (Playfair 1814, 491).

Playfair's anti-Catholicism is especially evident in two of his portraits. The first is of Patrick Duigenan, an Irish MP who was born a Catholic, but converted to the Protestant Church of Ireland at a young age, while the second is of Henry Grattan, another Irish MP. Grattan was in favour of Catholic emancipation, whereas Playfair was not. Playfair's anti-Catholicism is pervasive through many of his publications, but not vehement.

Among his subjects in *Political Portraits*, Playfair heaped praise on those who, beyond the usual political arena, directly served Playfair's interests. Early in the alphabetical order of the political portraits, Playfair liked Lord Ellenborough, who had been appointed chief justice of the King's Bench in 1802. Dismissing the controversy over Ellenborough's decision to join the cabinet in the Ministry of All the Talents in 1806 (he was the last chief justice ever to serve in cabinet), Playfair praised Ellenborough for two things that were close to Playfair's heart. The first was his treatment of patent law, which Playfair saw as better than former chief justices. The second was a new law, supported by Ellenborough, on the treatment of insolvent debtors, Playfair being one of them. Playfair rightfully pointed out that the new legislation removed some of the gross inequities of the past. Previously, a debtor could be held in prison indefinitely until the debt was paid, while a felon might be given a sentence for a fixed term. Instead of treating the insolvent debtors' law in detail under Ellenborough, Playfair waited until later in the alphabet to expound on his views. It was John Freeman-Mitford, 1st Baron Redesdale, who introduced the legislation in Parliament. Redesdale's biography is completely taken up with this legislation, including several of Playfair's suggestions to improve it.

Also related to debt was Playfair's high opinion of Lord Eldon, who presided over the High Court of Chancery. This was the court that heard bankruptcy cases, again a topic close to Playfair's heart. According to Playfair, Eldon was a good judge, but overworked; the number of bankruptcy cases had doubled in the previous twenty years and had increased five-fold over the previous century. Playfair's portrait of Eldon allowed him to ignore Eldon for a bit and go into a tirade about bankruptcy and the lawyers who handled such cases. Among these lawyers were what Playfair called "bankruptcy-hunters" on whom he commented: "There are numbers of attorneys who, with the aid of certain tradesmen, their confederates, sue out commissions, and act as solicitors and assignees. It would be well to have a list of those who appear frequently in such affairs, for frequency is a proof of connivauce [*sic*]" (Playfair 1813–14, 1, 291). A search of the *London Gazette* reveals that Samuel Wadeson, the lawyer most closely involved in Playfair's bankruptcy case, handled several bankruptcy cases over a twenty-year period. Perhaps Playfair was taking a shot at Wadeson without naming names.

Playfair definitely used another portrait to settle an old score – his entry for Joel Barlow, with whom Playfair had worked selling lands in Ohio through the Compagnie du Scioto. Playfair minced no words, describing Barlow as "taciturn and selfish," writer of a "long, tedious poem," "ignorant of business," "an enemy to all religion," and having a "total want of a principle of justice." His "cunning ingenuity" and his "being unrestrained by principal, and accompanied with a grave exterior" allowed him to be useful to revolutionary France and popular in America. Playfair took his shots while Barlow was lying in his grave, having died in December 1812.

Both *British Family Antiquity* and *Political Portraits* were a spawning ground for some of Playfair's later scams. The first scam to emerge was just before *Political Portraits* went to press. Probably still needing money, Playfair wrote to George Canning on 19 March 1813. Canning had held several government offices and had been especially close to Pitt. At the time of Playfair's letter, Canning was still an MP, but out of office. As a supporter of Pitt and his dealings with revolutionary France, one would think that Playfair would be positive towards Canning. Such was not the case. Playfair's letter to Canning is not extant; the reply is. It was a typical Playfair scam. Playfair thought he had some dirt on Canning, put it on paper, and sent it to Canning for his comment. Canning replied, "A public man must of course make up his mind to the seeing his character torne to pieces by any writer who may be disposed to seek either reputation or profit at his expence. But I am not aware that he is under any obligation to entertain the private communication of the abuse or calumny which it may be intended to publish against him, or to enter into a controversy with his revilers. I therefore take the liberty of returning your ms. Without availing myself of your obliging offer to add to, or to curtail it & must decline seeing the gentleman who you say is to wait upon me in your behalf."[14] What the dirt was, Canning did not mention in his reply. It may be guessed that it is tied to the activities of the French diplomat, spy, and political adventurer Louis-Emmanuel-Henri-Alexandre de Launay, Comte d'Antraigues.

When Canning became foreign secretary in March 1807, much to his dislike he was constantly courted through letters by d'Antraigues (Munch-Petersen 2008). Canning put his dislike aside, as d'Antraigues was seen as having secret connections to the Russian government of Tsar Alexander I. Although Canning never trusted d'Antraigues enough to put anything in writing to him during this time, what eventually cemented their relationship were events surrounding the Treaty of Tilsit signed on 7 July 1807 between Napoleon and Alexander. In the treaty, Napoleon and Alexander were no longer enemies but allies. One outcome of the treaty was an attempted continental blockade of British trade to Europe. This was Napoleon's attempt to isolate Britain economically. On 21 July 1807, d'Antraigues sent Canning a letter claiming to have secret information about the Tilsit agreement from Alexander's aide-de-camp. Part of d'Antraigues's new information supplemented the information on Napoleon's continental blockade. According to d'Antraigues's source, Denmark, Portugal, and Sweden would be forced to join the blockade. Although the letter was mostly the product of d'Antraigues's imagination, it led to the British view that the British had a line of communication to someone in the Russian court. It also led to the British bombarding Copenhagen from mid-August to early September 1807 and seizing the Danish fleet (Munch-Petersen (2013).

The timing of the Canning-Playfair correspondence and the publication of *Political Portraits* fit together nicely. The correspondence occurred in March 1813 and the first volume of *Political Portraits* was probably printed late in 1813, since the first advertisement for it, along with the second volume, appear on 1 March 1814, with the first review of both volumes appearing in April 1814 (*Morning Post*, 1 March 1814; *Critical Review*. 1814, 425–32). In a footnote to his article on Canning in *Political Portraits*, Playfair tells the same story of Canning and d'Antraigues with some added spice, claiming that d'Antraigues paid the Russian aide-de-camp £5,000 for the information, presumably put up by the British government. This may be the dirt that Playfair thought he had on Canning. When Canning did not bite, Playfair published, perhaps in a more measured version to avoid a lawsuit.

Chapter Seventeen

Playfair Tries to Pull a Major Scam on Lord Bathurst about Bonaparte

Playfair was back in Fleet Prison on 4 September 1813 for debt, his fourth incarceration there since 1801.[1] He was released within about a month. The editors for his *British Family Antiquity* had recently gone bankrupt, so there was no money to be had from that project.[2] His *Political Portraits* was not advertised until 1 March 1814 (*Morning Post*, 1 March 1814). From that project, money may have trickled in through the remainder of the year, but the trickle may have turned to a drip if his creditors still had a call on his debts. It was his continuing need for money, with little visible means to obtain it, that probably led him to try to extort money from the British government.

To set the stage for his scam, the allied troops of Austria, Prussia, and Russia took Paris on 31 March 1814 and within a week restored the Bourbon monarchy. By the time Napoleon's forces had reached Fontainebleau to move on Paris, about seventy kilometres away, it was too late. Bonaparte's senior military staff advised him to surrender, which he did. By the Treaty of Fontainebleau, Napoleon and all his family were required to renounce all claims to the throne of France. Further, all of Napoleon's property in France was to revert to the Crown. He retained his title of emperor, but he was emperor of a much reduced territory. It was Elba, an island off the west coast of Italy on approximately the same latitude as the northern tip of Corsica, Napoleon's birthplace. In his new principality of Elba, he was allowed to maintain a guard of 400 men. A detachment of troops escorted Napoleon from Fontainebleau to Saint-Tropez on the French Riviera, where he boarded a ship for Elba, arriving at the main port, Portoferraio, on 4 May 1814 (Thompson 1949, 1950). A British Army officer, Lieutenant-Colonel Neil Campbell, accompanied Bonaparte to Elba. Campbell was promoted to the rank of colonel soon thereafter, 4 June 1814 (Stephens and Fraser (2006).

It was late October 1814 when Playfair wrote to Henry Bathurst, 3rd Earl Bathurst, about Napoleon's planned escape from Elba (Playfair 1815, 14–15). Bathurst was the secretary of state for war and the colonies. Playfair also contacted the French ambassador, Claude-Louis-Raoul de La Châtre. According to Playfair, he had heard of the plot as early as 10 September but sat on the information to protect his informant. Further, there was no immediate danger of escape, Playfair thought, until after the peace conference, the Congress of Vienna, was concluded.

Here is a pertinent excerpt from the paper that Playfair enclosed with his letter to Bathurst, explaining the need to thwart Napoleon, and Playfair's proposed role in doing so:

> As Buonaparte is plotting, and the fact is certain, the best way will be to detect him, and then the two Emperors [Austria and Russia] will see the necessity of securing his person; at the same time that it will remove every scruple in regard to changing the mode of treating him.
>
> What is done must be in silence, and quickly when begun; for, though Buonaparte would naturally wait for a fit time, yet he will act the moment he sees danger.
>
> The information I have obtained is sufficient to enable me, if I had a trusty and confidential person with me, to bring proofs against Buonaparte, such as would convince the Allied Powers of the necessity of changing the treatment of him: but this should be done while Congress is sitting. (Playfair 1815, 13)

The subtext of this quotation is that Playfair wanted to go to the Continent, Paris in particular. It is made more explicit in the covering letter to Bathurst for the paper that Playfair wrote: "If your Lordship chuses to give me letters to Paris that I may be safe I shall soon be able to furnish your Lordship well."[3] Later Playfair claimed that both Bathurst and La Châtre ignored his warnings and offer of help.

The idea of Bonaparte escaping Elba, before he actually did so four months later, was not completely fanciful, and Playfair could easily have been inspired by reports in the London newspapers. On 9 October 1814, one London newspaper, *Bell's Weekly Messenger*, reported two conflicting rumours about Bonaparte: that he was working quietly in Elba and that the British had whisked him away to Malta or Gibraltar (*Bell's Weekly Messenger*, 9 October 1814). Less than a week later, another London newspaper, the *Morning Post*, reported that Bonaparte still had strong support in France (*Morning Post*, 14 October 1814). Five days after that article, the same newspaper reported that, after the Congress of Vienna concluded, Bonaparte would be moved from Elba to the Caribbean, probably St. Lucia (*Morning Post*, 19 October 1814). The *Morning Chronicle* reported on 13 October that during celebrations on Elba earlier in August Napoleon's troops shouted "Vive l'empereur!" (*Morning Chronicle*, 13 October 1814). Most telling was another news report. A man who had visited the iron mines in Elba reported that there were now 700 Italian and French troops on the island who were "passionately devoted to the person and fortunes of Napoleon" (*Morning Post*, 11 October 1814). Further, there was no appearance of personal restraint on Bonaparte's movements, and Colonel Campbell had left the island without armed vessels nearby. In that atmosphere, it would have been easy for Playfair to concoct a believable story about Napoleon planning to escape Elba.

Playfair's full story for Bathurst was spiked with intrigue and a clandestine meeting. And it involved his old crony Thomas Byerley. The story goes that in early September 1814, Playfair met an Italian named Caraman, who had been part of Bonaparte's guard on Elba (Playfair 1815, 12–13). For private reasons, Caraman had left Elba and was on his way to Sweden. On 10 September, Caraman, Playfair, and Byerley met in Pagliano's, a dining establishment near Leicester Square. Caraman and Playfair spoke in French, which Byerley did not understand. Caraman gave some documents to Playfair, who passed them to Byerley to copy. Playfair was required to return the documents to Caraman the next day. One document was in code and the other was the key to it. After dinner, Playfair informed Byerley of the gist of his conversation with Caraman. Playfair and Byerley then went away from the Pagliano's together, sat down, and deciphered the code. The coded message was an address to be given by Napoleon to the people of France. Byerley wrote down his recollections of the meeting in a memorandum 14 October 1814 (8–9). On thinking more about these revelations, Byerley advised Playfair to inform the French ambassador, La Châtre. He put this in a memorandum dated 20 October 1814.

Through November and for the next few months, Playfair let the matter rest. It was not until after Napoleon had escaped Elba on 26 February 1815 that Playfair played his next card. It took time for word to reach Britain. The *Morning Post* reported on 11 March 1815 that Napoleon had landed in France accompanied by about 1,000 men. Soon afterwards, 50,000 French troops joined him (*Morning Post*, 11 March 1815). Two weeks later, Playfair wrote to his publisher, John Stockdale, informing him of the conspiracy that he had revealed to Bathurst five months before (Playfair 1815, iii–vi). By early April, Stockdale published the letters and accompanying documents as a pamphlet. On 3 April, the *Morning Post* ran the pamphlet, nearly in its entirety, and the next day the *Times* ran it as well (*Morning Post*, 3 April 1815; *Times*, 4 April 1815). The publication caused a stir. Several London and provincial newspapers included the following item: "A Mr. Wm. Playfair has published a pamphlet, imputing to ministers negligence and incredibility, in not paying attention to his communications, made since October last, of Bonaparte's plan to escape and regain the crown of France."[4] The pamphlet was then reprinted in other newspapers.[5]

Then *London Courier and Evening Gazette* investigated, reporting on 10 April 1815:

> We have hitherto abstained from noticing a statement which appeared some [time] since in a respectable Morning Paper, relative to the discovery by a Mr. Playfair of a correspondence between Buonaparte and his adherents; because the statement seemed to carry with it the evidence of its imposture, and were therefore unwilling to give it importance which it did not deserve. Finding, however, that some of our readers are desirous of knowing the facts of this case, we now briefly lay before them the result of our investigation. We can distinctly assure them that Mr. Playfair received as far back as November last the means of proceeding to Paris, where he himself stated his proofs were to be adduced. He obtained from the French Ambassador in this country a letter of recommendation to the French Minister at Paris, and a promise of liberal remuneration in the event of his statement proving true. Of this Mr. Playfair never availed himself.
>
> We can also assure our readers that Mr. Playfair deserves as little credit for the discovery of a cypher as for his knowledge of Buonaparte's plot. Without adverting to the language of the Proclamation, which is not correct French, or to the circumstance of a *w* (which is not a French letter) being introduced into a cypher alphabet, intended solely for correspondence between two French persons, we can state that the cypher in question has been long known in all the public offices. We have moreover a pamphlet in our possession, entitled *Langue Diplomatique,* published in 1794, by a French Gentleman of the name of Cameron, and purporting to be a translation from German, in which this cypher is explained in great detail almost in the words used by Mr. Playfair. We mention this latter circumstance as it accounts for the introduction of the suspected letter *w* (which is a German letter), and as it may help our Readers, as well as Mr. Playfair himself, to discover his Ingenious Italian friend, Mr. Caraman, of whom Mr. Playfair professes himself unable to afford any satisfactory information. We will not detain our readers by a detail of the minor circumstances of this attempt to impose upon the public, because we conceive we have stated enough to prove his folly and absurdity.

One would think that after this report the matter would be closed. Such was not to be the case.

Playfair responded to the *Courier* in the *Morning Post* (11 April 1815). He challenged the editor of the *Courier* to come forward with the internal evidence from his pamphlet showing his claims to be fraudulent. As for the code, it was meant for all languages, not just French, Playfair claimed. Further, Playfair stated that he had received no funds to go to Paris. And finally, Playfair wrote, "The *Courier* says the Proclamation is bad French. Why does he not point out the errors?"

The controversy reached the floor of the House of Lords. In a debate about Napoleon's escape from Elba and how to proceed, Lord Bathurst stated on 12 April 1815, as reported in *Hansard*,

> The noble earl then alluded to a subject, on which, he said, he was almost ashamed to trouble the House – the information which a Mr. Playfair had stated he was able to communicate to government of the plan of Buonaparté. When Mr. Playfair was asked whether the person from whom he received his information, was a friend to Buonaparté or the Bourbons – how he became acquainted with him – what had induced him to make the communication – where he could be seen – to all these questions no answer could be returned. He had assured Mr. Playfair, that if he was able to substantiate his statement by proof, he should be rewarded. Mr. Playfair knew where to find him if he had any such proofs; and if he did not bring them forward, no blame could attach to any other person. At his solicitation, the French ambassador gave him a passport, and a letter of introduction to M. de Blacas; but he never heard of Mr. Playfair after; and when Mr. Playfair was asked why he did not avail himself of the passport and the letter, the reason he gave was, that the letter was sealed. (Great Britain 1815a, 571)

This speech was reported widely in the London and provincial newspapers. On 20 April MP James Abercrombie spoke in the House of Commons, saying "that he totally condemned the trash published by a person of the name of Playfair, which he conceived to be altogether unworthy of notice" (723).

Playfair had lost all credibility with the government, but there is still the question of how much of what Playfair wrote contained a grain of truth. At this point, it is apparent that Playfair was dealing at least in half-truths. There are two things left to check: the code and Bonaparte's alleged speech.

Figures 17.1 and 17.2 show Cameron's cipher from the 1794 *Langue diplomatique* and Playfair's cipher respectively (Cameron 1794, 13; Playfair 1815, 21). On the surface they appear to be different. The layouts of the tables are different. Further, Cameron omits the letter *j* while Playfair omits the letters *j* and *v*. In support of the editor of the *Courier*, to be French Playfair should have deleted *w* instead of *v*. When examined closely, however, the two tables are based on the same principles. Cameron's table is an exact replica of what is known as the Vigenère cipher, named for its supposed inventor, the sixteenth-century French cryptographer Blaise de Vigenère. Playfair's table is merely a variation of the Vigenère cipher. Consequently, the jury is out on whether Playfair copied Cameron. Even Vigenère's cipher had forebears. It was based on a synthesis of the earlier work of three others: Abbot Trithemius, Giovanni Battista Bellaso, and Giovanni Battista Porta (Mendelsohn 1940).

Napoleon's alleged proclamation in French is a different story. To help the editor of the *Courier*, Christian Genest of McGill University has translated many articles from English to French and vice versa. He is a stickler for accuracy and correct grammar. His conclusion was that "the French is rather dubious. It sounds to be as if it was originally written in English and then translated into French" (private communication with Christian Genest).

Before going into detail, this is the proclamation in French, followed by the English translation. The problematic words are underlined.

> Français! votre pays <u>était</u> trahi, votre Empereur seul peut vous remettre dans la position splendide <u>que</u> convient à la France. Donnez toute votre confiance à celui qui vous a toujours <u>conduite</u> à la gloire.
>
> Ses aigles planeront encore en l'air et étonneront les nations.

> Frenchmen! your country was betrayed; your emperor alone can replace you in the splendid state suitable for France. Give your entire confidence to him who has always led you to glory.
>
> His eagles will again soar on high and strike the nations with astonishment. (Playfair 1815, 22)

Here is Christian's analysis:

1 The word "était" should be replaced by a "été." No francophone would ever write "était trahi," even in the nineteenth century, but one can observe that "était trahi" is a literal translation of "was betrayed."
2 The word "que" should be replaced by "qui." The original says "que," which refers to a complement, while "qui" refers to a subject. Example: La femme *que* j'aime (whom I love) vs. la femme *qui* m'aime (who loves me). Now if you start from English and do not know the distinction, you might translate "that" into que or qui.

GRANDE TABLE

P. 13.

✱	a	b	c	d	e	f	g	h	i	k	l	m	n	o	p	q	r	s	t	u	v	w	x	y	z
a	b	c	d	e	f	g	h	i	k	l	m	n	o	p	q	r	s	t	u	v	w	x	y	z	a
b	c	d	e	f	g	h	i	k	l	m	n	o	p	q	r	s	t	u	v	w	x	y	z	a	b
c	d	e	f	g	h	i	k	l	m	n	o	p	q	r	s	t	u	v	w	x	y	z	a	b	c
d	e	f	g	h	i	k	l	m	n	o	p	q	r	s	t	u	v	w	x	y	z	a	b	c	d
e	f	g	h	i	k	l	m	n	o	p	q	r	s	t	u	v	w	x	y	z	a	b	c	d	e
f	g	h	i	k	l	m	n	o	p	q	r	s	t	u	v	w	x	y	z	a	b	c	d	e	f
g	h	i	k	l	m	n	o	p	q	r	s	t	u	v	w	x	y	z	a	b	c	d	e	f	g
h	i	k	l	m	n	o	p	q	r	s	t	u	v	w	x	y	z	a	b	c	d	e	f	g	h
i	k	l	m	n	o	p	q	r	s	t	u	v	w	x	y	z	a	b	c	d	e	f	g	h	i
k	l	m	n	o	p	q	r	s	t	u	v	w	x	y	z	a	b	c	d	e	f	g	h	i	k
l	m	n	o	p	q	r	s	t	u	v	w	x	y	z	a	b	c	d	e	f	g	h	i	k	l
m	n	o	p	q	r	s	t	u	v	w	x	y	z	a	b	c	d	e	f	g	h	i	k	l	m
n	o	p	q	r	s	t	u	v	w	x	y	z	a	b	c	d	e	f	g	h	i	k	l	m	n
o	p	q	r	s	t	u	v	w	x	y	z	a	b	c	d	e	f	g	h	i	k	l	m	n	o
p	q	r	s	t	u	v	w	x	y	z	a	b	c	d	e	f	g	h	i	k	l	m	n	o	p
q	r	s	t	u	v	w	x	y	z	a	b	c	d	e	f	g	h	i	k	l	m	n	o	p	q
r	s	t	u	v	w	x	y	z	a	b	c	d	e	f	g	h	i	k	l	m	n	o	p	q	r
s	t	u	v	w	x	y	z	a	b	c	d	e	f	g	h	i	k	l	m	n	o	p	q	r	s
t	u	v	w	x	y	z	a	b	c	d	e	f	g	h	i	k	l	m	n	o	p	q	r	s	t
u	v	w	x	y	z	a	b	c	d	e	f	g	h	i	k	l	m	n	o	p	q	r	s	t	u
v	w	x	y	z	a	b	c	d	e	f	g	h	i	k	l	m	n	o	p	q	r	s	t	u	v
w	x	y	z	a	b	c	d	e	f	g	h	i	k	l	m	n	o	p	q	r	s	t	u	v	w
x	y	z	a	b	c	d	e	f	g	h	i	k	l	m	n	o	p	q	r	s	t	u	v	w	x
y	z	a	b	c	d	e	f	g	h	i	k	l	m	n	o	p	q	r	s	t	u	v	w	x	y
z	a	b	c	d	e	f	g	h	i	k	l	m	n	o	p	q	r	s	t	u	v	w	x	y	z

Figure 17.1. Cameron's cipher table. Courtesy of Falvey Memorial Library, Villanova University.

Example: "the book *that* I am currently reading" vs. "the book *that* explains this concept."

3 The word "conduit" should be replaced by "conduits." Past participles are not bent in English but they are in French according to rules that even francophones do not always master. Here, "conduits" takes a final *s* because it refers to "vous." The question a high-school teacher would encourage you to ask is: "He has driven whom?" If

The Cipher in which Buonaparte corresponded.

A	a	b	c	d	e	f	g	h	i	k	l	m
B	n	o	p	q	r	s	t	u	w	x	y	z
C	a	b	c	d	e	f	g	h	i	k	l	m
D	z	n	o	p	q	r	s	t	u	w	x	y
E	a	b	c	d	e	f	g	h	i	k	l	m
F	y	z	n	o	p	q	r	s	t	u	w	x
G	a	b	c	d	e	f	g	h	i	k	l	m
H	x	y	z	n	o	p	q	r	s	t	u	w
I	a	b	c	d	e	f	g	h	i	k	l	m
K	w	x	y	z	n	o	p	q	r	s	t	u
L	a	b	c	d	e	f	g	h	i	k	l	m
M	u	w	x	y	z	n	o	p	q	r	s	t
N	a	b	c	d	e	f	g	h	i	k	l	m
O	t	u	w	x	y	z	n	o	p	q	r	s
P	a	b	c	d	e	f	g	h	i	k	l	m
Q	s	t	u	w	x	y	z	n	o	p	q	r
R	a	b	c	d	e	f	g	h	i	k	l	m
S	r	s	t	u	w	x	y	z	n	o	p	q
T	a	b	c	d	e	f	g	h	i	k	l	m
U	q	r	s	t	u	w	x	y	z	n	o	p
W	a	b	c	d	e	f	g	h	i	k	l	m
X	p	q	r	s	t	u	w	x	y	z	n	o
Y	a	b	c	d	e	f	g	h	i	k	l	m
Z	o	p	q	r	s	t	u	w	x	y	z	n

Figure 17.2. Playfair's cipher table. Public domain, Google digitized.

> the answer is singular, then write "conduit." If the answer is plural, then write "conduits."

Christian concludes, "It is possible that even an educated man such as Bonaparte could have made the third mistake if writing in a hurry. But certainly not the first two (unless rules changed over the past 200 years). The corrected text is far from educated anyway." The jury is in. Playfair made up the message in English, translated it into French, then coded it using a common cipher, not necessarily exactly the one given by Cameron. This puts into doubt his whole story about Caraman.

For his own benefit, at least, Playfair did obtain a passport to go to France. On the downside for Playfair, he did not receive any funds to get there, only the promise of funds should he ever come up with something more substantial.[6]

Chapter Eighteen

Playfair Continues Writing and Tries a Few More Scams to Get to Paris

The Bonaparte fiasco had played itself out by the end of April 1815, resulting in Playfair having in his hand only a passport to go to France, but no financial support from the British government. He could not immediately embark for France, for lack of funds. A few months later, he was still trying to raise money. After Napoleon's defeat at Waterloo on 18 June 1815, Playfair waited until August to try his luck again with funding. On 2 August 1815, he wrote to John Beckett, undersecretary of the Home Office. Enclosing something that is no longer extant, he asked Beckett to give the enclosure to Henry Addington, 1st Viscount Sidmouth. Sidmouth had been home secretary for less than two months.[1] The enclosure might be guessed at, on the basis of next two letters Playfair sent, to Beckett on 16 September and to Sidmouth on 28 September 1815.[2] Playfair claimed that a plot was hatching in Paris, and wanted the prince regent and others informed of it. Beckett wanted particulars, and Playfair responded that he needed to get to Paris to get firm information. Beckett would not give him the money. When he heard that Beckett had gone to Paris, Playfair wrote to Sidmouth that he felt Beckett's trip would not amount to anything: "I do not believe that any known public servant can get at the information wanted."[3] Alas for Playfair, it was another scam without a successful monetary conclusion for him.

Instead of travelling to Paris, Playfair was writing. His major activity until the end of the year was probably researching and writing the supplement to his earlier two-volume *Political Portraits*. His *Supplementary Volume to Political Portraits in the New Æra* was available for sale by mid-April 1816 (Playfair 1816a; *Morning Post*, 17 April 1816). On the surface, it was to be a collection of biographical sketches of Playfair's living heroes and villains, and monarchs and other royals whom Napoleon had deposed and were reinstated after his defeat, along with stray cases seemingly unrelated. There are no reviews of the book in the literary journals of the day, so it is difficult to gauge the public reaction to it.

On close reading, the *Supplementary Volume to Political Portraits* is a misnomer. It resembles more a political landscape strewn with royalty, military men, politicians, and diplomats. Some appear in the first two volumes of *Political Portraits*. Their revised treatments reflect Playfair's perception of the change in the political scene. Like *Political Portraits*, the supplementary volume is structured alphabetically by the characters whose portraits Playfair has drawn. Here Playfair often gives spotty treatment to the individual, then meanders off on a tangent close to his heart.

Several themes in the *Supplementary Volume to Political Portraits* continue from Playfair's earlier writing, reaching as far back as publications prior to the French Revolution: *The Commercial and Political Atlas*, *An Essay on the National Debt*, *Letters of Albanicus*, and *Joseph and Benjamin*. Then and now, Playfair was concerned about the size of the national debt, the military and economic strength of France, the vulnerability of Britain's possessions in India to encroachment from other European powers (previously just France, now France and Russia), and the form of government that countries should have. Tied to the national debt was the issue of a Corn Law passed in 1815, which imposed new tariffs and raised the price of foreign grain imported to Britain, thus protecting British agriculture. The main beneficiaries were the landed interests. According to Playfair, the Corn Law could lead to national bankruptcy and the possibility of revolution. Also tied to the national debt was the discontinuance of the income tax that had been imposed by William Pitt in 1798 to help pay for the war and its accoutrements. Playfair favoured continuing the tax, as it would contribute to public revenue and thereby reduce, or at least control, the debt.

In 1816, Britain was quickly sinking into a post-war depression that lasted until 1821 (for a succinct description, see Gash 1978, 152–4). One aspect contributing to the post–Napoleonic War depression was that the labour market was flooded with up to 300,000 returning soldiers. It played havoc with Britain's finances. And some of the havoc was self-inflicted, as Playfair noticed. In his portrait of the chancellor of the exchequer, Nicholas Vansittart, Playfair takes up the issue of national finances, thus illustrating that many of the portraits are really landscapes. Very little of the article on Vansittart is about the man himself. After some lengthy discussion, Playfair rightly points out two mistakes that Vansittart had made, which Playfair predicted would make the debt situation much worse: allowing the discontinuance of the income tax and the introduction of the Corn Laws. Playfair then returned to an old theme with an addition: "At present, the greatest part of the debt is owing to persons of independent fortune: the lower classes, public bodies, &c. have little to do with it, one consequence of which is, that the whole of the interest goes to persons who are mere idle consumers of the fruits of the earth, and that the industrious working classes naturally would be benefited by a bankruptcy" (Playfair 1816, 282). The old theme is that Britain's creditors, those invested in consols, were "idle consumers" who did not benefit the British economy. The addition is the misguided conclusion that the working class would benefit from Britain defaulting on its debt. Playfair had a new solution to the national debt problem. He suggested lowering the interest paid on the consols and using the savings to start paying down the debt. This would make the idle consumers responsible in helping to bring down the debt. Though Playfair was definitely not a radical, such a proposition was made in England twenty years later by the Chartist movement, which argued that the national debt was a subsidy to middle-class *rentiers*.

Playfair supported a constitutional monarchy like that enjoyed by Britain and felt that France should adopt one as well. And Playfair was very much a monarchist. Monarchs were the rightful heads of the executive branch of government. He gave his full support to George as prince regent, dismissing out of hand, for example, rumours and negative gossip about George's lifestyle. As for George's financial excesses, they happened because George was ill advised, so Playfair claimed. Playfair did not confine himself to Britain's regent. In addition to Louis XVIII of France, he commented on four other monarchs whose fortunes had been restored on Napoleon's defeat. Three of the four were a problem: Ferdinand VII of Spain, Ferdinand IV of Naples and Sicily, and Victor Emmanuel of Sardinia. He called Ferdinand VII an imbecile and praised Spain's parliament, the Cortes, for bringing the country together against Napoleon. On his restoration, Ferdinand VII dismissed the Cortes, punished several of its members, and imprisoned about 50,000 people. Supported by Britain during the Napoleonic Wars, Ferdinand IV, on the other hand, was just a waste of money, who should be left to his own devices. Victor Emmanuel was merely a despot. The passage of time tends to support Playfair's general views on the three. All three rejected liberal constitutions and became oppressive rulers. Playfair looked more favourably on the fourth monarch, whom he called Frederick, king of Belgium, known today as William I, king of the Netherlands. William had fought against Napoleon, a definite plus in Playfair's ledger. When William took the throne, his kingdom consisted of modern-day Belgium and the Netherlands. Playfair's well-founded concern for William was that his kingdom would be difficult to rule as the result of a "difference of religion, of interest, and of character" between what today are two countries. Belgium eventually broke away to become a separate country during William's reign.

Sprinkled throughout *Supplementary Volume to Political Portraits*, Playfair continued to express his anti-Catholicism, which was sometimes quite severe. One reason Louis XVIII was a weak sovereign was that he was a devout Catholic controlled by priests. Ferdinand VII of Spain was another dominated by priests. Playfair saw Pope Pius VII, pontiff during and after the Napoleonic Wars, displaying "an intollerant, as well as a revengeful spirit" (Playfair 1816a, 227) and as a "fanatic" (230). He commented that Pius "might moderate the absurd zeal of the frantic Ferdinand of Spain, but he seems rather to approve of it" (232). Veering off the topic of the Napoleonic Wars and its aftermath, under Pius's portrait Playfair revealed himself to be against Catholic emancipation in Ireland, which was still another dozen years away. He criticized Pius for not helping "to tranquillize a great part of the Roman Catholic population of Ireland." His anti-Irish bias also showed through, calling the Irish "mad-headed" (230). He described Catholicism elsewhere as "gloomy" (123). Taken as a whole, Playfair's comments were far from a ringing endorsement of that branch of Christianity.

While Louis XVIII was a problem (weak and unloved by his people), so was the country over which he ruled. Playfair was concerned about the balance of power in Europe, which had been disturbed by the French Revolution. He was critical of the Treaty of Paris, signed 30 May 1814, and of some of the decisions taken later at the Congress of Vienna, the peace conference that followed the Napoleonic Wars. Lord Castlereagh, the English diplomat at the congress, was the subject of Playfair's ire on the subject of the peace conditions. Playfair's position is summed up in a single sentence in Castlereagh's portrait: "It will not be forgotten, that by degrees the public mind was much altered with respect to the treaty of Paris; and it was discovered that what had been gained by the sword, had been given away by the pen" (Playfair 1816, 109–10). Playfair returned to the same topic, and similar phrase, under the portrait of a different diplomat, Baron Wilhelm von Humboldt. On British diplomacy, Playfair said, "They destroy with the pen what they have achieved by the sword" (161–2). To give flesh to this diplomatic failure, Playfair claimed that France was still too powerful. It should have paid more in war reparations, and should have returned the plunder taken during the Napoleonic Wars. The only way to keep France in check was to make Germany more powerful. By "Germany," Playfair meant Prussia and Austria. The decisions made at the congress did not go far enough. Saxony had been partitioned and part of it given to Prussia; according to Playfair, it should have come entirely under the control of Prussia.

There were three French villains in *Supplementary Volume to Political Portraits*. Napoleon, of course, was included for causing the war, and Charles-Maurice de Talleyrand for getting a good deal for France at the close of the war as the French representative at the Congress of Vienna. Playfair dismissed Napoleon as vain, cunning, and cowardly. Napoleon was a coward because, after Waterloo, he quietly left the field of battle in haste without telling his generals. Then he planned to seek refuge in the United States but was afraid of capture by the British navy. As for Talleyrand, his bad traits were succinctly described in the second sentence of Playfair's portrait of him: "Without honour or fixed principles, he is able at intrigue and deception, and displays an uncommon talent for taking hold of circumstances, as they occur, for his own advantage" (244). The rest of Playfair's portrait of Talleyrand fills in the details. Indeed, Talleyrand was a very talented political survivor, the Teflon man of his day. He successfully served several French governments, from Louis XVI to Louis-Philippe. His ability to transform himself as the political winds shifted may have aroused distrust among others including Playfair, but it was outweighed by the usefulness of his diplomatic skills and his political acumen. The final Frenchman was Joseph Fouché, who, similar to Talleyrand, supported the revolutionary republic, Napoleon, the Bourbon restoration, then Napoleon again during the Hundred Days, and finally the Bourbons again after Waterloo. In 1816, the French government proscribed Fouché as a regicide for his role in the execution of Louis XVI. Playfair dismissed him as having "grown old in iniquity and crime" (132).

Playfair's tendency to meander off political topics and onto other subjects close to his heart can be seen once more in his treatment of the MP Samuel Whitbread. Whitbread, whom Playfair considered a villain, had vocally supported Napoleon in England. Along other political lines, one of Whitbread's sins was supporting the Corn Laws. To give Whitbread some credit on that issue, he did have conflicting interests on the matter; he was both a brewer and a landowner. Completely off topic, Whitbread's other sin was running the Drury Lane Theatre badly. Although the theatre burned down in 1809, the new version of the Drury Lane Theatre may have been an issue close to Playfair's heart. The old theatre had connections to Playfair's Original Security Bank and to Richard Brinsley Sheridan.

Totally beyond the realm of politics and the Napoleonic Wars, Playfair described another individual with connections to the Drury Lane Theatre – Lord Byron, whose poetry Playfair described in the original

Political Portraits as "not of the light sort." He admired Byron's early poetry but not his most recent work (probably *Childe Harold's Pilgrimage*), which was dark, describing life as a burden. By the time the *Supplementary Volume to Political Portraits* was written, he had changed his mind about Byron, perhaps because Byron became a member of the managing committee of the Drury Lane Theatre. As Playfair says, "A play-house cannot be managed by mere men of business, like a brewhouse; men of taste and genius are necessary both to its success as an enterprise, and its being conducive of amusement to the public" (Playfair 1816a, 92).

The *Supplementary Volume to Political Portraits* must not have been a financial success. By July 1816 Playfair was once again desperate for money. And once again in these circumstances, he resorted to a variation on his usual extortion scam. As he had done with the concocted Bonaparte story only a year before, he worked with Thomas Byerley. This time he went after a prominent and highly successful engineer, John Rennie. The difference was that he knew the mark. Rennie's and Playfair's apprenticeships with Andrew Meikle had overlapped.[4] In about 1807, Rennie sent his eldest son, George, to study at the University of Edinburgh. While there, George lodged with John Playfair, William's brother (*Gentleman's Magazine* 1866, 749–51). Why William would be so stupid as to pick someone so close probably attests to the level of his desperation. The scheme backfired, but it did give Playfair some money and breathing space.[5]

It is impossible to say whether Playfair and Byerley met to devise their scheme before 11 July 1816, when Byerley wrote his first letter to Playfair. Byerley outlined a plan, which was a little vague, as should be the case in all scams that work. He claimed to have several unspecified documents about corrupt practices in public works projects such as bridge and canal building. These practices included such things as submitting a low bid and changing it once the project was underway; or the designer of the project (canal, bridge, etc.) taking money from the contractor to secure the job; or the designer taking a percentage of the total (and probably rising) cost of the project instead of a flat fee. Byerley told Playfair that he was going to write an exposé about this situation. Six days later, Playfair wrote to John Rennie saying that he had seen materials in which there were "allegations against you which appear to be supported." He also enclosed a hand-written prospectus for Byerley's proposed book, which had an ominous and lengthy title: *A Warning Address to the Monied Interest concerning the Abuse and Frauds Committed by the Making of False Estimates of Canals, Bridges, Mills, Machinery & Buildings by Civil Engineers with a Variety of Curious Facts.*

Playing good cop in the good cop/bad cop game, Playfair wrote, "As no man can be indifferent to such allegations if you should wish to see what is said of you before it appears I shall try to get you a sight of it & what is more get whatever is found wrong altered or left out."[6]

Rennie made a lengthy reply to Playfair, saying essentially that he was innocent of any wrongdoing, but acknowledging that others had carried out the alleged nefarious practices. Moreover, if there was any wrongdoing on his part, it was completely unintentional. Rennie asked for a meeting, and Playfair replied that he would meet him at a tavern at 9:00 in the morning of 25 July. After the meeting, in which Rennie had defended his own work ethic, Playfair wrote to Rennie, saying that Rennie was above reproach and that he would ask Byerley to insert in his proposed book that were some exceptions to Byerley's general observations and Rennie was one of them.

Then Rennie complicated things for himself. He was an honest man and checked his books to see if he had done anything amiss. He found a couple of very minor problems and wrote to Playfair explaining the situation. On 2 August, Playfair wrote to Rennie, enclosing a letter from Byerley to Playfair, written the previous day that challenged Rennie's defence of his professional conduct as an engineer. Nothing seemed to happen, at least in the surviving record, for about ten days. On 12 August, Byerley tried to play his trump card. He wrote a threatening letter to Playfair, but directed at Rennie. There were phrases in it such as "that he [Rennie] never took any percentage. I know the contrary & can prove it." Byerley also raised the spectre of Alexander Davison who, as procurer of coal for the government a few years before, was caught claiming extra fees and spent twenty-one months in Newgate Prison.

At this point, it appears that Rennie had had enough, and was onto the scam. The threatening letters from Byerley and Playfair end on 12 August, and the conciliatory letters from Playfair to Rennie begin on 23 August. It is no coincidence that all the correspondence came into the hands of Rennie; he probably demanded to be given all copies, both letters and any prospectuses. What Playfair got was a loan of £25 from Rennie. The general goodness of Rennie is seen in the size of the loan. Playfair sent Rennie two promissory notes, one for £20, and the other for £25. He could not remember which one was the agreed amount of the loan and so sent two notes. Only the promissory note for £25 survives. There is no evidence that the loan was ever repaid, and Rennie probably knew that was going to be the case. On 23 August, Playfair wrote a memorandum to Rennie stating that it was Byerley who had written the accusatory prospectus, that Playfair showed it to Rennie to prevent any accusation against Rennie, that essentially any statements against Rennie were wrong, and that Rennie was at the top of his profession. Further, the loan he was receiving from Rennie had nothing to do with the accusations against Rennie, but instead was an application for financial help from Rennie as an old friend.

The money from Rennie was apparently not enough. In September 1816, Playfair launched another scam. Like the Rennie scam, Playfair claimed to have a document that would be prejudicial, in this case to Archibald Douglas, 1st Baron Douglas. The circumstances of Douglas's birth had been clouded in controversy and became the centrepiece of the Douglas Cause, one of the most famous legal cases of the eighteenth century (Lowe 2004). Archibald Stewart's mother was the sister of the Duke of Douglas, whose estates were worth more than £12,000 a year. Archibald and a twin brother, who died young, were born in France when their mother was fifty years old. When the Duke of Douglas died, Archibald Stewart was recognized as the heir and changed his surname to Douglas. The Duke of Hamilton, who was next in line for the inheritance, went to court in 1762 claiming that Archibald was actually Jacques Louis Mignon, a child who had disappeared in July 1748 at the time when Archibald was born. The Stewarts, Douglas's alleged parents, had purchased the twins, so it was claimed. After great expense, said to total £54,000 for both sides, the case was settled in 1769 in the House of Lords in favour of Stewart/Douglas. Nearly fifty years later, Playfair claimed to have access to new documentary evidence that Archibald Douglas was indeed Jacques Louis Mignon.[7]

Playfair began his scam by writing to Archibald Douglas on 23 September 1816 with a teaser. He had a friend who had recently obtained papers related to the Douglas Cause, but had been kept back when the case was before the courts in the 1760s. The papers had been sequestered since December 1792, but could now be obtained at a reasonable price. Douglas wrote back on 28 September asking for more specifics about the papers and the cost to obtain them. Playfair obliged on 2 October 1816, when he informed Douglas that the papers were about a child obtained in France for an English couple and the subsequent enquiries that were made afterward. Playfair thought that the papers, which were then in France, could be obtained for £500 or less. He then advised Douglas to move quickly as a number of English intriguers in Paris might purchase the papers first. He finished on a personal note, trying to win Douglas's trust: "My brother James so often mentioned to me the honourable way in which your Lord^{sp} acted to him [In 1787 James Playfair worked on remodelling Bothwell Castle for Archibald Douglas], that in return for the obligation he received I shall be happy to do any thing that can give pleasure to your Lordship."

Douglas was away from home when Playfair's letter arrived and did not reply until 12 October. In his reply, Douglas dismissed the papers offered to him as of no consequence.

In the meantime, Playfair was becoming a little impatient and wrote again to Douglas on 14 October. Their letters crossed in the mail. In his latest letter, Playfair gave more detail. When Douglas's parents had arrived in France in 1748, they stayed at the house of a man named Michelle. There they purchased the infant son of a glasscutter named Mignon. When the enquiry was made for the court case in 1762, money exchanged hands in order to falsify this story. When Michelle's brother-in-law heard of this, he received 18,000 livres to keep quiet but put together a memorandum outlining the true story. This was the nature of the papers that sat in Paris.

When Playfair received Douglas's letter of 12 October, he replied with hurt and indignation, with veiled threats of the consequences of not buying the papers. When he heard nothing further from Douglas, Playfair wrote again on 7 December 1816 with an open threat. Playfair had a prospectus printed for a subscription to a publication on the Douglas Cause showing the "true facts" exposing instances of bribery, false witnesses, and contradictory statements. He threatened to go public.

Between Playfair's October and December letters, Douglas consulted with his solicitor, David Wemyss, writing to him on 22 October 1816. Douglas was convinced that the letters would have no effect on the lawsuit from the 1760s and had no intention of paying Playfair. He was concerned about Playfair going to the press and wanted to know from Wemyss what legal steps could be taken "to repel the attack in the most serious way the law will permit." Wemyss replied two days later. He agreed that Playfair's intention was to extort money and that the information that Playfair purported to have had "no legal consequences." He thought that Douglas could sue Playfair over his threatening letters, but advised against it. His general advice was to let the matter drop.

As an indication generally of how far Playfair had fallen by this time, after describing Playfair's three brothers in one of his letters to Archibald Douglas, Wemyss commented in his letter of 24 October 1816: "I think I have heard of another brother in London who I presume is your correspondent; and I think I have heard, altho' I cannot at present recollect the source from which I derived my information, that this Brother in London was a person held in no respect. His letters upon this occasion I think shew, that like many others in London he wishes to live upon his shifts."[8] Later in the letter Wemyss called Playfair "a daring worthless fellow."

After Playfair's December letter, Douglas consulted with his son Charles, who was a fellow of All Souls College, Oxford, at the time (Foster 1888, 1:379). Charles was also a barrister, having been admitted to Lincoln's Inn in 1796 and called to the bar in 1802 (Baildon 1896, 1:558). The son consulted the attorney general, William Garrow, on the matter. In his law career, Garrow had specialized in criminal law and was at Lincoln's Inn at the time that Charles Douglas was there studying for the bar. Charles wrote to his father on 3 January 1817: "I had only time yesterday in the cover of a letter to Archy [Charles's brother, later 2nd Baron Douglas], to give you the result of my conversation with the Attorney General. I now inclose you his written opinion upon the point, or rather two, as put to him. He admits the possibility of a prosecution for what has passed, but doubts the prudence of such a step at present."[9] Playfair's scam did not work this time; and luckily for him he did not wind up in Newgate again, had the attorney general pursued the case. Douglas did not pay; Playfair did not publish; and Douglas did not pursue the matter in the courts.

A year after Playfair's unsuccessful attempt to obtain money from Viscount Sidmouth, Playfair wrote again to Sidmouth on a different topic, but again in search of money. The issue that Playfair raised was frame breaking. By 1816, the knitting industry in Nottingham was experiencing a crisis as the result variously of the post-war depression, the decline of the apprenticeship system in the industry, and an oversupply of workers (Patterson 1948). This resulted in violence in which knitting machinery was smashed. On 12 October 1816, Playfair wrote to Sidmouth saying that he had information on a plan concerning frame breaking from a Mr. Wheatly, an old acquaintance. He suggested to Sidmouth that he meet his acquaintance soon, as Wheatly would be in town only a few days.[10] Ten days later, Playfair wrote again to Sidmouth with a specific plan to prevent frame breaking.[11] As usual, the plan was short on specifics. Of course, Playfair wanted money in return. He expressed it modestly as "a reward as to any two gentlemen shall be thought proportioned to the service performed." He also stated the reason that he wanted the money much more explicitly than he had expressed in his plan: "I am no longer young & have a daughter who is blind & unprovided for & I cannot think but that if I prevent expence & bloodshed I shall deserve some remuneration."[12] Sidmouth passed the case on to Sir Nathaniel Conant, the chief magistrate at Bow Street. Conant wrote to Sidmouth on 24 October saying that he had requested a meeting with Playfair. Nothing more seems to have come of it.

During the rest of 1816 and into 1817 until he left for France, Playfair wrote five articles on the British government's finances, which appear in *Monthly Magazine*, owned and edited by Sir Richard Phillips (Playfair 1816b, 28–9, 129–30, 292–3; Playfair 1817, 6–8, 123–4). In these articles, Playfair reiterates the points that he made in his *Supplementary Volume to Political Portraits*, especially his article on Nicholas Vansittart. The one difference is that Playfair supports his arguments with more extensive statistics not given in *Political Portraits*. The choice of this venue for publishing seems odd for Playfair. Although he had known Phillips for perhaps a decade and a half, Playfair and Phillips had widely differing political views. Phillips had "strong radical and republican views" and was once prosecuted and jailed for publishing Thomas Paine's *Rights of Man*, a book that Playfair wrote against in 1796 (Seccombe and Loughlin-Chow 2008). Even Playfair acknowledged the apparent contradiction and, after commenting on his longstanding support for William Pitt, offered justification: "Having never supported ministers, but measures, there is no inconsistency in my writing against them; and I chose the Monthly Magazine, as being full of useful and important matter, and widely circulated. I do not agree with all its political principles, but that is nothing to the purpose; you wish well to your country and that is enough. All men are not obliged to be of one opinion in politics" (Playfair 1816b, 129). It can be speculated that Playfair received some payment for his articles, money that he, as usual, desperately needed.

Playfair remained in London until at least 10 March 1817. On that date, he wrote again to Sidmouth.[13] As in the case of frame breaking, political unrest was on the rise in Britain. In March, Parliament had before it a bill to prevent seditious meetings (*Morning Post*, 2 and 8 March 1817). Already at the beginning of the year, a magistrate had published a letter in the newspapers, anonymously written, to Viscount Sidmouth noting that seditious meetings were occurring in public houses (*Morning Post*, 3 January 1817). The publican should be responsible to suppress these meetings, the writer demanded, on penalty of forfeiture of the publican's licence. Playfair's letter to Sidmouth is similar, but went further. He wanted pubs to close at 11 o'clock, in part to prevent such meetings from occurring. Further, all seditious papers should be removed from public houses and, instead, papers sympathetic to the government should be offered free of charge. Once again, his proposal appears to have fallen on deaf ears.

Soon after 10 March, Playfair returned to Paris, perhaps on the proceeds from his articles in *Monthly Magazine*.

Chapter Nineteen

Playfair Returns to Paris

Despite Playfair's distain for the Treaty of Paris, evident in his portrait of Castlereagh, he stood to gain from it. By Article XXV of the treaty, corporations and other public establishments could be reimbursed in instalments for funds confiscated by the government during the French Revolution. In the treaty, it was put more diplomatically as "deposited with the government." Back in 1788, when Playfair went into partnership with Joseph Gerentet, he had been required to deposit 3,200 French livres with the Paris banker James Carrey.[1] According to an audit done by the French government on 9 January 1794 (or 20 nivôse an II by revolutionary dating), Playfair still had the 3,200 livres on account with Carrey.[2] In 1815, Carrey requested and obtained the papers that had been seized from his bank in 1792. For some reason, it was Mary Playfair who wrote to the French government in December 1816 requesting the return of her husband's papers from Carrey's bank. This was fulfilled in January 1817. As mentioned in the previous chapter, Playfair was in London until 10 March 1817.[3] He likely embarked for Paris shortly thereafter, probably in pursuit of some or all of the 3,200 livres.

In Paris, Playfair came into contact with Giovanni Antonio Galignani, who had come to France from Brescia in northern Italy in about 1790, but soon left for England. In England, Galignani married Anne Parsons, whose father was a London printer. After Napoleon had taken power in France, thus making the political situation relatively stable, he returned to France in about 1800, where he established a printing and bookselling business with his father-in-law. In 1814, Galignani started an English-language newspaper called *Galignani's Messenger*. Popular from the very beginning, the paper printed "English, American, and continental news from all sources" (Barber 1961). Sometime during the first half of 1817, Playfair began working as the editor of *Galignani's Messenger*.

During the summer of 1817, one of the items that crossed Playfair's desk was the book *France* by Sydney, Lady Morgan. Born Sydney Owenson, she became a very successful Irish novelist, whose work contains themes of women's rights and freedom, as well as the treatment of issues surrounding the social and political condition of Ireland. *France* was her first travel book, published in 1817 after a trip to France the year before.

Many of Sydney Morgan's novels have elements of her own personal experience in them with heroines modelled on herself. The writing of *France* is similar, but without a heroine. Annetta Bishop describes the background to *France*: "They [Lady Morgan and her husband, Sir Charles Morgan] took with them letters of introduction, and they met all the noted men of literature and public affairs, as well as all the women whose beauty, fashion, or talents for intrigue, had made them leaders of society" (Bishop 1938). *France* was both praised and panned in London. Around the time of its publication, the *Morning Post* inserted a teaser in one of its issues: "Public curiosity is strongly excited by the announcement of the speedy publication of Lady Morgan's *France*. Extracts have been read in private circles, and reports speak of it as exhibiting the most brilliant view of society and manners in Paris ever produced. It is to appear next week at Paris, as well as London" (*Morning Post*, 17 May 1817).

At the other end of the spectrum was a reviewer for the *Quarterly Review*, thought to be John Wilson Croker. He was a fellow Irishman and author, as well as a politician, sitting as an MP in Westminster. He, or the reviewer, did not like Lady Morgan's earlier work, nor did he like her current effort. After a more than three-page preamble, he summed up his criticisms of *France* in one paragraph: "Our charges (to omit minor faults) fall readily under the heads of – Bad taste – Bombast and Nonsense – Blunders – Ignorance of the French Language and Manners – General Ignorance – Jacobinism – Falsehood – Licentiousness, and Impiety. – These, we admit, are not light accusations of the work; but we undertake, as we have said, to prove them from Lady Morgan's own mouth" (*Quarterly Review* 17 [1817]: 264). The reviewer ran through a withering attack of *France* over the next twenty-three pages. All his charges are treated individually in minute detail.

The earliest concrete evidence of Playfair's editorship comes from his book review of *France* in *Galignani's Messenger* (8 August 1817). Sitting in Paris, Playfair was also not amused by *France*. In a brief article, he found her knowledge of French society and history uninformed, misleading, and inadequate, her praise of the French Revolution offensive and inaccurate. A year or two after his return to England, more would come from Playfair's pen, turning the short review into a full-length book.

All was going smoothly for Playfair in Paris until a ghost from his past rekindled his indignation, dormant since his banking days in London. Initially, he did not react to news of the ghost. Near the end of July, a short notice appeared in *Galignani's Messenger*, reporting that the funeral of the Comte de St. Morys had taken place and was well-attended by officers and other troops of the French army; St. Morys, son of the St. Morys who had died at Quiberon, had held the rank of Maréchal de camp in the army (*Galignani's Messenger*, 26 July 1817). His demise came about from a duel with Colonel Barbier-Dubay, who had served with the revolutionary army and with Napoleon. After St. Morys's family estate, Houdainville had been confiscated during the French Revolution, Barbier-Dufay had bought it in order to tear down the buildings and sell the materials. When St. Morys returned to France after the Bourbon restoration, he was able to live on a small part of the estate that had not been confiscated. Barbier-Dufay had supported Napoleon during the Hundred Days, but was pardoned on the second Bourbon restoration. With this as background, the neighbours were definitely not on friendly terms. After Barbier-Dufay made several defamatory statements against St. Morys, the latter challenged the other to a duel. Barbier-Dufay proposed that they hold pistols at each other's chest; only one would be loaded, causing certain death for one of the combatants. St. Morys refused this form of duel. Instead, they went through four rounds of pistols (both loaded and fired at a distance) and then turned to swords, at which point St. Morys was run through and died (*Gentleman's Magazine* 1817, 116–19).

It would have ended there for Playfair had not St. Morys's widow and daughter accused Barbier-Dufay and the Duc de Mouchy of murdering the comte. She took the case to court, lost, appealed to a higher court, and lost again. Afterwards, she and her daughter wrote a pamphlet about the case. Playfair saw the pamphlet and wrote a review for *Galignani's Messenger*, which appeared in late March 1818. Since what Playfair wrote had great impact on his remaining days in Paris, it is included here in its entirety:

> Madame la Comtesse de St. Morys has published a Memoir of twenty-seven sheets, against the Duke of Mouchy and Colonel de la Faye. We conceive this Lady, to say no more of it, to be extremely ill-advised – His Majesty, the Marshalls of France, the Chamber of Peers, all have decided against his – "Flower of Chivalry and Honour." The Public are acquainted with the real facts of the case, hence the former Memoir, filled with the

most violent accusations against Col de la Faye, and the most extravagant praises of the conduct, honour and virtue of the Count St. Morys, required no answer in justification from Colonel de la Faye, for sophistry can never remove the knowledge of a fact, and the Public regarded it as the angry ebullitions of sorrow, and retained it opinion. We confess it is a great misfortune for Madame de St. Morys to have lost her husband in a duel, but it is still greater, that he should die with the universal impression of being less brave than his sword – want of bravery is so rare in the French character, that an instance is noticed as remarkable, and consequently prudence should have dictated to the Countess de St. Morys, to have let her husband's memory sink into oblivion as fast as possible, rather than to seek to keep alive sentiments and recollections which do his memory no honour. (*Galignani's Messenger*, 17 March 1818)

Why would Playfair write such as thing about St. Morys? A contemporary article, in *Gentleman's Magazine*, was full of praise for the man. Likely he was being true to an old friend, in this case Colonel John Gordon Sinclair. St. Morys had been involved in a late 1790s lawsuit between Charles-Alexandre de Calonne and Sinclair, and had distinctly sided with Calonne (Sinclair 1798, 10). This was natural. St. Morys's father was colluding with Calonne and his brother Abbé de Calonne. Through his wife, the younger St. Morys had family ties to the Calonnes. Sinclair had spent a year in Newgate as a result of the lawsuit, so the review in *Galignani's Messenger* was probably Playfair's attempt at a bit of revenge.

In England, Playfair may have been safe. Freedom of the press was established there. He had made strong statements about a few of his subjects, who were very much alive, in *Supplementary Volume to Political Portraits in the New Æra* and had come to no harm by it. Such was not the case in France, at least in the early days of the Bourbon restoration. At times the press had been tightly controlled by the state, and, prior to a reform in 1881, a maze of laws had grown regarding what could and could not be published (Kuhn 1995, 47–8). Playfair had been caught in it. The Comtesse de St. Morys and her daughter sued Playfair and the proprietor of the paper, Galignani, for defaming the deceased husband and father. At the same time, Barbier-Dufay sued the comtesse and her daughter for the strong accusations they made against him, including cowardice, in what they had written in their pamphlet.

The case was heard before the Tribunal de police correctionnelle, with the plaintiff's case presented on 7 July 1818 and the defence's response on 15 July.[4] On the first day of the trial, the offending article was read in court and several military witnesses were called to attest to

St. Morys's bravery. One testified that Barbier-Dufay's first proposal for the duel was contrary to practice, that it amounted to a form of assassination, and that Saint-Morys was right in refusing to duel in that way. With that established, the comtesse's attorney accused Barbier-Dufay of dictating the offending article to Playfair. Barbier-Dufay, who was in court, denied the charge vehemently, saying that he had not even seen the article. On the second day of the trial, it was established that Playfair had written the article and had received no assistance from Galignani in writing it. Galignani was then off the hook. Playfair also could not provide evidence to support his statements about St. Morys. And he went further, making ironical and sarcastic remarks in the course of his defence. Judgment was rendered at a hearing of the court on 28 July 1818. Playfair was found guilty. Because of the statements he had made during his defence, his sentence was more severe than originally proposed. He was sentenced to three months imprisonment, given a 300-franc fine (about £12 6d.), 1,000 francs (about £42) in damages, and a five-year suspension from the rights allowed by Article 42 of the Code Penal (the right to vote, to hold any public office, and to bear arms, among others) (France 1819, 2:8–9). The amount of the fine and damages was moderate, amounting to about £4,800 in today's money. Playfair appealed his sentence to the Royal Court, but lost his appeal near the end of the year.[5]

The comtesse and her daughter got off lightly in Barbier-Dufay's defamation suit against them. The court was convinced by arguments that the two acted in a state of grief. They paid only a fifty-franc fine, much less than Playfair. They also had to pay for 150 printed copies of the judgment against them and the cost of placarding the judgment in public (*Galignani's Messenger*, 16 August 1818).

It is not known whether Playfair actually served time in a French prison, or if he fled the country to escape his incarceration and fines before judgement was rendered on his appeal. Whatever happened, he was back in England in 1819. The earliest firm evidence of his presence there is October 1819.

Chapter Twenty

Playfair Spends His Last Few Years in England in Poverty

Once the final court judgment went against him on the St. Morys affair in late 1818 or early 1819, Playfair returned to England. He probably wanted to avoid his prison sentence, as well as the fines that were imposed. By October 1819, Playfair was firmly ensconced in London in Castle Street near Covent Garden, writing an obituary of his former employer, James Watt, and writing letters to newspapers about French politics (*Monthly Magazine, or British Register* 48, pt. 2, 1 October 1819, 230–9; *Morning Post*, 4 November 1819).

On 20 July 1819, Playfair's brother John died, and a month later, on 25 August, James Watt died. Both received substantial death notices in the newspapers (see, for example, *Morning Chronicle*, 27 July 1819, for John Playfair; *Morning Chronicle*, 1 September 1819, for James Watt). Playfair was hoping to inherit something from his brother, but he was disappointed. John Playfair left an estate in excess of £3,000, the bulk of it going to his two sisters, with minor legacies of £100 each to each of two nephews, the sons of James Playfair the architect.[1] Strapped for money as usual, William Playfair tried to capitalize on Watt's death in another way. He had better luck with this than with his brother, but with mixed results.

For some monetary compensation, Playfair wrote two memoirs of James Watt, the first for the *New Monthly Magazine* and the second for the *Monthly Magazine*. Playfair had longstanding connections with Sir Richard Phillips, owner and editor of the *Monthly Magazine* (Clemit and McAuley 2020). Its younger competitor, the *New Monthly Magazine*, began publication in 1814, nearly two decades after the *Monthly Magazine* began, under the editorship of Henry Colburn (Garside 2006). Under Colburn, the *New Monthly Magazine* was known as "a virulently Tory publication, which opposed liberal journals such as the *Edinburgh Review* and Phillips's *Monthly Magazine*" (Higgins 2006).

Playfair's political sympathies leaning towards Colburn may have been the reason that Playfair went to Colburn first. Money was undoubtedly the major motivating factor for the second choice.

With two memoirs on James Watt near publication, in the hopes of further money Playfair wrote to Watt's son, James Watt Jr., on 25 September 1819, enclosing the proof copy of the article for the *Monthly Magazine*.[2] He used the enclosure to point out first-hand knowledge of James Watt's early career, including industrial espionage that had been carried out against Watt. Probably knowing of the general speculation that Watt Jr. was going to write a biography of his father (which never happened), Playfair offered his services for information about Watt's early career. He further bolstered his credentials by telling Watt Jr. that he had written another memorial of Watt, slightly varied in content, for the *New Monthly Magazine*. He had expected that it would have been in print by now, but was still waiting for the proofs of the article. He concluded his letter by begging for money. Expecting but receiving nothing from John Playfair's estate, he needed a loan of £10 or £20 to carry him through until December.

Unknown to Playfair, Henry Colborn had sent the proofs of Playfair's article to Watt Jr. on 11 September 1819.[3] Acting on advice from John Rennie, Colburn asked Watt Jr. to review the proofs and make any corrections he thought necessary. Rennie wrote to Watt the same day informing Watt of his meeting with Colburn. Also unknown to Playfair, Watt Jr. was busy with business, as well as his father's affairs, and was daily making trips from his residence in Aston Hall into town. Becoming impatient, Playfair wrote to Watt Jr. on 12 October 1819 with some petulance: "If I was wrong in expecting you would do any act of friendship I had at least a right to expect that from mere civility you would have acknowledged the receipt of the memoir."[4] To the letter, Playfair attached an abridged copy of a letter he claimed was written to him by a J. Smith, but probably written by Playfair himself. Smith claimed that in his memoir of James Watt, Playfair had missed key damning details about how Matthew Boulton had originally obtained his money. Smith also claimed that Watt was stingy. Playing the hero, Playfair claimed that he had prevented Smith from taking his information to the press. Had this been a typical Playfair scam, Playfair would have asked Watt Jr. for money to suppress the publication. This time it is likely that Playfair was clumsily trying to impress upon Watt Jr. his loyalty to the family in order to obtain the monetary support he needed.

Watt finally wrote to Playfair on 20 October 1819 replying to both letters. His major concern was that Playfair had written a biography of Watt's father, encompassing his whole life and work, without consulting any of his father's friends or associates. Had he done so, he would have obtained an accurate description of his father's work, so Watt claimed. Watt Jr. considered Playfair's action indelicate and improper. Moreover, Watt Jr. was convinced that some of Playfair's claims about his father's early life were incorrect, although he admitted that other aspects were accurate. One example was singled out. Playfair claimed that James Watt had been apprenticed in his youth. The son checked his father's papers and corresponded with some of his father's friends, but found no such evidence. Playfair responded to Watt Jr. on 22 October 1819 in his last letter to Watt. He questioned some of the corrections that Watt had sent to him on 20 October and accepted others. Then he defended his interaction with the supposed J. Smith: "As to Mr Boulton I did what you would have done – Told the man I did not believe it & it is for him to do as he pleases."[5] This does not quite jibe with what Playfair's earlier report that he had intervened in some way to prevent publication of Smith's claims.

On 25 October 1819, John Rennie wrote to Watt Jr. Rennie interpreted Playfair's request for money and the letter from Smith as the same kind of scam that he had experienced from Playfair in 1816. For Rennie, the fictitious letter was from a non-existent Thomas Byerley, not realizing that Byerley was real and Smith was probably not. Watt Jr. closed the whole affair with a letter to Playfair on 16 January 1820. He informed Playfair that he had submitted Smith's letter to Boulton, Watt, and Co., thanked him for it, and let the matter rest, as he saw no evidence of illegality in the contents of it.

The second memoir of Watt that Playfair had written appeared in the October issue of *Monthly Magazine*, two months before his first. When the two are compared, Playfair was true to his word. The *Monthly Magazine* memoir may have had nearly the same information as for the *New Monthly Magazine*, but there were significant differences in layout and wording. As a quick example, Playfair began his article for the *Monthly Magazine* with a more extended history of the development of the steam engine than what appeared later in the *New Monthly Magazine*. One big difference between the two articles is that tacked on to the end of the *Monthly Magazine* memoir is an obituary of Watt that first appeared on 4 September 1819 in the *Scotsman*. Whether this was part of the proofs that Playfair sent to Watt Jr. or added later to the proofs by Richard Phillips is impossible to tell. Without the addition from the *Scotsman*, both memoirs have natural but different endings. What is possible to tell is that Playfair did not follow all the corrections Watt Jr. had sent to him. The memoir in the *Monthly Magazine* continued to claim that James Watt had been apprenticed to a mathematical instrument maker, despite the son's contrary claim. For his part, Henry Colburn was angered by the fact that Playfair had written a second memoir of Watt. As Colburn prefaced the memoir in the *New Monthly Magazine*, "The following Memoir was expressly written for the proprietors of this Magazine by Mr. Playfair, who received a handsome remuneration for the same; while, however, it was in the hands of a friend of the late Mr. Watt, for revision, the writer thought proper to dispose of a copy of the same to the old Monthly Magazine. The Editor was felt compelled to make this statement, to account for it having already appeared in that journal, and leaves the public to form their own opinion on the conduct of Mr. Playfair." Colburn gives the impression that the articles were identical, but, though similar in content, they were not. Perhaps Colburn was just kicking himself for taking the time to check with Watt Jr. through John Rennie on the accuracy of Playfair's memoir.

Just prior to John Playfair's death, the fourth edition of Lady Morgan's *France* was advertised for sale (*Morning Post*, 12 July 1819). It was obviously a popular book. This was a red flag that prompted Playfair to turn his attention to expanding his August 1817 short review of *France* in *Galignani's Messenger* into a book. It grew into a two-volume 731-page tome: *France As It Is, Not Lady Morgan's France* (Playfair 1819). Advertised at the end of October 1819 as soon to be published, it was available for sale early in December (*Morning Chronicle*, 29 October 1819; *Morning Post*, 2 December 1819). Playfair was highly critical of Morgan's book, sometimes vituperatively so. The nature of the red flag that sent Playfair on the rampage might be summed up in two quotes from modern scholars of Lady Morgan's work: "Owenson [Lady Morgan] sympathized with the ideals of the French Revolution and expressed immense enthusiasm for Britain's former nemesis, Napoleon Bonaparte." "Her liberal sympathies led her to exaggerate both the oppression of the *ancien régime* and the beneficial effects of the revolution" (Donovan 2009, 9; Suddaby and Yarrow 1971, 36). In the past, Playfair had spilt gallons of ink writing about the excesses of the French Revolution that he had seen and experienced first-hand. This new book that had crossed his desk in Paris had incensed him. Seeing the book's popularity must have rankled.

There were at least four contemporary reviews of *France* (Suddaby and Yarrow 1971, 300). All point to errors in fact made by Lady Morgan in her book. The earliest review, appearing in *Quarterly Review* in 1817 soon after *France* was published, has been described as "a rabid denunciation of the work and of the writer." In the same discussion of these reviews, Playfair's was dismissed as "a dull, malicious, and confused work" (302). Playfair's review also came late to the scene, at least two years after the initial publication. Earlier, the reviewer in the *Quarterly Review* had torn apart *France* under eight headings: Bad Taste, Bombast and Nonsense, Blunders, Ignorance of the French Language and Manners, General Ignorance, Jacobinism, Falsehood, and Licentiousness and Impiety. Playfair attacked Lady Morgan under only three general headings: Indecency, Falsehood, and Ignorance. Playfair and the *Quarterly Review* generally overlap.

Playfair's book was a lengthy refutation of what Morgan had written. His comments were not always the mark. One probable reason that it was described as dull is that Playfair presents statistics, as he had in his economic and political works, to make his point. That is typically not palatable to the general reading public. It was at times confused. Playfair misinterprets or misrepresents what Morgan has written. A contemporary review of Playfair's *France As It Is, Not Lady Morgan's France* best describes Playfair's shortcomings:

> His book is, as far as it goes, a picture of "*France as it is*," and certainly *not* Lady Morgan's "*France*." Hers is a vivid picture of manners of the people; his a dull dissertation on the finances of the government. She depicts, with exquisite skill, society, literature, and the fine arts; he dilates, with dry detail, on manufactures, commerce, and agriculture. Lady's Morgan's "France" is a Parisian *elegante* – gay, lively, and fascinating; in the exuberance of her spirits, too careless, perhaps, on serious subjects, but meaning nothing wrong, abounding with vivacity and wit, delightful to hear, and lovely to look upon. Mr. Playfair's "France" is a *roturier*, dull, heavy and vulgar; useful perhaps, in the way of his business, but rude in his manners and coarse in his conversation; sometimes amusing by an attempt at clumsy humour, but often abusive to his competitors in trade; and though he now and then appears decent and respectable in his Sunday clothes, his general habits are dirty and disagreeable. (*Freeman's Journal and Daily Commercial Advertiser*, 23 June 1820)

An example of one of Playfair's general habits that detracts from his critique of Lady Morgan's *France* is that he takes a few gratuitous shots at the Irish and Lady Morgan's Irish background. Of course, it was a Dublin newspaper that disliked what Playfair had written.

As 1819 turned into 1820, things once again started to go badly for Playfair. He had ambitions of publishing a statistical account of the various départements of France, which never came to fruition (Playfair 1820, 16). He tagged a little advertisement for this ambition to the end of a short pamphlet promoting the advantages of emigration to France. As the result of what Playfair reported were oppressive taxes and dearness of living in Britain, and as a result of the post-war depression, many English families had emigrated. Playfair considered several locations for immigration and rejected them all disparagingly, except France. His son Andrew William Playfair had stayed in Canada after the War of 1812. Father dismissed this as living in the woods and herding with the "savages of North America." What France had to offer was relatively cheap land that was close to Britain. Taxes under Bonaparte that had fallen on property owners, as well as the law forbidding Bonaparte's family and his closest supporters from living in France, had brought much property onto the market, which Playfair claimed was discounted by 25 per cent of its value. He expected that the price of land would soon rebound so that there was a speculative opportunity as well as produce and rents to be earned from the land. Playfair did not take his own advice and return once again to France, nor did any of his immediate family.

Playfair's ambitions for a major statistical publication on France with some cash flow coming his way came to nought. On 6 March 1820, he was once again taken to prison for debt, this time at Whitecross Street Debtors' Prison. His friends thought the incarceration unjust, dismissing the debt as a seemingly trifling matter that Playfair incurred while helping others. In 1829 Robert Stuart described the situation: "In the autumn of 1822 [it was actually 1820], he was released from the prison in which he had been confined many months for a debt (for which he had been surety for a friend) of a few guineas" (Stuart 1829, 549). The debt actually amounted to £798 13s, a much larger amount than a few guineas (Great Britain 1822a, 85). It is possible that the small debt mentioned by Stuart triggered the initial legal action against Playfair, which then cascaded into several other creditors wanting their money. Playfair also thought that his imprisonment was unjust, but for different reasons. He was imprisoned during a short interval when old legislation on insolvent debtors had lapsed. This was a combination of legislation whose benefits he had enjoyed initially in 1801 and of newer legislation he had praised in his political portraits of Lords Ellenborough and Eldon in 1813.

A month into his imprisonment, Playfair wrote to the directors of the Literary Fund on 8 April 1820 asking for monetary assistance.[6] He was in poor health, he said, as he wrote from the sick ward of the debtors' prison, and needed to support his wife and blind child. He justified his request by giving a one- or two-sentence summary of his literary career and then tugged on the directors' moral and political heartstrings: "That he has on all occasions without a single exception supported the cause of religion and morality & written various tracts to counteract the fatal principles that bad, or mistaken men promulgate unfortunately with too much success." There was no immediate response to his request, so Playfair wrote again on 29 April. It took over a month and a half from his initial request for the directors to respond; Playfair was given £10, the maximum available. Still in prison and unlikely satisfied with the extent of the help he had received, he wrote to the directors on 2 July 1820 suggesting that the Literary Fund provide "literary men" with regular support, not just support up to the maximum amount in times of emergency. By mid-August, his money had run out at a time when release was possible, but at a cost of £5. On 22 August he wrote again to the Literary Fund requesting the £5, but was turned down.[7] With money from somewhere, finally Playfair's petition for release was heard at Justice Hall, in the Old Bailey, on 16 October 1820 (*London Gazette*, 23 September 1820, 1812–13). He was released the following month. But Playfair was not done with the Literary Fund. He wrote to the directors on 15 December 1820, requesting information on the finances of the Literary Fund. He wanted to submit a plan to the directors on the establishment of an "accumulating fund" that would support regular financial aid to literary men like him. His request was turned down.[8]

The article on James Watt published in *Monthly Magazine* in 1819 brought Playfair back into contact with Sir Richard Phillips, the editor and owner of the magazine. Phillips commissioned Playfair to draw a graph for his *Chronology of Public Events and Remarkable Occurrences within the Last Fifty Years*. Annual editions appeared over the years 1820 to 1824. (The graph shown earlier in figure 4.1 is from the 1821 edition.) Playfair is acknowledged in the foreword: "The scale of the progressive variation of certain prices and amounts, connected with public finance, for which the Editor is indebted to the ingenious Mr. WILLIAM PLAYFAIR, will serve to check any errors of the press in regard to those subjects; while, as a picture, it cannot fail to interest the politician and philosopher."

Phillips was also interested in science (Seccombe and Loughlin-Chow 2008). And it was through Phillips that Playfair, in his last hurrah in engineering, attempted to become involved in a steam carriage newly patented by Julius Griffith in 1821. The carriage is shown in figure 20.1. Griffith had travelled extensively in Europe; he had patents for his steam carriage in France and Austria as well (Whittaker 1825, xx–xxiv). The invention caught the attention of the Bramah brothers, Timothy and Francis, who had taken over the engineering works of their father, the renowned engineer Joseph Bramah (McNeil 1968, 178–9). The Bramahs built the steam carriage according to Griffith's design in their

Figure 20.1. Griffith steam carriage 1821. Alamy Stock Photos.

Pimlico workshop. The carriage is described in detail in *Jackson's Oxford Journal* (14 September 1822). In the article it is claimed that the carriage could carry three tons of merchandise or passengers, and travel at ten miles per hour. An article like this may have piqued Playfair's interest.

Through Sir Richard Phillips, Playfair met Griffith in early December 1822. Phillips introduced Playfair as having been "bred" with Boulton and Watt, thus validating Playfair's engineering credentials to Griffith. After their meeting, Playfair gave thought to this new invention and made suggestions for its improvement. He visited the Bramah workshop on 24 December, but did not meet with anyone. On 1 January 1823, Playfair wrote a letter addressed "To the Patentee of the Steam Carriage at Messrs Bramah Engineers Pimlico."[9] In it he compared the steam-powered carriage to a steam-powered vessel on water. On water, because of the nature of the water resistance, the vessel would initially move easily but would have an upper limit of ten miles per hour. On land, experiencing a different type of resistance, the carriage would have a slow start, but could reach a speed of fifteen or twenty miles per hour. Playfair suggested that the engine should be made of rolled and hammered copper to reduce the weight and, at the same time, maintain strength. He also recommended having the flywheel run at high velocity, but again with reduced weight."

Playfair's anti-Catholic sentiments emerged once again late in his life. Although there had been a softening of anti-Catholic feelings in

England during the French Revolution, they did not disappear, even with the emancipation of Catholics in Great Britain through an act of Parliament in 1829. Anti-Catholic opinion was still flourishing in 1821 when Playfair was rubbing shoulders with members of Orange lodges in London. On 5 November 1821, the day after the birthday of William III, eight Orangemen met at the Café Royal in London's Regent Street to celebrate. (*Morning Chronicle*, 6 November 1821). Jeremiah Stockdale (probably a relation to the bookseller John Stockdale, Playfair's old publisher), deputy grand master of the Metropolitan Lodge of Orangemen, presided. It must have been a jolly affair. Over twenty toasts were drunk (*Oxford University and City Herald*, 17 November 1821). It was not so jolly for the proprietor, Thomas Rose. He had not realized who his guests were and was very uncomfortable with their behaviour. He asked for a written apology and wrote about it to the *Morning Chronicle*: "I called, and warmly remonstrated with him [Stockdale] on the shameless effrontery of *eight* persons, *only eight*, sitting in a small room on my 2d floor, and daring to assume a political character so justly odious. To their unconscious host they appeared to be a party of tradesmen met together for the purpose of enjoying a few luxuries, of which they were not in the habit of taking" (*Morning Chronicle*, 28 November 1821). Playfair may not have been present at the dinner. But he was there when Rose demanded an apology from Stockdale. On the basis of that interaction, Playfair wrote letters to two newspapers in defence of Stockdale (*Morning Post*, 24 November 1821; *Morning Chronicle*, 1 December 1821). The editor of the *Morning Chronicle* decided to print only an extract from one, because he "was unwilling to make our paper a vehicle for personal abuse."

Needing money once again, Playfair continued to write at a furious pace. Part of his late work might be described as fighting old battles with discredited ideas. Most of it was re-spinning old work with a new twist. His use of statistical graphics was creative and novel. He had found success with it in the past, and he was reinvesting it in the promotion of his latest literary efforts. He found a new publisher, William Sams. Little is known about Sams other than that he produced a number of books with engravings done by aquatint. The surviving publications that Playfair printed through Sams all have coloured graphs.

In 1821, Playfair published *A Chronological Chart of the Contemporary Sovereigns of Europe from A.D. 1060 to 1820*, and a review of it appears in a literary magazine (*Eclectic Review* 15, 15 June 1821). It was a single chart that cost five shillings in monochrome and seven shillings in colour. From the description in the review, it may have been patterned on, or inspired by, his charts in *British Family Antiquity*. The sovereigns covered during the time period are England, Scotland, Spain, Portugal, Germany (more formally the Holy Roman Empire of the German Nation), Hungary, Russia, Poland, Prussia, Denmark, Sweden, and the Popes. The reviewer commented that, in general, a graph is only as good as the data going into it. He was happy to give "our unqualified approbation to the very creditable compilation before us."

During the depression following the Napoleonic Wars, unemployment rose as a result of stagnation in the manufacturing industries. There was lack of demand for manufactured goods as the government reduced its purchases and as European countries grappled with weak purchasing power due to the effects of war. When unemployment rose, wages of the factory workers fell. During the wars, rents on land had risen. Not wanting to see a loss in their incomes, landowners petitioned Parliament once peace was established. The result was the Corn Laws of 1815, which restricted the importation of foreign grain, thus keeping the price of grain in England high. This, in turn made it difficult for the ordinary worker, now living at a subsistence level, to afford bread, a dietary staple. There was unrest among the general populace and fear of revolution among the ruling classes. That unrest played out in mass meetings, such as the most famous one held at St. Peter's Field in Manchester on 16 August 1819. Fear of revolution translated into the authorities' overreaction to the meeting. After sending in cavalry, many of them inexperienced, to break up the meeting, the result was the Peterloo Massacre in which eighteen civilians were killed and between 400 to 700 were injured.

In his *A Letter on Our Agricultural Distresses, Their Causes and Remedies*, Playfair tried to tackle one aspect of the problem of this post-war depression – the price of wheat and hence bread (Playfair 1821, 1822a, 1822b). The pamphlet, which was first published in July 1821, was popular enough that it went through two more editions in 1822.[10] It was an elaboration of what he had published in 1810 as a series of letters he had written to the editor of the *Morning Advertiser* and to criticism of him in responding letters in the same paper.[11]

As solutions to reducing the price of wheat, Playfair dismissed reducing taxes or altering the Corn Laws. He did have the poor in mind, but with a completely wrong-headed solution. Decreasing taxes would result in decreasing relief to the poor, which amounted, according to Playfair, to little more that 40 per cent of the taxes. He also doubted that any reduction in rents on land would have great effect since many occupiers of land currently "do not pay their rents." The real problem, according to Playfair, rested with the role of the corn factor, the recently emerged middleman who bought grain from the farmers and sold it to the millers. The millers, in turn, sold their flour to the bakers. Playfair argued, "*Between the sale of wheat by the farmer, and the purchase of flour by the baker*, a large sum of money is LOST: I say LOST because it does not go to any productive purpose." After a lengthy argument using historical examples, and especially historical data from 1565 to 1821, Playfair concluded that the solution was to eliminate the practice of bakers paying on credit. Rather, it should be ready money, i.e., cash. It might not be practicable to implement such a rule, so he suggested passing a law that "would make any debt from being legally recovered after an expiration of two months." Then those with money would pay cash and those without it would not get credit.

Playfair's pages read like a diatribe, and his pamphlet was attacked by some reviewers. One pointed out a major problem in his solution: "Mr. Playfair treats the subject with much novelty and acuteness, and certainly throws lights, which ought to be known to every body, though we fear, that the evils exposed are remediless; unless the Farmer himself turns Factor, and directly supplies the Baker, which many might do, if of sufficient capital" (*Gentleman's Magazine* 1822a, 59). On the positive side for the reviewer were three graphs that Playfair constructed; the graphs do not appear in the newspaper articles from 1810. After being critical of Playfair's recommendations, another reviewer commented, "Mr. P. has given us three charts, ingeniously constructed, perspicuous, and useful: the *first* displays at one view the price of the quarter of wheat [eight bushels], and the wages of labor by the week [shown in figure 20.2 and labelled "No. 1"], from the year 1565 to 1821; the *second* gives the yearly average prices of the quartern loaf [four-pound loaf] of best wheaten bread, from 1740 to 1821; and the *third* shews the value of the quarter of wheat in shillings, and *in days' wages of a good mechanic*, from 1565 to 1821" (*Monthly Review* 1822, 88).

Playfair's argument rests on the evidence presented in his charts. In the reviewer's first chart labelled "N^{0}.1," shown in figure 20.2, the five-year average of the price of a quarter of wheat, given in shillings, is shown with vertical bars. With some regular variation, the price decreased between the sixteenth and seventeenth centuries, remained stable until about 1795, and then shot up in the early nineteenth century. At the bottom of the graph are the weekly wages in shillings of a "good mechanic," again a five-year average. What Playfair meant by the phrase "good mechanic" was "artisan." Generally, wages for this group had been rising. Then Playfair manipulates the data a little to obtain the chart labelled "N^{0}.3," shown in figure 20.3. In this case, Playfair has taken twenty-five-year averages for the price of a quarter of wheat in shillings. This can be calculated by averaging the heights of

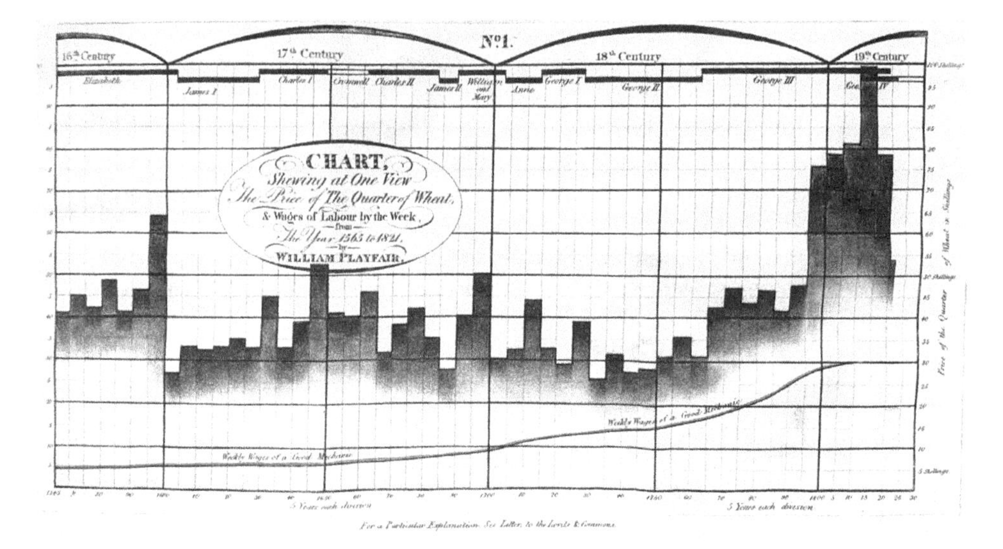

Figure 20.2. The price of wheat compared to labourers' wages. Courtesy of Stephen M. Stigler, University of Chicago.

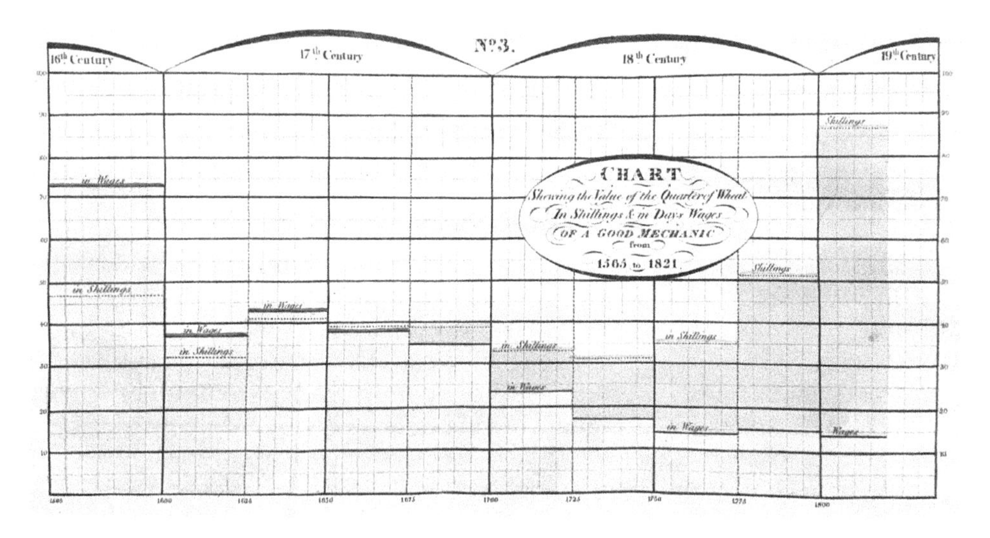

Figure 20.3. The price of wheat and the number of days' wages needed to purchase the wheat. Courtesy of Stephen M. Stigler, University of Chicago.

the appropriate five bars in chart N⁰.1. The height of the resulting bar is shown by the dotted line in each bar. On the basis of the wages of a good mechanic, Playfair then calculates how many days of wages it would take to purchase a quarter of wheat, shown by the solid line in each bar. The appropriate part of the bar is coloured blue. As can be seen from this graph, although the cost of the wheat in shillings is generally increasing over the time period, the number of days in wages required to buy that wheat is decreasing. The positive gap is shown in red. Playfair concludes, "The wages of labour never bore so favourable a proportion to the price of wheat as they do at this time." What he observed was contrary to many who argued during these difficult economic times that the price of wheat should be regulated by the price of labour.

Playfair had a very nice set of graphs to prove his point. What he missed was that graphs are only as good as the data going into them. He had used the wages of "good mechanics," whereas the real issue was among the factory workers, whose wages were decreasing drastically after the war. In 1805 he had ignored the emerging factory system when updating Smith's *Wealth of Nations*. Sixteen years later he continued to ignore a system that was now more firmly in place. A reduction in the tariffs on wheat brought by a change in the Corn Laws would have helped these factory workers now in dire straits. Although Playfair takes some swipes at some landowners, his arguments against changes to the Corn Laws directly support the landed classes, whom Playfair saw in earlier publications as the natural ruling class for Britain. At heart, *A Letter on Our Agricultural Distresses* is a piece of political propaganda relying on "lies, damned lies, and statistics" to make its point. Through some well-constructed graphs it does so beautifully, although incorrectly.

Playfair's pamphlet *Can This Continue?*, probably written in 1822, falls in line with what he expressed as early as 1786 and more formally in his 1805 book *An Inquiry into the Permanent Causes of the Decline and Fall of Powerful and Wealthy Nations* (Playfair 1822?). Once again, he expressed concern about the national debt and how it was creating a class of people living off the interest created by the debt. These investors made no contribution to the economic well-being of the nation, according to Playfair. He supported his argument with the graph shown in figure 20.4. The graph shows three time series: rental income from landed property, the same rental income with mortgages and other expenses deducted, and the national debt, which provides the income for those living off their investments. Rental income rose substantially during the Napoleonic Wars. Further, the gap between rental income

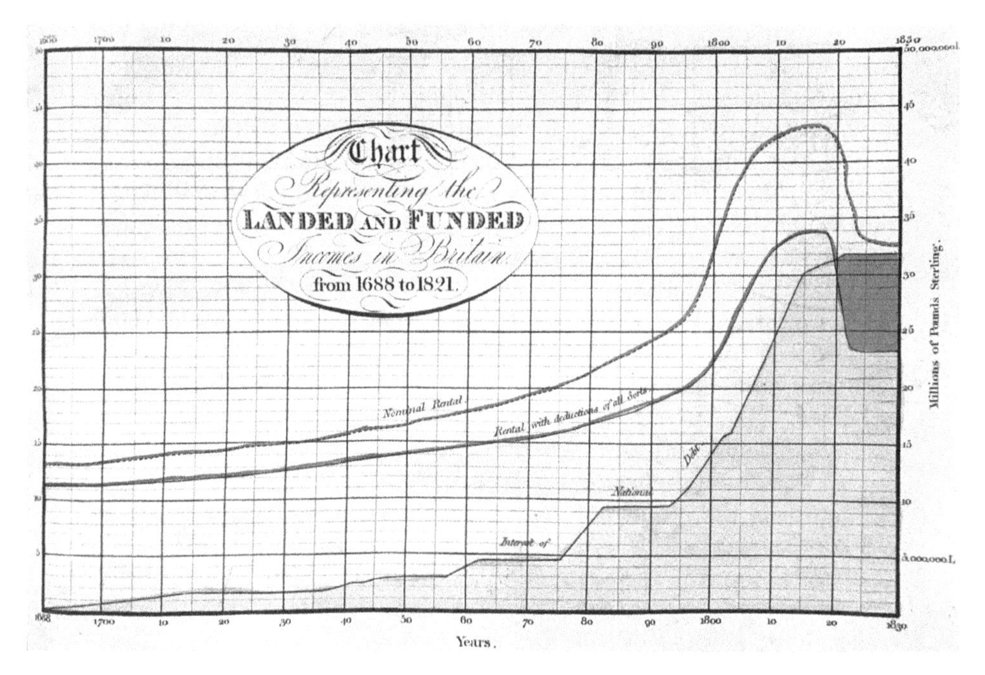

Figure 20.4. Landed income and income from investment in the national debt. Courtesy of Stephen M. Stigler, University of Chicago.

and net rental income increased on account of the ease of borrowing money so that the purchase of many estates was done through mortgages. The concern, which Playfair underlined by the graph, was that when the pamphlet was published the national debt now exceeded net rental income. And to sound the alarm even louder, his projection of the trends to 1830 showed that the gap was widening substantially. Playfair's warnings were a false alarm. He neglected to take into account British manufacturing, which soon recovered and went on to power the British economy through the rest of the century.

Playfair continued to try to publish political pamphlets, although as he noted in his *Political Portraits*, pamphlets were on the wane from publishers' perspective. In his last pamphlet, he attacked Lord Castlereagh. On 12 August 1822, Castlereagh, who had served as foreign secretary since 1812, committed suicide. He had represented Britain at the Congress of Vienna while it met from September 1814 to June 1815. As government leader in the House of Commons, Castlereagh had to defend repressive and unpopular responses to the depression following the Napoleonic Wars, along with unrest among the general populace and fear of revolution among the ruling classes. To make matters worse for Castlereagh, he was criticized for his post-war association with Britain's allies against Napoleon. He came to be seen as the personification of repression and reaction (Bartlett 1966, 181–90).[12] A month after Castlereagh's death, George Canning was made foreign secretary.

Playfair had disagreed with Castlereagh's approach to foreign affairs, and about a month after Canning was in office, he wrote a pamphlet advocating a change in approach.[13] No copy of the pamphlet seems to have survived. On 20 November 1822, Playfair sent his pamphlet to Canning. Perhaps he thought that he would be sympathetic, since Canning and Castlereagh had fought a duel in 1809 in which Canning received a pistol shot to the thigh. On the contrary, Canning was not at all sympathetic. On receipt of the letter and pamphlet, Canning's secretary wrote to Playfair the same day: "Mr. Canning cannot help expressing his extreme regret that Mr Playfair should have thought it expedient, at the present moment, to revive discussion on subjects which have been long buried in oblivion, and to which (after the melancholy event of August) it is no less painful to Mr. Canning than it must be distressing to the friends of Lord Londonderry, that public attention should be recalled."[14]

Playfair continued to show his lack of judgment when he replied to Canning's letter on 22 November 1822: "British interests have suffered so much since 1814 when one blunder has followed another."[15] He continued to ask for a change in approach to foreign affairs. In the hopes of mollifying Canning, Playfair added that if a second edition of the pamphlet were to come out, it would state that the original pamphlet was written without Canning having knowledge of it. There is no further correspondence to Playfair from Canning's office.

Playfair had suffered from problems with his legs since at least June 1807. At that time, he mentioned in a letter to the Literary Fund that he had had an attack of "rheumatic gout."[16] When he was incarcerated in Newgate in 1809, he complained that he was unable to go out in the courtyard of the prison because of a bad leg.[17] During his later incarceration for debt in 1820, Playfair was placed in the sick ward.[18] Despite being active enough to meet Julius Griffith in early December 1822 and possibly to visit the Bramah workshop in late December, Playfair's condition had seriously deteriorated by mid- to late January 1823. On 21 January, he wrote to the Literary Fund for financial help: "I have not strength sufficient to write to any serious purpose."[19] Within a week or two he was carried from his home in Bedford Street, Covent Garden, to St. George's Hospital to undergo an operation on his legs since they were "showing symptoms of mortification" (Stuart 1829, 2:549–50). He died at the hospital on 11 February 1823. Ever active to the last trying to make ends meet, Playfair was writing a book at the time of his death. Carrying the title, *A Vindication of the King*, it was to be published by William Sams.[20]

To give medical context to this information about Playfair's death, there seemed to be a similar case in 1825 of a young man named John Hammond brought to St. George's Hospital with a cut on his leg that perhaps had hit an artery (*Lancet* 7, no. 4 (1825): 113–15; and no. 5, 130–50.[21] He was treated and the wound was tightly bound. Several days later Hammond died. There were complaints by Hammond's friends about the cleanliness of the hospital linens on the bed, the cleanliness of Hammond himself and his bedclothes, and the lack of attention given to the patient after the initial binding of the wound by the surgeon.[22] It was also claimed that Hammond's wound was bound too tightly, which resulted in swelling of the leg above the knee. A coroner's jury decided that Hammond died from lack of proper surgical treatment, but a hospital investigation of the incident exonerated the surgeon. Dr. Vivian McAlister, a highly respected surgeon who has an interest in the history of medicine, noted that gout could be complicated by diabetes and heart disease.[23] The "mortification" mentioned in the sources probably means ulceration or gangrene in the legs – symptoms that would most likely accompany diabetes. Playfair's condition was terminal and the only treatment was cutting away the damaged tissue or amputation of the legs, which, in either case, would have been excruciatingly painful. Knowing the terminal nature of Playfair's condition, the surgeon at St. George's probably did not operate, and instead gave him laudanum to ease his pain. Playfair likely died of heart failure from the heart disease associated with gout, from respiratory failure from sleep apnoea also associated with the disease, or from the laudanum.

Playfair was buried three days later on 14 February 1823 in the cemetery attached to the church of St. George's Hanover Square.[24] He appears to have been buried in the well-heeled area of the cemetery rather than the section given over to paupers (Spence, Fenn, and Klein 2017). This is a bit of a conundrum, as Playfair had little or no money. One possible explanation might be conjectured from his final letter to the Literary Fund three weeks before his death. Written on the back of the letter in someone's handwriting (not Playfair's) is "Miss Playfair at Alexander Johnstons Esq. Hornsey Lane Highgate." This probably refers to Zenobia Playfair, who worked as a governess. She, or her brother

John, may have supplied the money to ensure that their father did not wind up in a pauper's grave. Or her likely employer Alexander Johnston put up the money.

Playfair's late plea to the Literary Fund did not fall on deaf ears. Slightly more than a month after his death, the Literary Fund sent £20 to Mary, Playfair's widow.[25] With that, an eventful, sometimes highly successful, sometimes disastrously unsuccessful life came to a close. His living legacy is his graphs.

Afterword

Playfair Avoids a Shakespearian Epitaph

The evil that men do lives after them;
The good is oft interred with their bones.

Julius Caesar, 3.1

When I first started to research the life and work of William Playfair, one historian colleague, knowledgeable in the history of mathematics, wanted to know why. To this colleague he was simply crazy. I have already mentioned his bad reputation among American historians in the early twentieth century. Some of his roguish behaviour has been described in more recent publications by aficionados of statistical graphs and their development. I have not swept Playfair's faults under the carpet. Instead I have tried to give context to his scams and other questionable activities. For the most part, it was due to financial desperation. He was also ambitious, wanting to hobnob with the upper reaches of English and French society. There is another side to Playfair that is not seen in the archival material that I found in my research. Playfair's friend Thomas Byerley summed it up in a single sentence in his obituary of Playfair: "In private life Mr. Playfair was amiable and firm in his friendship, as he was loyal to his principles" ([Byerley] 1823, 172). The problem with historical research is that the sources may provide only a partial view of the subject. Without more source material attesting to the truth of Byerley's assessment, Playfair's roguishness has clouded his reputation today, and even in his own day.

Playfair's positive legacy is in economics and statistics. To repeat Edgeworth's assessment of Playfair, "[He] evinces some acumen as an economist as well as some originality as a statistician" (Edgeworth 1899, 117). Playfair's contributions to economic theory were small but significant. He did see the problems with Adam Smith's approach to a legislated maximum for the legal interest rate at a low level; and, although overlooked, he was a pioneer in the theory of capitalist development. Playfair's main contribution to economics was on the applications side. His graphical work was inspired mainly by trying to illustrate issues using economic data.

Playfair's lasting contribution has been to statistical graphics. It has not always been recognized. His graphs have come in and out of focus in the historical lens over the past 200 years. After his death in England in 1823, little was heard about Playfair's graphs in England until William Stanley Jevons unearthed them about thirty-five years later while working on economic statistics in the early 1860s. With the exception of Edgeworth's brief notice at the turn of the century and an entry in the *Dictionary of National Biography* at about the same time, Playfair disappeared from view until the 1930s. He resurfaced once again in the 1990s and has remained above the surface since. In France, the graphical groundwork that Playfair laid was better built upon, at least in the nineteenth century. In the 1840s, Charles-Joseph Minard began producing many new forms of graphs that took Playfair's contributions to new levels (Palsky 1996, 112–38; Friendly 2002). Like Playfair, Minard was trained as an engineer. Unlike Playfair, Minard was very successful as an engineer throughout his working life and retired comfortably to devote himself to his graphs.

I have mentioned some of the graphical elements that Playfair used and that remain standard to this day: hachure, stippling, and colour. He also used other innovative elements that are apparent in the graphs produced throughout the book (discussed in Playfair 2005, 23–7; Friendly and Wainer 2021, 99–100) and are also standard to this day. These elements include the use of grid lines with differences between major and minor grid lines, labels for axes, dotted lines emanating from solid lines to show projected values in the data represented by the solid line, and event markers to show what the effect might be on the data of a historical event. Playfair also framed his graphs to offer space for labels and to provide a more pleasing appearance to the graph. Generally, Playfair had an excellent grasp of a well-constructed, pleasing, and informative graph, which is still held by a few today as evidenced, for example, by the continuing popularity of Darrell Huff's *How to Lie with Statistics*.

I leave you with a quote that has been attributed, most likely incorrectly, to Oscar Wilde.[1] Whatever its source, I think it best captures major aspects of Playfair's career. "Anyone who lives within their means suffers from a lack of imagination." Playfair had plenty of imagination and could never live within his means.

Appendix

Assignat Forging by French Émigrés in England

Rather than giving a chronological list of what happened in forging operations for assignats, which occurred mainly in England, I will peel back layers of information almost as I found them. Consequently, I may jump around in time over the years 1791 when forging operations started, through approximately 1795, when they ended.

The idea of forging assignats did not originate with Playfair. The London press reported a few instances of it over 1791–2. The earliest reports mention operations carried out in London (*World*, 11 August, 17 September 1791), and the London police (at the time, the Bow Street Runners) were pursuing these cases. Forgery or counterfeiting was illegal; the magistrate in charge of the case, Sir Sampson Wright, offered 50 guineas (or about £6,600 in today's money) for information about one of these operations. The next year, there was a report of a large-scale importation of forged assignats into France from Italy, totalling about 30,000 livres (*World*, 12 April 1792). There was also illicit activity in France itself. Another report from 1792 mentions that four persons were imprisoned in Paris for forging assignats (*World*, 21 November 1792).

Under Pitt, the British government did indeed forge assignats as part of its war effort, or so the London newspapers reported in 1794. In early January 1794, it was reported that Pierre-Joseph Cambon, a member of the French National Convention in Paris, denounced Pitt, claiming that the British government was responsible for forging assignats "to ruin the National Credit of France" (*Sun*, 3 January 1794). Cambon reported that the forgeries were so bad "that the least discerning eye could discover their real value." It was only a few months later, in March 1794, that Pitt's purported scheme was exposed in Parliament. In the Commons, the MP (and playwright) Richard Sheridan claimed to have knowledge of paper mills employed to make forged assignats

(*Morning Advertiser* and *Sun* 22 March 1794). Another MP, Michael Taylor, supported Sheridan, saying that he could name the mills involved. Pitt dodged Sheridan's accusation by saying, "The Hon. Gentleman ought not to rely much on his information, if it came from those concerned in forging Assignats." Both Sheridan and Taylor were members of the Whig opposition led by Charles James Fox and Charles Grey. Nothing more came of the accusations in Britain. Within a couple of weeks, Sheridan's comments in the House of Commons were known in France (*Oracle and Public Advertiser*, 24 April 1794). The French soon responded by altering the appearance of the paper used and changing the watermark (*Oracle and Public Advertiser*, 10 May 1794).

The scheme in Britain soon went awry. French republican forces, which had invaded the Netherlands on 16 February 1793, were making progress. By June 1794, the British were in retreat and in late June they abandoned Ostend. French troops moved in and found 2,270,00 French livres in forged assignats left behind by the British (*Oracle and Public Advertiser*, 30 July 1794). It was claimed in the National Convention in Paris that the forgery was clumsy and easy to detect (*Oracle and Public Advertiser*, 31 July 1794). In mid-November, to indicate that they were no longer involved in forging assignats, when they probably still continued to do so, the British government arrested eighty foreigners "concerned in the traffic of forged assignats" and expelled them from the country (*Sun*, 19 November 1794). British spying operations were run out of the alien office within the Home Office, which was why the British were able to round up and to quickly expel so many people (Sparrow 1999). The roundup could have been a smokescreen for continued activity. In a letter of 10 July 1811 to William Wyndham Grenville (at the time leader of the Whigs in Parliament), William Playfair stated that the British government was involved with French émigrés in London in forging assignats during 1794 and 1795.[1] Forging continued in France well into 1795 (*Oracle and Public Advertiser* 19 May 1795; *Telegraph* 27 June 1795).

Supported by the British navy and some British troops, on 23 June 1795 French royalist troops landed at Quiberon on the south coast of Brittany. Though they captured and held the fort at Quiberon, the whole operation was disorganized and fraught with division. The republican forces counterattacked on 20 July, and the royalist forces surrendered the next day. It was reported, perhaps a propaganda exaggeration, that the republican commander found 10 billion French livres in forged assignats (*Oracle and Public Advertiser*, 10 August 1795). Either the British were continuing to forge assignats or the French émigrés, unknown to the British (which is unlikely), had their own independent operation.

Some republican troops pocketed some of the captured counterfeit assignats. Their theft was soon discovered and the knapsacks of all republican troops were searched (*Oracle and Public Advertiser*, 28 August 1795).

After the fiasco at Quiberon, back in England a lawsuit in which forged assignats feature prominently had wound its way to the Court of King's Bench. The case was heard in the latter part of 1795 and reported on in January 1796 (*Morning Chronicle*, 8 January 1796; legal report of the case in Espinasse 1801, 388–91). The suit was over payment of a promissory note drawn by a Mr. Lukyn and payable to a Mr. Caslon. The note had been endorsed to a Mr. Strongitharm, who wanted his money. Only surnames are given in the published reports of the case.

Although the participants are not completely identified in the lawsuit, they can be identified through an online paper chase based on their professions related to the book trade.[2] Paul Lukyn was a London stationer. In 1799, he committed suicide by slitting his throat with a razor.[3] The act was attributed to insanity, so he was given a Christian burial.[4] John Strongitharm was a London engraver working in Pall Mall. From 1799 to 1820, he was seal engraver to the Prince of Wales (Exeter Working Papers in Book History). William Caslon had already gained some prominence at the time of the trial in 1795; he was letter founder to the king from 1785 to 1819 (Exeter Working Papers in Book History). All were active in London from at least 1785 on (Pollard 1953, 5, 7, 15). It is evident that Lukyn was a junior player trying to use the services of more prominent individuals for the operation.

Reports of the trial indicate that someone connected to an unnamed secretary of state contacted Lukyn, who was in the business of selling paper. Lukyn asked Caslon for help in obtaining engraved plates to forge some assignats and sealed the deal with a promissory note. Caslon passed the work on to Strongitharm and signed the promissory note over to him. Initially, Strongitharm refused the job. When he was told that the forged assignats were to be used by the Duke of York's army against the French and that the secretary of state had approved the operation, he relented and engraved the plates. Then, for some reason, Lukyn balked at paying up. Lukyn's defence counsel contended that Lukyn did not need to pay; because of the fraud involved, the whole transaction was "contaminated." The jury found in favour of Strongitharm.

Giving substance to the London newspaper reports are the memoirs of an Englishman and a Frenchman, both serving in the royalist cause against the French Republic. These memoirs flesh out the newspaper reports, telling the story in different and biased ways, as each tried to justify his own actions. They all point to the forging of assignats initially by French émigrés in continental Europe and later by the same men in England with the connivance of the British government. The main actors in this play are five French émigrés (Charles-Alexandre de Calonne; his brother, the Abbé de Calonne; the Comte de St. Morys; his brother, the Chevalier de Moligny; and Joseph-Geneviève de Puisaye) and one English soldier (John Gordon Sinclair). Calonne had been comptroller-general of finances in France from 1783 to 1787, when he was dismissed from office. He soon went to England. The Abbé de Calonne was elected to the Estates General. He was arrested in July 1789, released, and then followed his brother to England (Galarneau 1987). For brevity, I will refer to Charles-Alexandre as Calonne and his brother as the Abbé de Calonne. St. Morys was the commissary-general to the French royalist army in 1795 for the invasion of Quiberon. There were further family connections. St. Morys's son married a niece of Calonne. Puisaye initially favoured the revolution, but had a falling out with the Jacobins in 1793. The next year he went to England where he convinced Pitt to back the invasion of Quiberon. The British recognized Puisaye as the commander of the troops for the invasion. French royalist forces in France did not, which was one of the reasons why the expedition failed. St. Morys died on the expedition in 1795 and his son assumed the title of comte de St. Morys. John Gordon Sinclair was commissioned a colonel in the Royal Rangers of the French royalist army (Armée des princes) in 1791 when it was raised at Coblenz under the Comte d'Artois. Calonne and St. Morys were involved in the regiment. There was a falling out between Calonne and Sinclair, at which point St. Morys's son was made colonel of the regiment at the age of eighteen.

Back in England, Sinclair was convinced that Calonne still owed him money from his time in Coblenz. He went to court over it. Not only did he lose, but also Calonne went after Sinclair for perjury. Sinclair was found guilty and sent to prison in 1798. During the dispute, Sinclair wrote a tell-all memoir (Sinclair 1796). After he was sent to prison, he wrote another memoir exonerating himself, as well as a petition asking for release from prison.[5]

In his memoirs, Sinclair claimed that Calonne sent him to England in 1791 to equip the regiment. As part of his trip, he was to bring back two Bank of England notes of £20 each, one dated 1790 and the other dated 1791, as well as a £10 note and implements that Sinclair later understood could be used for printing and engraving. Further, Sinclair claimed that St. Morys was an excellent engraver (which seems unlikely) and that Moligny could forge signatures. He heard about the operation to counterfeit assignats taking place in Coblenz when he was in

England and verified it when he returned to Coblenz early in 1792. Less than twenty kilometres from Coblenz, the two brothers, St. Morys and Moligny, had set up a forging operation in the town of Neuwied. According to Sinclair, the operation was guarded by troops from the regiment. The forgeries were justified on the grounds that the republican assignats were spurious, a theme that was to recur. The money raised through the forgeries was to be used to defray the expenses of raising the regiment. The regiment joined the army of Prussian and Austrian troops that the Duke of Brunswick had assembled, also in Coblenz, for an invasion of France. Brunswick's troops were defeated at Valmy in Champagne by republican forces in September 1792 and the invasion eventually came to nothing. With their regiment disbanded after the Battle of Valmy, Sinclair, the Calonnes, St. Morys, and Moligny all went to England.

In his own memoir, Puisaye backs up some of Sinclair's claims. Making no mention of Bank of England notes, Puisaye writes that in 1791 or 1792 Calonne proposed issuing forged assignats and that the whole operation (in translation from the French): "was entrusted to M. de St. Morys, and all the necessary plates and utensils were still in the possession of the latter, when I arrived in London" (Puisaye 1804, 376). Puisaye also claims that when he arrived in London in 1794, the forging of assignats seemed to be commonplace in the city. He states (in translation from the French), "Several parts of Germany and Switzerland were covered with clandestine manufactories of false assignats; and London contained seventeen or eighteen when I arrived there" (378). One of Puisaye's sources of information on the French émigré forging operation was the Abbé de Calonne. The abbé claimed not to be involved in the forging operation, but only got his information about it from St. Morys, with whom he lodged.[6] After this conversation, Puisaye went to St. Morys and talked to him about the multitude of forging operations, claiming that it was actually detrimental to their cause. He suggested that instead they have one very large operation to produce assignats. The forgeries would contain secret marks on them to identify them as royalist. It needed the approval of the French king and his regent. When the monarchy was restored, the secret marks would identify those assignats as legal tender. Since it was the French government in exile doing this, to Puisaye the only legitimate government of France, it was legal for them to issue these assignats. The British government had no such legal basis, Puisaye claimed. Puisaye finishes his description of the situation by quoting from an order of 20 November 1794 issued by the French royalist military council in Brittany. The order follows Puisaye's suggestions.

That the émigrés were printing some form of paper money in Coblenz near the end of 1791 was already known in Paris. An article in *Revolutions de Paris* in December reported such activity (*Révolutions de Paris, dédiée à la Nation*, 3 December 1791, 445). The émigrés were printing bills in twelve-, six-, and three-livre denominations carrying a picture of Louis XVI with the motto *Fidélité au roi & à la religion*. These could be exchanged at par with assignats. It would be a short step to forging assignats.

Work by nineteenth-century historians pinpoints a paper mill near Haughton Castle as a major source of paper for forged assignats. Haughton Castle is about thirty miles from Newcastle upon Tyne. These credible claims were made in 1856 by Thomas Doubleday, an English politician and author, later in 1858 by W.C. Trevelyan, cousin of the more famous G.M. Trevelyan, and finally by The Rev. G. Rome Hall in 1882. They all seem to have used the same source of information: direct descendants (son and grandson) of the owner of the paper mill that produced the paper for forged assignats. Doubleday met with the son during a fishing trip in 1852. With one exception, Trevelyan provides a more succinct description than Doubleday of what happened. Trevelyan writes,

> The paper for the assignats was manufactured at Haughton paper-mill (built in 1788), a few miles from Hexham, in a very picturesque part of Northumberland. The transaction was managed for Mr. Pitt by Mr. (afterwards Alderman) Magnay, whose family was and is connected with that part of the county. One of the moulds in which the paper was made is still in the possession of the proprietor of the mill, in whose family some of the assignats were also long preserved, but they have now been lost. The assignats were probably printed, in London, but on this and other questions information might probably be obtained from the successors of the alderman, who might, perhaps, also be able to tell what number, and in what year they were circulated. (*Notes and Queries*, 25 September 1858, 2nd ser., 6 (143): 255)

The one exception is a fuller description of Magnay by Doubleday:

> Towards the end of the American war, and during his father's lifetime, one of the farms of the estate of Haughton Castle was held by a farmer named Magnay, a married man, with more than one son. In due time, one of the young Magnays found his way to London in search of fortune; and being probably favoured by the partners in the paper mill, of whom the then owner of the estate was one, he became a wholesale stationer, and being

> steady, clever, and active, he soon rose to eminence as a stationer, and became an alderman of the city of London. (Doubleday 1856, 41)

At the Annual Meeting of the Newcastle Society of Antiquaries in 1882, Hall brought to the meeting a mould used to forge assignats. A newspaper report of the meeting states,

> It had been lent to him by Sir William Smith, grandson of Mr Smith of Haughton Castle, who was the then owner of the paper mill there, at which the assignats were made. It was thought of no interest, and placed in one of the lumber rooms, and after having been found it was restored by Mrs Smith herself. It had upon it the date of the forging of the assignats. He understood that a Mr Magnay was the then Court stationer in Newcastle, and his father was the foreman of the paper mill at Haughton Castle. It was said the assignats were sent out in connection with the Duke of York's army in expedition to Flanders 1793 – not to pay the troops, but cheat the French. The assignat was meant to depreciate the currency which was then very fast down. (*Newcastle Journal*, 26 June 1882)

The Magnay in question is Christopher Magnay, who was a wholesale stationer and rag merchant. Between 1793 and 1796, he worked out of 72 Upper Thames Street in London (Exeter Working Papers in Book History).

There is some evidence that at least two other paper mills were used to forge assignats. The first, Albury Park Mill in Surrey, was operated by Charles Ball and his sons from about 1790 to 1810. Recollecting in 1869 events of 1793 or 1794 told to him by his grandfather, Charles Ashby Ball said that the mill produced paper for forged assignats at the personal request of the Comte d'Artois, then living in exile in England (*Stationer and Fancy Trades' Register*, 5 April 1869). D'Artois arrived anonymously one day at the mill, gave the senior Mr. Ball notes with a certain watermark, and asked Ball to make paper with the same watermark. Ball successfully carried out the work and d'Artois returned on several occasions with more orders for paper. It was only later that Ball discovered that his anonymous client was d'Artois. The second paper mill was in Dartford, southeast of London. It has been reported that, in 1790 or 1791, John Finch produced paper for forged assignats at Dartford under contract from a stationer working out of St. Paul's churchyard in London (Dunkin 1844, 310; see also *Notes and Queries*, 1 April 1865, 2nd ser. 7:270; 7 (157): 16).

Pulling back the next layer deeper reveals contemporary manuscripts related to these memoirs and other seemingly random manuscripts.

Some of the random bits turn out not to be random, but connected. In these manuscripts there is overwhelming evidence that connects the British government to French émigrés forging assignats. The major cache of manuscripts related to forged assignats is in volume 101 of the Puisaye Papers, held by the British Library.[7]

The first thing that the Puisaye Papers reveal is that Puisaye was protecting the Abbé de Calonne. Puisaye's *Mémoires* were published in 1804. By that time, the abbé was serving very successfully as a parish priest in Canada. Rather than being merely an onlooker as Puisaye reports in his *Mémoires*, the Abbé de Calonne was one of the London organizers for the forging of assignats. This comes out in the different versions of the order of 20 November 1794 issued by the French royalist military council in Brittany. The *Mémoires* has four articles attached to the order, numbered I through IV. A printed version of the order in the Puisaye Papers also has four articles, but numbered I, II, V, and VII. Then there is a handwritten version of the order. This has eight articles, four of which correspond the numbers I, II, V, and VII in the printed version. Article III in the handwritten version, missing in printed versions, states that the Abbé de Calonne and St. Morys would be responsible for manufacturing assignats.

Also, among the Puisaye Papers on assignats is a letter that directly connects William Windham to the forging operation. Pitt had made Windham secretary at war in July 1794 (Wilkinson 2008). In the letter, a daughter of St. Morys asks the Marquis de Ménilles (Puisaye's father-in-law) to obtain an order from Windham for forty pounds of ink. Ménilles replied positively and recommended that she be careful in her choice of printers.[8] In view of these letters, it is highly likely that Windham was the secretary of state mentioned in the case of Strongitharm v Lukyn.

Windham was well known for hiring his relatives and close friends into positions that he controlled through his ministry. Salaries for these positions totalled about £8,000 annually or about £950,000 in today's money (Thorne 1986, 5:628). Two of these relatives were his half-brother Robert Lukin (no connection to Paul Lukyn) and another having the odd name of Emperor John Alexander Woodford. Woodford was from the Thellusson (also spelled Thelusson) family of bankers and businessmen; Windham's nephew and heir married a Thellusson. Lukin had the job title "Agent to the Foreign Corps" and Woodford's title was "Inspector-General of Foreign Forces in British Pay." Both dealt extensively with the French royalist regiments raised to battle the French republican troops. Over the years 1794 through 1801, Lukin spent over £1,500,000 on the French regiments (Great Britain 1809). Woodford's name crops up in the Puisaye Papers on assignats, but with nothing specific tying him to the forgeries. What is more revealing is in the papers of William Huskisson, undersecretary at war beginning in 1795.[9] It is evident that Huskisson was Puisaye's direct paymaster. There are several receipts from Puisaye, one for £4,000 for "support of the army," despite the fact that Robert Lukin was in charge of the accounts for supporting the French royalist army. There are also several receipts from Woodford. Tantalizingly, but inconclusively as far as assignats are concerned, are two receipts, one dated 25 November 1795 and the other 22 July 1796, each for £3,000 and described as "for secret service."

Brushes with assignat forging also turn up in Birmingham, as well as the Newcastle area. The two cities are loosely connected in this enterprise although they are about 230 miles apart. On 24 October 1793, Obadiah Westwood of Birmingham wrote to Thomas Bewick, an engraver living in Newcastle. Bewick responded on 30 October; only the response is extant (Bewick 1887, 378). Westwood wanted Bewick to engrave some plates "as to pass for real Assignats." Bewick declined to do the job, saying "that no temptation however great will ever induce me to depart from the line of rectitude which I have mark'd out to steer my course thro' life." Westwood, along with his brother, had a reputation for forging trade tokens in Birmingham. Bewick's letter pushes back concrete references to the forging operations in England to at least the last quarter of 1793, just prior to the claims in London newspapers that the British government was behind the forging operations.

John Gregory Hancock, a die maker and a Westwood associate, was also involved in trade token forgeries, as well as in the assignat operation (Dykes 1999). Samuel Garbett, a prominent Birmingham manufacturer and merchant, and a founder of the Birmingham Assay Office, wrote to Matthew Boulton on 9 December 1794: "Blocks to print assignats were brought to a workman of our assay master's to cast, and which were to be repaired by Hancox. I saw the blocks and acquainted Lord Hawkesbury with the particulars; what farther passed must be the subject of conversation."[10] A member of Pitt's ministry, Lord Hawkesbury was Charles Jenkinson. He was made Earl of Liverpool in 1796.

It is probably no coincidence that St. Morys was arrested in late February or early March 1795 for forging assignats. This may have been at the tail end of the general crackdown reported in the press on 19 November 1794. St. Morys's son wrote to Puisaye asking him to enlist the help of the Marquis de Spinola. For the previous two years or so, Spinola had represented the Republic of Genoa in London. He also had strong French connections. He had been the Genoese minister in France, married a French woman, and, after of the violent events of 10 August 1792, left France to live in London. With the efforts of Spinola, St. Morys was soon released. Spinola wrote to William Wyndham Grenville, the foreign secretary, that St. Morys was Spinola's secretary. Spinola reported to Puisaye on 3 March 1795 that he had essentially claimed diplomatic immunity for St. Morys. The next day, St. Morys was released. Someone found out about the affair and threatened to go public naming St. Morys and Puisaye as the heads of a ring to forge assignats. The writer also claimed that St. Morys was arrested for a 3,000-livre debt. The same person also threatened to write to Lord Hawkesbury. The letter, written in French, in which the threats were made, is dated 4 March 1795, and the intended recipient was the *Morning Post* newspaper. It appears that the letter was never published.[11]

Puisaye had claimed in his *Mémoires* that there were secret marks on the forged assignats, so that when the monarchy was restored these assignats could be identified as legal tender. The Puisaye Papers has a list of seventeen or more assignats that were forged. For each there was a description of the secret marks. One of the assignats, for fifty sols, supposed to have been forged is shown in figure A.1. On the top left, the assignat has the date of the law under which it was produced (Loi du 23 Mai 1793). On the top right is the year of the republic (L'An 2me de la République). On his list, Puisaye wrote the same date for the top left, but L'An 2me de la Liberté for the top right. Puisaye got the wrong assignat. "Liberté" appears on an assignat printed according to the law of 4 January 1792. Peter Bower has displayed a genuine assignat in this series for fifty sols alongside a modern printing from an eighteenth-century woodblock made to forge assignats. On the top right is L'An Quatrième de la Liberté (Bower 2018).

The forging operation may have had residual consequences for the St. Morys family. After the Comte de St. Morys died on the expedition to Quiberon, his son inherited his estate and title. The deceased comte's brother, Moligny, proved the will while the son was off in a regiment in continental Europe. The Comte de St. Morys had lent the abbé £500 and the method of repayment of the debt was not clear. When the young Comte de St. Morys returned to England, he sued the Abbé de Calonne for repayment of the £500 debt and included his uncle Moligny as a defendant in the suit (Vesey 1827, 577–92). At the time the suit came to trial in 1799, the abbé had run out of money and could not even afford legal representation. He had spent his money writing and publishing the counter-revolutionary newspaper *Courier de l'Europe*. The repercussions from this episode may have been felt over a decade and a half later when Playfair returned to France and the young Comte de St. Morys died in a duel.

Figure A.1. A forged assignat. Courtesy of Welcome Collection. Attribution 4.0 International (CC BY 4.0).

By 1795, the forging of assignats seems to have come more into the open. Those involved, who needed to find paper, engravers, and printers, seem to have been much less secretive and less careful than earlier, as seen in a letter dated 15 October 1795 from Matthew Boulton to his agent in London, Richard Chippindall.

> I am sorry to find you are about to lose Mr. Brooks. I don't love changing; however, what can't be cured must be endured. In respect to the business you last wrote to me upon, let it suffice in this and in all other similar cases that I answer: I will never be tempted by any pecuniary advantages to do a thing that is contrary to all principles of justice, honor, and the laws that do or ought to exist between nations, and, what is more, is contrary to the laws of the land we live in.

> I have had a similar offer from two foreigners within these three days, and I gave them a similar answer, in consequence of which they have applied themselves to a paper-man – I mean a man who has been accustomed to forge bills, assignats, and money, as well as to appear at the bar and to reside in Newgate: I have no doubt they will succeed with him, but I advise Mr. P. … to have nothing to do with so desperate and hazardous a business, which the law calls a misprision of treason and is punishable by heavy fines and imprisonment.[12]

The foreigners' interest in Boulton, and perhaps Chippindall as well, most likely arose because Boulton had established a mint at his Soho factory, which had been producing overseas coinage since 1787. He employed highly skilled French and German engravers (Srinivasan 2005, 8). The most intriguing part of the quotation is the reference to "Mr. P." From the context of the letter, it appears that Mr. P. was contemplating getting involved in the foreigners' scheme. It could be a reference to Playfair – or not.

All the evidence presented here points to a major assignat forging operation in England centred on French émigrés beginning as early as 1793 or even before. The British government under Pitt turned a blind eye to the operation and surreptitiously provided the funds. Had Pitt and his associates been more directly involved, the government would have been in danger of falling and Pitt himself impeached. According to the letter of the law, Puisaye was correct. The British government had no legal basis to forge assignats; the French government in exile did.

Notes

Preface

1 Funkhouser (1937) labelled the first half of the nineteenth century the "Expansion of the Graphic Method" and the second half "The Age of Enthusiasm in Graphics."

1 Playfair Is Sent to Newgate Prison

1 National Archives TS11/457/1520.
2 National Archives HO 77/16.
3 National Archives TS11/457/1520.
4 National Archives HO 26/6.
5 National Archives TS11/457/1520.

2 Playfair Goes to Birmingham to Work for Boulton and Watt

1 Library of Birmingham MS 3782/12/65/16.
2 National Archives HO 26/6. Criminal Registers. Middlesex. Newgate, etc.
3 Library of Birmingham MS 3782/12/65/44.
4 Library of Birmingham MS 3147/5/1458.
5 A biography of Boulton may be found in Dickinson (1937). His work as a toymaker and as a manufacturer of silverware and silver plate may be found in Robinson (1963), and Tann and Burton (2013, 23–32, 67–83).
6 Library of Birmingham MS 3782/12/72/45.
7 Library of Birmingham MS 3782/12/74/51.
8 Library of Birmingham MS 3782/12/65/16 and MS 3782/12/24/116.
9 Library of Birmingham MS 3782/12/24/116.
10 Library of Birmingham MS 3782/12/24/157 and MS 3782/12/65/44.

11 Library of Birmingham Library of Birmingham MS 3147/3/2 and MS 3782/12/65/31. How Boulton and Watt charged for their steam engines is described in Tann and Burton (2013, 98).
12 Library of Birmingham MS 3782/12/76/37.
13 I have not been able to find the original source. Watt's assessment of Hall is quoted by Dickinson and Jenkins (1981, 285), who also describe the production problems and the problem with Hall.
14 Library of Birmingham MS 3782/12/65/20.
15 For example, John Playfair's grave is located in the Old Burying Ground, Perth, Lanark County, Ontario, Canada. On his gravestone is written that he died 22 April 1853, aged 73. Further, an early twentieth-century history of the Playfair family provides a pedigree of the Playfair family in Canada (Playfair 1913, between 12 and 13). In the pedigree, the head of this branch of the Playfair family is given as William Playfair who married Mary Morris, with no date or even year of marriage. Most likely, the information for this pedigree was obtained from family members.
16 *Register of Banns of Marriage Published in the Parish Church of St. Mary le Strand*, 20 January 1771–3 January 1808, SML/PR/2/2, City of Westminster Archives Centre..
17 Ian Spence did an extensive search for the Playfairs' marriage record and found nothing. My assessment that the marriage likely took place comes from an e-mail from Rebecca Probert, author of *Marriage Law and Practice in the Long Eighteenth Century: A Reassessment* (Probert 2009) and the leading expert in her field related to the title of her book. Her e-mail to me says, "It is possible that the marriage was not registered. While it was meant to be registered immediately after it took place, there were inevitably omissions and errors. The Commission that examined Anglican marriage registers in the 1830s was very unimpressed with the clergy's record-keeping."
18 Library of Birmingham MS 3219/4/123/146. See also Muirhead (1854, 3:27–35) for a copy of the patent and a description of its development.
19 Information on the copier and its invention may be found in Dickinson (1937, 107–9) and Dickinson and Jenkins (1981, 50–1). Playfair's involvement in this work can be gleaned from letters from William Playfair to Matthew Boulton, Library of Birmingham MS 3782/12/25/58, 63 and MS 3782/12/26/106.
20 Playfair to Boulton, Library of Birmingham MS 3782/12/25/63. The 150 copying machines produced is found in Dickinson (1937, 108).
21 Microfilm: *Industrial Revolution: A Documentary History, Series One*, Part 15. Smeaton to Playfair, 1 November 1780. Adam Matthew Digital. Marlborough, Wiltshire. The orginal letter is in Library of Birmingham.
22 Microfilm: *Industrial Revolution: A Documentary History, Series One*, Part 12. Buchanan to Watt, 8 July 1781. Adam Matthew Digital. Marlborough, Wiltshire. The orginal letter is in Library of Birmingham.

23 Library of Birmingham MS 3782/12/26/106.
24 Library of Birmingham PF 629.
25 Library of Birmingham MS 3219/4/90.

3 Playfair Goes to London to Set Up His Own Business

1 Microfilm: *Industrial Revolution: A Documentary History, Series One*, Part 7. Walker to Boulton, 24 December 1781. Adam Matthew Digital. Marlborough, Wiltshire. The original letter is in Library of Birmingham.
2 London Metropolitan Archives MDR/1785/2/535, 537.
3 V&A/Wedgwood Collection MS E11/30436.
4 Library of Birmingham MS 3782/12/27.
5 Library of Birmingham MS 3782/12/28/8.
6 Library of Birmingham MS 3782/12/5/268.
7 V&A/Wedgwood Collection MS L84/14906 – 14913.
8 The Royal Society Journal Book records one Playfair as a guest at 11 meetings between 20 February 1782 and 13 June 1782. A second Playfair is a guest at the meeting of 9 May 1782. On that day, the Playfairs in attendance were "Rev Dr Playfair" and "Mr Playfair." At the other meetings, the Playfair in attendance was only described as "Mr Playfair," "Dr Playfair," or "Rev Mr Playfair." It is a good assumption that these were all John Playfair.
9 V&A/Wedgwood Collection MS E55/30060 – 3006a. The letter, but not the attachment to the letter, is reprinted in Smiles (1895, 266–7).
10 Library of Birmingham MS 3782/12/28/38.
11 Library of Birmingham MS 3728/12/30.
12 Library of Birmingham MS 3219/4/94.
13 I have been unable to see the original letter. Lord's reference is Tew MSS, which is an early reference system from the Library of Birmingham.
14 Royal Archives GEO/MAIN/25075.
15 The partnership was dissolved 25 June 1785. *London Gazette*, 25–8 June 1785.
16 V&A/Wedgwood Collection MS L84/14906–14913.
17 See the advertisements in *Morning Post and Daily Advertiser*, 9 and 21 March 1795, and *Public Advertiser*, 13 April 1785, compared to the advertisements in *Morning Post and Daily Advertiser*, 3 June 1785 and *Public Advertiser*, 7 July 1785.
18 London Metropolitan Archives CLC/B/192/F/001/MS11936/321/492137, MS11936/328/505247, and MS11936/347/538138. Later insurance policies, such as one in 1789 for James Playfair, make no mention of William, who by that time was in France. See MS11936/361/559048.
19 Patrick Colquhoun to William Playfair. Copy letter book of Patrick Colquhoun, Murray 551, 109r–110r, University of Glasgow Archives and Special Collections.

4 Playfair Evolves into a Writer by Profession

1 See Griffin (2014) and articles in the *Bee, or Literary Intelligencer*, 12 January 1791, 63; and 11 May 1791, 13.
2 The standard reference is now Smith et al. (1976, bk. 2, chap. 3, para. 28).
3 National Archives HO 42/9 fols. 581–2.
4 Seccombe and Loughlin-Chow (2008).
5 British Library, Loan 96 RLF 1/121/6.
6 London Metropolitan Archives MJ/SD/038.

5 Playfair Expresses His Early Political Views

1 The name of the group is also reported as "Chamber of Commerce." The meeting is reported in several London and provincial newspapers, such as *World and Fashionable Advertiser*, 12 February 1787; *Derby Mercury*, 8–15 February 1787; *Saunders's News-Letter*, 19 February 1787; and *Felix Farley's Bristol Journal*, 17 February 1787.
2 Mention of Playfair seconding the motion, along with a Mr. Bell, is in *World and Fashionable Advertiser*, 12 February 1787. Only Mr. Bell is given as the seconder in, for example, *Whitehall Evening Post*, 10–13 February 1787.
3 Patrick Colquhoun to William Playfair. Copy letter book of Patrick Colquhoun, Murray 551, 109r–110r, University of Glasgow Archives and Special Collections.
4 Patrick Colquhoun to William Playfair. Copy letter book of Patrick Colquhoun, Murray 551, 109r–110r, University of Glasgow Archives and Special Collections.
5 The attribution is made, for example, in the electronic databases WorldCat and ECCO (Eighteenth Century Collections Online).
6 The first advertisement that I can find for the complete version of the *Commercial and Political Atlas* is in *Morning Post and Daily Advertiser*, 14 July 1786.

6 Playfair Makes His Mark on Statistical Graphics

1 Sir Adam Ferguson, 3rd Baronet of Kilkerran (1733–1813) is not to be confused with Adam Ferguson (1723–1816), the philosopher and historian who was made joint professor of mathematics with John Playfair at the University of Edinburgh in 1785.
2 University of Glasgow, Murray 551, 52v–53r and 109r–110r.
3 Advertisements for the first instalment of *The Commercial and Political Atlas* appear in *General Evening Post*, 10–13 and 17–20 September 1785; and *Public Advertiser*, 14 September 1785. The advertisement contains the table of contents for the instalment.

4 Library of Birmingham MS 3728/12/30.
5 Library of Birmingham MS 3219/4/94.
6 Library of Birmingham MS 3219/4/23.
7 Patrick Colquhoun to George Delmpster. Copy letter book of Patrick Colquhoun, Murray 551, 52v–53r, University of Glasgow Archives and Special Collections.
8 The Irish House of Commons operated separately from the Parliament at Westminster until the Act of Union of 1800, which created the United Kingdom of Great Britain and Ireland, and merged the parliaments in Dublin and Westminster.
9 Information on James Corry, John Foster, and William Burgh may be found Geoghegan (2008) and Malcomson (2009).
10 Patrick Colquhoun to William Playfair. Copy letter book of Patrick Colquhoun, Murray 551, 109r–110r, University of Glasgow Archives and Special Collections.
11 There could be some minor quibbles with this statement. For example, according to Macpherson, imports from Flanders and Scottish exports to Flanders between the Christmases of 1780 and 1781 amount to £45,804 and £56,452 respectively, while Playfair's graphs show slightly less than £45,000 for imports and slightly less than £55,000 for exports. It is a quibble since one has to examine the graph carefully to notice the difference.
12 The author of the review is identified in Nangee (1934, 1 and 175); the review itself may be found in the *Monthly Review* (1789, 414–17).

7 Playfair Goes to Paris

1 Playfair, *Memoirs*. Memoir 1.
2 Playfair, *Memoirs*. Memoir 1.
3 A list of the holdings is held in the Wolfson Reading Room of the Special Collections Department of the University of Aberdeen. https://www.abdn.ac.uk/collections/documents/SLA_PDF/ca001.pdf.
4 Archives de Paris DQ/10/1443/3128.
5 Archives nationales de France MC/ET/LXV/502, 504 and 506.
6 Archives nationales de France T/1610; Pierre Robin (1929, 347).
7 The location is mentioned in the articles of association. Archives nationales de France MC/ET/XXXVIII/702.
8 The use of furnaces in rolling mills in the late eighteenth century is mentioned in Bekasova (1958).
9 Archives nationales de France MC/ET/XXXVIII/702.
10 Archives nationales de France O/1/1498.
11 Playfair (1793c, 43–4).
12 Archives nationales de France O/1/1293 and O/1/1498.

13 Playfair (1798, Vol. 1, 92–3).
14 Playfair (1798, Vol. 2, 472).
15 Playfair, *Memoirs*, Memoir 2.
16 The information is taken from four sources: (1) James Swan to Thomas Jefferson, 5 October 1788, Founders Online, National Archives, https://founders.archives.gov/documents/Jefferson/01-13-02-0537; (2) Richard Claiborne to Thomas Jefferson, 16 December 1788, Founders Online, National Archives, https://founders.archives.gov/documents/Jefferson/01-14-02-0141; (3) Samuel Blackden to George Washington, 28 May 1789, Founders Online, National Archives, https://founders.archives.gov/documents/Washington/05-02-02-0289; (4) Louis-Philippe Gallot de Lormerie to George Washington, 28 May 1789, Founders Online, National Archives, https://founders.archives.gov/documents/Washington/05-02-02-0291.
17 William Playfair to Alexander Hamilton, 30 March 1791, Founders Online, National Archives, https://founders.archives.gov/documents/Hamilton/01-08-02-0165.
18 The events of the French Revolution are described in detail in Schama (1989).
19 William Playfair to Thomas Jefferson 13 March 1789, Founders Online, National Archives, https://founders.archives.gov/documents/Jefferson/01-14-02-0404.

8 Playfair Tries to Take Advantage of the French Revolution

1 Taken from an advertisement, appearing 25 May 1792 in *Morning Herald*, for his *Letter to the People of England, on the Revolution in France.*
2 Barlow to Walker 21 December 1790 (Belote 1907b, 72–3).
3 Archives nationales de France. MC/ET/LXXI/90.
4 Archives nationales de France. MC/ET/LXXI/90.
5 The agreement is translated into English and reprinted in Belote (1907b, 48–54).
6 Playfair *Memoirs*, Memoir 4.
7 William Playfair to Thomas Jefferson, 13 March 1789. Founders Online, National Archives.
8 The names of the five individuals with their occupations are given in Bouchary (1940, 131). The names do not appear in the list of land purchasers that appear in Moreau-Zanelli (2000).
9 Archives nationales de France. MC/RE/CXVI/9.
10 Archives de la Préfecture de police de Paris. A/A 88 fols. 388 and 389.
11 National Library of Scotland. Adv.MS.33.5.25 "Journal of Architecture" by James Playfair.
12 Archives nationales de France. 158AP/12, dossier 4.

13 The claim that Caplazy was asking for a refund on his 600-acre purchase is pieced together from the 23 January 1790 entry in Farmain's notebook (Archives nationales de France. MC/RE/CXVI/9) and from Archives de la Préfecture de police de Paris. A/A 85 fols. 51–4.
14 "Joel Barlow William Playfair, July 22, 1790, Table of Accounts, in French." Thomas Jefferson Papers, Library of Congress. https://www.loc.gov/resource/mtj1.014_0145_0146/?sp=2&st=image.
15 William Playfair to Thomas Jefferson, 20 March 1791, Founders Online, National Archives, https://founders.archives.gov/documents/Jefferson/01-19-02-0154.
16 William Playfair to Thomas Jefferson, 30 March 1791, Founders Online, National Archives.
17 Archives nationales de France. 158AP/12, Dossier 4.
18 Playfair to Walker, 18 March 1791, VFM 2110, Ohio History Connection Archives. The letter comes with a typed transcription, which reads, "Vibert (?), Lalemond & the Nolaine." My interpretation of Playfair's sometimes indecipherable handwriting is that "Vibert" is correct while "Nolaine" should read "notaire."
19 Archives de la Préfecture de police de Paris. A/A 88 fols. 388 and 389.
20 Archives de la Préfecture de police de Paris. A/A 85 fol. 120.
21 Archives nationales de France F/7/4737.
22 The complete set of despatches for August 1792 is in Browning (1885, 206–15). Lord Gower was George Leveson-Gower at the time Earl Gower. Lord Grenville was William Wyndham Grenville, 1st Baron Grenville.
23 In his *Memoirs*, Playfair says only that the Jacobin was the "natural son of the Count de Lorigais." Ian Spence, in his transcription of the *Memoirs* thinks, most likely correctly, that this is Lauraguais; Playfair sometimes had idiosyncratic spellings. The son may be identified through Jones (2013, 30) and Douglas (1898, 57–9). Berkowitz (2018, 161) identifies the son as Antoine-Constant de Brancas, another son of Arnould and Lauraguais. He bases his identification on a list of Jacobin Club members in Aulard (1889, xxxix). However, Liévyns, Verdot, and Bégat (1844, 229–30) put Brancas campaigning with l'armée du Nord during 1792. Goncourt and Goncourt (1885, 149 and 194–5) say that Sophie Arnould registered both her sons in the Jacobin Club. Consequently, I think that Auguste-Camille is the better candidate. In his *Memoirs*, Playfair also says that Louis Léon Félicité de Brancas, Duc de Lauraguais, and Louis-Paul de Brancas, Duc de Céreste, were brothers. This incorrect; they were distant cousins (Gallian 2016).

9 Playfair Escapes from France and Returns to England

1 The Dieppe-Brighton route was a common one. See, for example, Young (1792, 100) and the discussion in Burney (2001, 897).

2 The Duke of York's military campaigns are described in detail in John Watkins (1827, 208–6).
3 An advertisement for the pamphlet appears in *Morning Herald*, 12 April 1793.
4 The earliest advertisement I have found is in the newspaper *Sun*, 17 May 1793.
5 National Archives HO 42/29/8 fols. 474–82.
6 British Library. Loan 96 RLF 1/121/6.
7 Playfair (1794). The title page gives the date 27 January 1794, and the first advertisement that I can find in the newspapers is in the *Sun*, 22 April 1794.
8 Playfair and the story of his telegraph can be pieced together from several sources: "Mechanical Inventions: The Telegraph," *English Review of Literature, Science, Discoveries, Inventions, and Practical Controversies and Contests* 27 (1796): 585–91; "Description of the Telegraph," *Britannic Magazine; or, Entertaining Repository of Heroic Adventures and Memorable Exploits* 4 4 (54) (1796): 360–5; "Telegraph," *Encyclopædia Britannica; or, a Dictionary of Arts, Sciences, and Miscellaneous Literature on a Plan Entirely New*, 18:334–9, 1797, Dublin: James Moore; Playfair (1814, 2:220); Farrow (1895, 1:692); Great Britain (1796, 317).
9 In the *Morning Post and Fashionable World*, 19 September 1794, there is an advertisement for a display of the French telegraph by a Mr. Astley, who claimed to have been well acquainted with it from his travels and connections with "some of the first Philosophers of the age." This may have been advertising hyperbole.
10 Royal Archives GEO/MAIN/17355-17480. My thanks to the archivist, Julie Crocker.
11 *St. James's Chronicle or British Evening Post*, 21–3 May 1795. The advertisement states that the book will be available at the end of May. An advertisement for the sale of the book is in *Morning Post*, 22 June 1795.
12 National Archives PRO 30/8/167.

10 Playfair Becomes an Avid Anti-Jacobin Propagandist

1 The earliest advertisement that I have seen for the proposal is in *Star and Evening Advertiser*, 11 April 1795.
2 Notice for the availability of the book as of May 30 appears in *St. James's Chronicle*, 21 May 1795.
3 University of Michigan, William L. Clements Library. Viscounts Melville Papers, Box 10.
4 National Archives HO 69/27/46.
5 British Library Add Mss 37875.

6 National Archives HO 42/29/8 fols. 474–82.
7 British Library Add Mss 37875.
8 British Library Add Mss 37876. Playfair to Windham, 26 March 1796.
9 British Library Add Mss 37876.
10 The year 1789 does not appear in the *Tomahawk*. I have inferred that from Miles's correspondence written from Frankfurt between July and December 1789. See Miles (1890, 121–43).
11 British Library Add Mss 37876.
12 Playfair, *Memoirs*, Memoir 8.
13 Anti-war motions presented in Parliament are listed in Newmarch (1855, 260–2).
14 British Library Add MS 37876.

11 Playfair Gets Involved with Forged Assignats

1 Temple University Library. H.A. Cochran Collection.
2 The Swinburne manuscript is transcribed in *Archeaologia Aeliana* and a photographic reproduction of it appears in *The Quarterly: The Journal of the British Association of Paper Historians*.
3 Northumberland Archives ZSW 590.
4 Historical Manuscripts Commission (1927, 155–7).
5 Reports of the lawsuit with Playfair as a witness appear in several London newspapers: *Morning Post and Gazetteer*, 1 March 1798; *Oracle and Public Advertiser*, 1 March 1798; *Sun*, 1 March 1798; *True Briton*, 1 March 1798; and *Star*, 10 March 1798.
6 Playfair to Sinclair, London Metropolitan Archives Add MS 403. Biographical information on Sinclair and Irwin may be found in Sinclair (1798) and Howard (1985).
7 British Library. Add MS 37875. fols. 227 and 232.
8 Temple University. H.A. Cochran Collection.
9 In late 1794, Playfair sent a different proposal to Pitt, which seems to have been ignored. This letter is held in the National Archives among Pitt's papers, but Playfair's 1793 assignat proposal is not there. The Clements Library of the University of Michigan holds Dundas's papers, but a copy of the plan is not there, although another Playfair plan from approximately the same time, which was a scam, is among Dundas's papers. The scam was originally sent to the MP Sir John Sinclair and wound up on Dundas's desk. The assignat plan is also not among the Home Office papers in the National Archives, although a 1794 Playfair proposal received by Dundas is there. This proposal was for a propaganda campaign against French republicans. National Archives HO 42/29/474–84.
10 British Library Add MS 37875. Windham Papers Vol. XXXV.

12 Playfair Starts a Bank and Goes Bankrupt

1 The notice for the new bank appeared in several newspapers, such as the *Times*, 2 December 1797.
2 *Statement &c by W Playfair*, in London Metropolitan Archives Add MS 403.
3 British Library Add Mss 37876.
4 Playfair to Mackintosh, London Metropolitan Archives Add MS 403.
5 The acts may be found in *Statutes of the Realm*: 15 George III c 51 (1775); 17 George III c 30 (1777); 31 George III c 25 (1791); George III c 32 (1797); and 37 George III c 61 (1797).
6 *Statement &c by W Playfair*, in London Metropolitan Archives Add MS 403.
7 The advertisement appeared in several newspapers, for example *Oracle and Public Advertiser*, 25 March 1797.
8 Playfair to Hutchinson, 23 August 1797, London Metropolitan Archives Add MS 403.
9 *Statement &c by W Playfair*, in London Metropolitan Archives Add MS 403.
10 The £14,000 is reported in *London Courier and Evening Gazette*, 23 November 1804, and the £8,000 is reported in the *Morning Chronicle*, 22 November 1804.
11 The notice appears in several London newspapers. See, for example, *Morning Post and Gazetteer*, 28 October 1797.
12 National Archives HO 26/7.
13 Playfair to Wadeson, 18 September 1805, London Metropolitan Archives Add MS 403.
14 Playfair to Hutchinson, 23 August 1797. London Metropolitan Archives Add MS 403.
15 Playfair to Vertaul, n.d., London Metropolitan Archives Add MS 403. My translation.
16 Playfair to Hutchinson, 23 August 1797; Playfair to Hutchinson, n.d., London Metropolitan Archives Add MS 403.
17 Somerset Archives DD/SAS/C795/TN/155/4.
18 Some items were sold from an address described as "late Union Bank."
19 Playfair to Wadeson, 3 February 1798, London Metropolitan Archives Add MS 403. Punctuation added.
20 *Statement &c by W Playfair*, London Metropolitan Archives Add MS 403. Punctuation added.
21 Playfair to Wadeson, 29 June 1798, London Metropolitan Archives Add MS 403.
22 Playfair to Wadeson Hardy & Barlow, 8 May 1798, London Metropolitan Archives Add MS 403.
23 National Archives HO 26/6.
24 Playfair to Hutchins, 17 June 1798, London Metropolitan Archives Add MS 403.

25 Playfair to Wadeson, 18 September 1805, London Metropolitan Archives Add MS 403.
26 *Statement &c by W Playfair*, London Metropolitan Archives Add MS 403.
27 The quotations come from *Statement &c by W Playfair*.
28 National Archives TS 25/1/18, fol. 27.
29 National Archives PROB 11/1525/134; *London Gazette*, 1 March 1825.
30 National Archives PROB 11/1854/29.
31 London Metropolitan Archives MJ/SD/038. Punctuation added.

13 Playfair Ekes Out a Living as a Bankrupt

1 The sale may be found at https://www.christies.com/lotfinder/Lot/playfair-william-lineal-arithmetic-applied-to-shew-5388577-details.aspx. The book sold for $7,500 US.
2 The third edition of the *Commercial and Political Atlas* and the *Statistical Breviary* have been reprinted in 2005. See Playfair (2005).
3 The most damning reviews of *Strictures on the Asiatic Establishments of Great Britain* are found in the *Critical Review* (1801, 438–42); and the *Monthly Review* (1800, 403–5). The accuracy of the data in the *Statistical Breviary* is questioned in the *Critical Review* (1802, 76–8).
4 National Records of Scotland GD51/3/646/1.
5 National Records of Scotland GD51/3/646/2.
6 National Archives HO 25/6. Criminal Registers, Newgate, 1797–8.
7 British Library Add MS 37868.
8 Cleveland made his comment with respect to another graph in the *Statistical Breviary* dealing with sizes of several cities rather than nations.
9 British Library Loan 96 RLF 1/121/1–3.
10 London Metropolitan Archives. CLA/047/LJ/084/030.
11 John and Andrew William are listed in the 1851 Census of Canada West, 97, sub-district 173, Lanark County, Canada West, RG31 Census of Canada . Zenobia is listed in the 1861 Census of England and Wales, 26, fol. 118, RG9 Middlesex Piece 95.
12 Elizabeth appears as Mrs. E. Bingley in the 1851 Census of Canada West, 93, sub-district 173, Lanark County, Canada West, RG31 Census of Canada. Louisa is mentioned in Rogers (1887, 55).
13 http://www.britishmuseum.org/research/collection_online/collection_object_details. aspx?objectId=742057& partId=1&searchText=thomas+rowlandson&page=5.
14 National Archives PRIS 10/52.
15 National Archives TS 11/457/1520. *Rex v Robert Kennett*; Henry Kennett; Brackley Kennett and William Playfair for a conspiracy in relation to bankruptcy of Robert Kennett 1805: Chancery and King's Bench.

14 Playfair Has a Good Year during 1805 with Hints of Ending Badly

1 The first advertisement that I can find for the book is in *Morning Chronicle*, 21 October 1805. The dedicatory letter in the book to Thomas Jefferson is dated 17 April 1805.
2 Temple University Library, H.A. Cochran Collection. Letter and proposal from William Playfair to Viscount Mellville, 9 October 1804.
3 William Playfair to Thomas Jefferson, 17 April 1805, Founders Online, National Archives, https://founders.archives.gov/documents/Jefferson/99-01-02-1534.
4 Smith's treatment of monopolies is described succinctly in Salvadori and Signorino (2014).
5 In this supplementary chapter, Playfair mentions "those persons who serve for only three years, or a short term as in Scotland." He went to Bolton and Watt at the age of eighteen immediately after he left Meikle.
6 Playfair's position is discussed more fully in Rothschild (2001, 91, 96–7).
7 The first advertisement for the availability of *Decline and Fall of Powerful and Wealthy Nations* is in the *Morning Chronicle*, 21 October 1805.
8 This is discussed in detail in Grossman (1948).
9 See, for example, Eisenstein (1979, 21).
10 National Archives HO 42/81 fol. 195–6.

15 Playfair Has Serious Legal and Other Problems

1 National Archives TS 11/457/1520. All the information and quotations about the case are taken from this source.
2 National Archives TS 11/457/1520.
3 National Archives TS 11/457/1520.
4 Gout has been known and diagnosed since ancient times. See, for example, Nuki and Simkin (2006).
5 British Library. Loan 96 RLF 1/121/4. Playfair to Baker-Holroyd, Baron Sheffield.
6 Windham's patronage is mentioned by [Byerley] (1823, 172).
7 Census of Great Britain 1841, Parish of Martham, Norfolk, National Archives HO107/766/6.
8 I am indebted to Michael Umpherson for information he obtained on Elizabeth Playfair at this school.
9 London Metropolitan Archives Add MS 403.
10 National Archives TS 11/457/1520.
11 National Archives TS 11/457/1520.
12 National Archives TS 11/106/318/11. Playfair to Perceval, 13 May 1809.

16 Playfair Dabbles Deeply in Family History and Political Biography

1 Royal Archives GEO/MAIN 28532–28533.
2 National Archives B/3/4197.
3 Royal Archives GEO/MAIN 28528.
4 Royal Archives GEO/MAIN 28528 and Sharpe (1888, 373).
5 A copy of the form letter is in British Library Add MS 13806.
6 He makes this point in a letter to Charles Kirkpatrick Sharpe. Sharpe (1888, 386).
7 Public Record Office of Northern Ireland. D607/I/127.
8 British Library Add MS 13806.
9 Public Record Office of Northern Ireland. D1503/11/143.
10 British Library Add MS 13806.
11 The immigration information was obtained from Playfair family records courtesy of Michael Umpherson.
12 The whole affair can be pieced together from manuscripts in the Royal Archives. GEO/MAIN/28527–8, 28532–4, 28537.
13 Letterbook, vol. 2, John W. Croker Papers, Clements Library, University of Michigan.
14 British Library Add MS 89143/1/2/26: 1816-1827.

17 Playfair Tries to Pull a Major Scam on Lord Bathurst about Bonaparte

1 National Archives PRIS 10/52, PRIS 10/ 53 and PRIS 10/156.
2 National Archives B 3/4197.
3 British Library, Loan. MS. 57/8.
4 See, for example, *London Chronicle*, 5 April 1815; *Leeds Mercury*, 8 April 1815; *Salisbury and Winchester Journal*, 10 April 1815; and *Bury and Norwich Post*, 12 April 1815.
5 For example, *Caledonian Mercury* 8 April 1815; *Bell's Weekly Messenger*, 9 April 1815.
6 *Morning Chronicle*, 13 April 1815.

18 Playfair Continues Writing and Tries a Few More Scams to Get to Paris

1 National Archives HO 42/145/145.
2 National Archives HO 42/146/354–355 and 443.
3 National Archives HO 42/146/443.
4 National Library of Scotland, MS 19828 fols. 46 and 49.
5 The entire exchange of letters is in National Library of Scotland, MS 19829 fols. 3, 14–16, 21–5, 29–32, 36–41, 46, 48.
6 National Library of Scotland, MS 19829 fol. 14.

7 The original documents pertaining to Playfair's scam are in NRAS of Scotland, NRAS859 box 44/bundle 10 and NRAS859 box 217/bundle 2. Transcriptions of some of the documents and a description of the affair are in Spence and Wainer (1997b).
8 NRAS859 box 217/bundle 2.
9 NRAS859 box 217/bundle 2.
10 National Archives HO 42/153/143.
11 National Archives HO 42/154/182.
12 National Archives HO 42/154/183.
13 National Archives HO 42/161/243.

19 Playfair Returns to Paris

1 Archives du Paris DQ10/1443/3128.
2 The amount of 3,500 livres is given in Archives nationales de France T1600; the amount of £3,500 is given in Archives du Paris DQ10/1443/3128.
3 National Archives HO 42/161/243.
4 The case may be pieced together from articles in *Galignani's Messenger*, 8, 16, and 29 July 1818; *Journal des debats: politques et littéraire*, 8 and 29 July 1818; and *Morning Post*, 3 August 1818 and 4 January 1819.
5 The case was also reported in Britain: *Morning Post*, 3 August 1818; *Morning Chronicle*, 5 August 1818; *Caledonian Mercury*, 29 August 1818; and *Morning Post*, 4 January 1819.

20 Playfair Spends His Last Few Years in England in Poverty

1 National Records of Scotland, SC70/1/21 658–64.
2 Playfair's letters to James Watt Jr. and Watt's replies are found in Library of Birmingham Archives MS 3219/6/3/3/1/1/2/1–8.
3 Watt's correspondence with Henry Colburn and John Rennie are found in Library of Birmingham Archives MS 3219/6/3/3/1/1/3/1–8.
4 Library of Birmingham Archives MS 3219/6/3/3/1/1/2/5.
5 Library of Birmingham Archives MS 3219/6/3/3/1/1/2/7.
6 British Library Loan 96 RLF 1/121/15, 17.
7 British Library Loan 96 RLF 1/121/11–18.
8 British Library Loan 96 RLF 1/121/19, 20.
9 Temple University (SPC) MSS 118 COCH.
10 A notice that the book would soon be published appears in the *Morning Post*, 9 March 1821. An advertisement for the sale of the first edition appears in the *Morning Post*, 10 July 1821. I could not find any advertisements for the second and third edtions.
11 Playfair's letters are in *Morning Advertiser*, 24 November, 5, 7, and 24 December 1810. The responses to the letters are in *Morning Advertiser*, 27 November, 6, 13, and 26 December 1810.

12 Bartlett (1966, 181–90).
13 British Library Add MS 89143/1/1/26.
14 British Library Add MS 89143/1/1/26.
15 British Library Add MS 89143/1/1/26.
16 British Library Loan 96 RLF 1/121/4.
17 National Archives TS 11/106/318/11. William Playfair to Spencer Perceval, 13 May 1809.
18 British Library Loan 96 RLF 1/121/15.
19 British Library Loan 96 RLF 1/121/21.
20 British Library Loan 96 RLF 1/121/21.
21 *Lancet* 7, no. 4 (1825): 113–15; and no. 5, 130–50.
22 The lack of cleanliness at St. George's, and other hospitals of the time, is described in Smith (1979, 262).
23 Communication with the author by e-mail, 17 September 2018.
24 City of Westminster Archives Centre, Westminster Burials, St George, Hanover Square, p. 171.
25 British Library Loan 96 RLF 1/121/22.

Afterword

1 The quote most likely comes from the American actor Lionel Jay Stander in an interview with *Esquire Magazine*. See Quotationize (n.d.).

Appendix

1 Manuscripts of J.B. Fortescue, Historical Manuscripts Commission (1927).
2 The individuals can be identified using the website Exeter Working Papers in Book History: Biographical and Bibliographical Information on the Book trades, https://bookhistory.blogspot.com/.
3 London Metropolitan Archives CLA/041/1Q/02/012/pt. 1 00-55 and pt. 2, 56–152.
4 See London Lives: 1690 to 1800, https://www.londonlives.org/browse.jsp?id=LMCLIC65012_n1480-4&div=LMCLIC65012IC650120731#highlight.
5 Sinclair (1798) and HO 42/45/174–93, 197–200.
6 The full description in French is in Puisaye (1804, 380–90).
7 British Library Add MS 8072. Puisaye Papers, vol. 101 on assignats.
8 British Library Add MS 8072. Dillaye (1877, 33), among others, misreads the marquis's name as Dumesnil.
9 British Library Add MS 38769. Huskisson Papers, vol. 36.
10 Library of Birmingham MS 3782/12/62/131.
11 British Library Add MS 8072.
12 Library of Birmingham MS 3782/12/59/81.

References

A.D. 1791. "On the History of Authors by Profession. II." *Bee* 3:13–15.

Académie française. n.d. "André Morellet." http://www.academie-francaise.fr/les-immortels/andre-morellet.

Adams, C.K. 1909. *A History of the United States*. Boston: Allyn and Bacon.

Alger, J.G. 1899. "British Visitors to Paris, 1802–1803." *English Historical Review* 14 (56): 739–41. https://doi.org/10.1093/ehr/XIV.LVI.739.

Anticipation, on Politics, Commerce, and Finance, during the Present Crisis. 1808. London: Glindon.

An Account of the Lodge of Nine Muses No. 235 1777–2012. 2012. London.

Analytical Review. 1793. "Thoughts on the Present State of French Politics, and the Necessity and Policy of Diminishing France, for Her Internal Peace, and to Secure the Tranquillity of Europe." Art. 34. 16:194–7.

– 1797. "A Letter to Sir William Pulteney, Bart. M.P. &c, on the Establishment of Another Public Bank in London." Art. 64. 26:188–9.

Antonetti, G. 2007. *Les ministres des Finances de la Révolution française au Second Empire (I): Dictionnaire biographique 1790–1814*. New edition. Vincennes: Institut de la gestion publique et du développement économique. https://www.economie.gouv.fr/igpde-editions-publications/ministres-des-finances-revolution-francaise-au-second-empire-0.

Ashton, T.S. 1955. *An Economic History of England: The 18th Century*. London: Methuen.

– 1969. *The Industrial Revolution: 1760–1830*. Oxford: Oxford University Press.

Aulard, F.A. 1889. *La Société des Jacobins: recueil de documents pour l'histoire du Club des Jacobins de Paris*. Tome 1. Paris: Jouaust, Noblet, and Quantin.

Baildon, W.P. 1896. *The Records of the Honourable Society of Lincoln's Inn: Admissions from A.D. 1420 to A.D. 1893, and Chapel Registers*. Vol. 1. London: Lincoln's Inn.

Bailey, W. 1781. *Bailey's Northern Directory, or, Merchant's and Tradesman's Useful Companion for the Year 1781*. Warrington, UK: William Ashton.

Bannerman, G., and C. Schonhardt-Bailey. 2016. *Battles over Free Trade*. Vol. 1, *The Advent of Free Trade, 1776–1846*. Abingdon, UK: Routledge.
Barber, G. 1961. "Galignani's and the Publication of English Books in France from 1800 to 1852." *Library* 5 (4): 267–86. https://doi.org/10.1093/library/s5-XVI.4.267.
Barnes, D.G. 1930. *A History of the English Corn Laws from 1660–1846*. London: George Routledge & Sons.
Bartlett, C.J. 1966. *Castlereagh*. London: Macmillan.
Barton, H.A. 1967. "The Origins of the Brunswick Manifesto." *French Historical Studies* 5 (2): 146–69. https://doi.org/10.2307/286173.
Bekasova, L.M. 1958. "Development of the Rolling Industry (until the 20th Century)." *Metallurgist* 2:201–5. https://doi.org/10.1007/BF00736177.
Belote, T.T. 1907a. *The Scioto Speculation and the French Settlement at Gallipolis*. Cinncinati: University Studies of Cincinnati Press. Series 2, vol. 3.
– 1907b. "Selections from the Gallipolis Papers." *Quarterly Publication of the Historical and Philosophical Society of Ohio* 2 (2): 37–92.
Bentham, J. 1788. *Defence of Usury: Shewing the Impolicy of the Present Legal Restraints on the Terms of Pecuniary Bargains*. Dublin: Williams, Colles, White, Byrne, Lewis, Jones, and Moore.
Berkowitz, B. 2018. *Playfair: The True Story of the British Secret Agent Who Changed How We See the World*. Fairfax, VA: George Mason University Press.
Berry, M. 1865. *Extracts of Journals and Correspondence of Miss Berry: From the Year 1783 to 1852*. Edited by Theresa Lewis. London: Longmans, Green.
Betham, W. 1805. *The Baronetage of England; Or, the History of the English Baronets, and Such Baronets of Scotland as Are of English Families*. London: E. Lloyd.
Bewick. T. 1887. *A Memoir of Thomas Bewick, Written by Himself*. Prefaced and annotated by Austin Dobson. London: Bernard Quaritch.
Bishop, A.C. 1938. "Lady Morgan and Her Circle." Master's thesis, McGill University.
Bishop, J.D. 1995. "Adam Smith's Invisible Hand Argument." *Journal of Business Ethics* 14 (3): 165–80. https://doi.org/10.1007/BF00881431.
Bötticher, J.G. 1789. *Statistische Uebersichts-Tabellen aller Europäischen Staaten nebst deren Münzen, Maaßen und Gewichten*. Königsberg: Hartung.
– 1800. *Statistical Tables Exhibiting a View of All the States of Europe: Showing, with the Greatest Accuracy, Their Population, Military and Marine Strength, Revenue and Expenditures, Form of Government*. Translated by William Playfair. London: Stockdale.
Bonnycastle, J. 1788. *The Scholar's Guide to Arithmetic; Or a Complete Exercise-Book for the Use of Schools*. London: J. Johnson.
Bonvarlet, A. 1868. "Notice de la commune de Pitgam au Ouest-Quartier de Flandre." *Annales du Comité flamand de France, 1867* 9:182–230.

Bötticher, J.G.I. 1789. *Statistische uebersichts-tabellen aller Europäischen staaten, nebst deren münzen, maassen, und gewichten*. Königsberg.

Bouchary, J. 1940. "Les compagnies financières à Paris à la fin du XVIIIe siècle: Compagnie du Scioto." *Annales historiques de la Révolution française* 99:129–51.

Bouchary, J. 1941. *Les compagnies financières à Paris à la fin du 18e siècle. Tome 2. La Caisse patriotique, la Maison de secours, la Caisse des billets de parchemin et les Caisses de confiance privées à Paris (1790–1793)*. Paris: Rivière.

Bowden, W. 1924. "The Influence of the Manufacturers on Some of the Early Policies of William Pitt." *American Historical Review* 29:655–74. https://doi.org/10.2307/1841230.

Bower, P. 2018. "Sir John Swinburne's 1793 Memorandum on the English Forgery of French Assignats at Haughton Castle Paper Mill, Northumberland in the 1790's." *Quarterly: The Journal of the British Association of Paper Historians* 106 (April): 1–19.

Boyer, A. 1797. *Boyer's Royal Dictionary Abridged*. London: Bathurst.

Brandstetter, T. 2005. "'The Most Wonderful Piece of Machinery the World Can Boast Of': The Water-Works at Marly, 1680–1830." *History and Technology: An International Journal* 21 (2): 205–20. https://doi.org/10.1080/07341510500103750.

Bret, P. 2017. "Du laboratoire de l'Académie de Dijon à celui de l'École polytechnique: Trente-six ans d'enseignement de la chimie." *Bulletin de la SABIX: Société des amis de la bibliothèque et de l'histoire de l'École polytechnique* 60:9–36. https://doi.org/10.4000/sabix.1841.

Brezis, E.S., and F.M. Crouzet. n.d. "The Role of the Assignats during the French Revolution: Evil or Rescuer?" https://econ.biu.ac.il/sites/econ/files/shared/staff/u46/french_revolution.pdf.

Brinton, W.C. 1914. *Graphic Methods for Presenting Facts*. New York: Engineering Magazine.

British Critic. 1793a. "General View of the Actual Force and Resources of France in January 1793." Art. 34. 1:107.

– 1793b. "Thoughts on the Present State of French Politics, and the Necessity and Policy of Diminishing France for Her Internal Peace, and to Secure the Tranquillity of Europe." 1:215.

Browning, O. 1885. *The Despatches of Earl Gower, English Ambassador at Paris from June 1790 to August 1792, to Which Are Added the Despatches of Mr Lindsay and Mr Monro, and the Diary of Viscount Palmerston in France during July and August 1791, Now Published for the First Time*. Cambridge: Cambridge University Press.

Burney, F. 2001. *The Wanderer; or Female Difficulties*. Edited by M.A. Doody, R.L. Mack, and P. Sabor. Oxford: Oxford University Press.

Burrows, S. 2000. *French Exile Journalism and European Politics, 1792–1814.* Woodbridge, UK: Boydell.
[Byerley, T.] 1823. "Biography, Mr. Wiliam Playfair." *Literary Chronicle and Weekly Review* 200 (15 March): 171–2.
Byerley, T. 1817. "An Account of the French Settlements on the Sioto Begun in the Year 1789." *Colonial Journal* 6:375–80.
Cameron. 1794. *Langue diplomatique. Nouvelle manière d'écrire sans que personne puisse lire votre écriture, que ceux à qui vous écrivez.* Liège: J.A. Latour.
Campbell, R.H., and A.S. Skinner. 1982. *Adam Smith.* London: Croom Helm.
Cassis, Y., and P.L. Cottrell. 2015. *Private Banking in Europe: Rise, Retreat and Resurgence.* Oxford: Oxford University Press.
Censer, J.R. 1994. *The French Press in the Age of Enlightenment.* London: Routledge.
Chaffers, W. 1883. *Gilda Aurifabrorum: A History of English Goldsmiths and Plateworkers, and Their Marks Stamped on Plate, Copied in Facsimile from Celebrated Examples.* London: W.H. Allen.
Chartrand, R. 1999. *Émigré and Foreign Troops in British Service (1): 1793–1802.* Oxford: Osprey Publishing.
Church of Scotland. 1783. *The Principal Acts of the General Assembly of the Church of Scotland. Convened at Edinburgh 22d May 1783.* Edinburgh: James Dickson.
Clemit, P., and J. McAuley. 2020. "Phillips, Sir Richard (1767–1840)." *Oxford Dictionary of National Biography.* Oxford: Oxford University Press. https://doi-org.proxy1.lib.uwo.ca/10.1093/ref:odnb/2216.
Cleveland, W.S. 1993. *Visualizing Data.* Murray Hill, NJ: AT&T Bell Laboratories.
Collins, A. 1806. *The Baronetage of England: Containing a New Genealogical History of the Existing English Baronets, and Baronets of Great Britain, and of the United Kingdom, from the Institution of the order in 1611 to the Last Creation.* London: John Stockdale.
Compagnie du Scioto. 1789. *Avis. La Compagnie du Scioto, établie à Paris pour l'exploitation & la vente de trois millions d'acres anglois de terres, situés dans l'Amérique Septentrionale.* Paris: Tardieu.
– 1790. *Nouveau prospectus de la Compagnie du Scioto avec plusieurs extraits de lettres, écrites du Scioto même en date du 12 Octobre 1790.* Paris: Clousier.
Cone, C.B. 1951. "Richard Price and Pitt's Sinking Fund of 1786." *Economic History Review* n.s. 4:243–51. https://doi.org/10.1111/j.1468-0289.1951.tb00613.x.
Cope, S.R. 1946. "The Original Security Bank." *Economica* n.s. 13:50–5. https://doi.org/10.2307/2549442.
Costigan-Eaves, P., and M. Macdonald Ross. 1990. "William Playfair (1759–1823)." *Statistical Science* 5 (3): 318–26.

Courcelles, J.B.P.J. 1826. *Histoire généalogique et héraldique des pairs de France, des grands dignitaires de la couronne, des principales familles nobles du royaume et des maisons princières de l'Europe, précédée de la généalogie de la maison de France.* Tome 7. Paris: L'auteur.

Critical Review. 1788. "Joseph and Benjamin. A Conversation." 65:395.

– 1794. "Peace with the Jacobins Impossible." Ser. 2, 11:466.

– 1795. "Letter to the Right Hon. the Earl Fitzwilliam Occasioned by His Two Letters to the Earl of Carlisle." Ser. 2, 14:446.

– 1801. "Strictures on the Asiatic Establishments of Great Britain; With a View to an Inquiry into the True Interest of the East-India Company." Ser. 2, 31:438–42.

– 1802. "The Statistical Breviary, Showing, on a Principle Entirely New, the Resources of Every State and Kingdom in Europe." Art. 8. Ser. 2, 35:76–8.

– 1806. "Playfair's Inquiry into the Permanent Causes into the Decline and Fall of Powerful and Wealthy Nations, &c. (Conclude from p. 12)." Art. 7. Ser. 3, 8:153–70.

– 1814. "Political Portraits in This New Æra, with Explanatory Notes. Historical and Biographical." Art. 9. Ser. 4, 5:425–32.

[Cutler, M.]. 1787. *An Explanation of the Map Which Delineates That Part of the Federal Lands, Comprehended between Pennsylvania West Line, the Rivers Ohio and Scioto, and Lake Erie; Confirmed to the United States by Sundry Tribes of Indians in the Treaties of 1784 and 1786 and Now Ready for Settlement*. Salem, MA: Dabney and Cushing.

Davis, M.T. 2009. "Walker, Thomas (1749–1817)." *Oxford Dictionary of National Biography*. Oxford: Oxford University Press. https://doi.org/10.1093/ref:odnb/63603.

Dawson, P.J. 1970. "Outline of the Story of the Lodge of the Nine Muses No. 235, during the First Fifty Years of Its Existence [1777–1827]." Typescript LF SN 937 Lodge File, Lodge of the Nine Muses. London.

Day, S.H. 1911. *Family Papers*. London: Printed by the author.

Dempster, G., and J. Ferguson. 1934. *Letters of George Dempster to Sir Adam Fergusson, 1756–1813: With Some Account of His Life*. London: Macmillan.

Desan, S., L. Hunt, and W.M. Nelson. 2013. *The French Revolution in Global Perspective*. Ithaca, NY: Cornell University Press.

Dickinson, H.W. 1937. *Matthew Boulton*. Cambridge: Cambridge University Press.

Dickinson, H.W., and R. Jenkins. 1981. *James Watt and the Steam Engine: The Memorial Volume Prepared for the Committee of the Watt Centenary Commemoration at Birmingham 1919*. Ashbourne, UK: Moorland Publishing.

Dillaye, S.D. 1877. *The Money and the Finances of the French Revolution of 1789: Assignats and Mandats*. Philadelphia: H.C. Baird.

Donnadieu, A. 1851. *Catalogue of Highly Interesting and Valuable Autograph Letters and Historical Manuscripts Being the Well-Known Collection of Monsr. Donnadieu*. London: Puttick and Simpson.
Donnant, D.F., and W. Playfair. 1805. *Statistical Account of the United States of America*. London: Greenland and Norris.
Donovan, J. 2009. *Sydney Owenson, Lady Morgan and the Politics of Style*. Bethesda, MD: Academica.
Doubleday, T. 1856. *The Political Life of Sir Robert Peel, Bart: An Analytical Biography*. Vol. 1. London: Smith, Elder.
Douglas, R.B. 1898. *Sophie Arnould: Actress and Wit*. Paris: Charles Carrington.
Doyle, W. 1967. "The Parlementaires of Bordeaux at the End of the Eighteenth Century 1775–1790." PhD diss., University of Oxford.
Dunkin, J. 1844. *The History and Antiquities of Dartford with Topographical Notices*. London: John Russell Smith.
Dunyach, J.-F. 2010. "Les réseaux d'un excentrique: vies et parcours de William Playfair (1759–1823)." In *Cultural Transfers: France and Britain in the Long Eighteenth Century*, edited by A. Thomson, S. Burrows, and E. Dziembowski, 115–27. Oxford: Voltaire Foundation.
– 2014. "William Playfair (1759–1823), Scottish Enlightenment from Below?" In *Jacobitism, Enlightenment and Empire, 1680–1820*, edited by D.J. Hamilton and A.I. Macinnes, 159–72. London: Pickering & Chatto.
– 2016. "Le *British Family Antiquity* de William Playfair (1811): une entreprise généalogique." In *L'entreprise généalogique en Europe (XVe–XXe* siècle), edited by S. Jettot and M. Lezowski, 319–40. Brussels: Peter Lang.
Dykes, D.W. 1999. "John Gregory Hancock and the Westwood Brothers: An Eighteenth-Century Token Consortium." *British Numismatic Journal and Proceedings of the British Numismatic Society* 69:173–86. https://www.britnumsoc.org/publications/Digital%20BNJ/pdfs/1999_BNJ_69_8.pdf.
Dziembowski, E. 2011. "Lord Shelburne's Constitutional Views in 1782–3." In *An Enlightenment Statesman in Whig Britain: Lord Shelburne in Context, 1737–1805*, edited by N. Aston and C.C. Orr, 215–32. Woodbridge, UK: Boydell.
Ëataux, L. 1885. *Généalogie de la maison de Brachet de Floressac*. Limoges: Dugourtieux.
Eclectic Review. 1821. "A Chronological Chart of the Cotemporary Sovereigns of Europe from A.D. 1060 to 1820." n.s. 15:555–7.
Edgeworth, F.Y. 1899. "Playfair, William (1759–1823)." In *Dictionary of Political Economy*, edited by R.H.I. Palgrave, 3:116–17. London: Macmillan.
Eisenstein, E.L. 1979. *The Printing Press as an Agent of Change: Communications and Cultural Transformations in Early-Modern Europe*. Cambridge: Cambridge University Press.
Electronic Enlightenment Project. 2008. "Jeremy Bentham to George Wilson Tuesday, 15 May 1787." Bodleian Libraries: Oxford University Press. https://www.e-enlightenment.com/.

Encyclopædia Britannica; or, a Dictionary of Arts, Sciences, and Miscellaneous Literature on a Plan Entirely New. 1797. Dublin: James Moore.
English Review. 1787. "Joseph and Benjamin: A Conversation." Translated from a French manuscript. Art. 4. 10:24–7.
– 1796. "Mr. Playfair's History of Jacobinism." Art. 9. 28:246–54.
Ermene, J.J. 1956. "The Machine of Marly." *French Review* 29:242–4.
Espinasse, I. 1801. *Reports of Cases Argued and Ruled at Nisi Prius, in the Courts of King's Bench and Common Pleas from Easter Term 33 George III. 1793, to Hilary Term 36 George III. 1796*. Vol. 1. London: J. Butterworth.
– 1804. *Reports of Cases Argued and Ruled at Nisi Prius, in the Courts of King's Bench and Common Pleas from Easter Term 41 George III. 1801, to Hilary Term 43 George III. 1803*. Vol. 4. London: J. Butterworth.
Farrow, E.S. 1895. *Farrow's Military Encyclopedia: A Dictionary of Military Knowledge*. New York: Military Naval Publishing.
Feltham, J. 1802. *The Picture of London, for 1802; Being a Correct Guide to All the Curiosities, Amusements, Exhibitions, Public Establishments, and Remarkable Objects, in and near London; With a Collection of Appropriate Tables*. London: Lewis.
Fierro, A. 1996. *Histoire et dictionnaire de Paris*. Paris: R. Laffont.
Flinn, M.W. 1966. *The Origins of the Industrial Revolution*. London: Longmans.
Foster, J. 1888. *Alumni Oxonienses: The Members of the University of Oxford, 1715–1886*. Oxford: Parker.
France. 1819. *The Penal Code of France: Trans. into English with a Prel. Dissertation and Notes*. Vol. 2. London: Butterworth.
Friendly, M. 2002. "Visions and Re-Visions of Charles Joseph Minard." *Journal of Educational and Behavioral Statistics* 27:31–51. https://doi.org/10.3102/10769986027001031.
– 2008. "The Golden Age of Statistical Graphs." *Statistical Science* 23 (4): 502–35. https://doi.org/10.1214/08-STS268.
Friendly, M., and H. Wainer. 2021. *A History of Data Visualization and Graphic Communication*. Cambridge, MA: Harvard University Press.
Fry, M. 2009. "Dundas, Henry, First Viscount Melville (1742–1811)." *Oxford Dictionary of National Biography*. Oxford: Oxford University Press. https://doi.org/10.1093/ref:odnb/8250.
Funkhouser, H.G. 1937. "Historical Development of the Graphical Representation of Statistical Data." *Osiris* 3:267–404. https://doi.org/10.1086/368480.
Funkhouser, H.G., and H.M. Walker. 1935. "Playfair and His Charts." Supplement, *Economic Journal* 45 (S1):S103–9. https://doi.org/10.1093/ej/45.Supplement_1.103.
Galarneau, C. 1987. "Calonne, Jacques-Ladislas-Joseph de." *Dictionary of Canadian Biography*. Vol. 6. Toronto and Quebec: University of Toronto and Université Laval.

Galignani, A., and W. Galignani. 1838. *Galignani's New Paris Guide*. 13th ed. Paris: A. & W. Galignani.
Gallian, J. 2016. "Généalogie de la famille BRANCAS." Coleméo. https://en.calameo.com/books/001513011077dc2135901.
Gamble, J. [1795?]. *Observations on Telegraphic Experiments. Or the Different Modes Which Have Been or May Be Adopted for the Purpose of Distant Communication.*
Garside, P. 2006. "Colburn, Henry (1784/5–1855)." *Oxford Dictionary of National Biography*. https://doi-org.proxy1.lib.uwo.ca/10.1093/ref:odnb/5836.
Gash, N. 1978. "After Waterloo: British Society and the Legacy of the Napoleonic Wars." *Transactions of the Royal Historical Society* 28:145–57. https://doi.org/10.2307/3679205.
Gauer, M. 2011. *Histoire et généalogie de la famille Julien de Vinezac et de ses alliances*. Cahiers ardéchois. https://en.calameo.com/books/00035755078a073bf214f.
Gentleman's Magazine. 1817. "The Character of the Count of St. Morys." 87 (2): 116–19.
– 1822a. "A Letter on Our Agricultural Distresses, Their Causes and Remedies Expounded with Tables and Copperplate Charts, Shewing and Comparing the Prices of Wheat, Bread and Labour from 1565 to 1821." 92 (1): 59.
– 1822b. "Mr. John Debrett." 92 (2): 474.
– 1866. "G. Rennie, Esq., C.E., F.R.S." Ser. 4. 1:747–51.
Geoghegan, P.M. 2008. "Howard, Frederick, Fifth Earl of Carlisle." *Oxford Dictionary of National Biography*. https://doi-org.proxy1.lib.uwo.ca/10.1093/ref:odnb/13899.
– 2009. "Corry, James." *Dictionary of Irish Biography*. http://dib.cambridge.org/viewReadPage.do?articleId=a2064.
Gifford, J. 1809. *A History of the Political Life of the Right Honourable William Pitt; Including Some Account of the Times in Which He Lived*. Vol. 1. London: Cadell and Davies.
Goncourt, E. de, and J. de Goncourt. 1885. *Sophie Arnould d'après sa correspondance et ses mémoires inédits*. Paris: Charpentier.
Great Britain. 1796. *A List of the Officers of the Army and Marines*. 44th ed. London: G.E. Eyre and W. Spottiswoode.
– 1797. *Parliamentary Register; Or, History of the Proceedings and Debates of the House of Commons*. Ser. 3, vol. 2. London: J. Debrett.
– 1801. "A Bill, Intituled, an Act for the Relief of Certain Insolvent Debtors." Parliamentary Papers. 1801[88]i.247.
– 1802. *Abstract of the Answers and Returns Made Pursuant to an Act, Passed in the Forty-First Year of His Majesty King George III. Intituled "An Act for Taking an Account of the Population of Great Britain, and the Increase or Diminution*

Thereof." Enumeration. Part I. England and Wales. Part II. Scotland. House of Lords Sessional Papers (1714–1805), Vol. 3.
– 1803. *Journals of the House of Commons. Vol. 41. From January the 24th, 1786, in the Twenty-Sixth Year of the Reign of King George the Third, to December the 14th, 1786, in the Twenty-Seventh Year of the Reign of King George the Third.* London: House of Commons.
– 1809. *Accounts, Estimates & Papers Relating to the Army.* 10:26–30.
– 1813. *Report from the Select Committee on Mr. Thomas Croggon's Imprisonment in Newgate.* Parliamentary Papers 1812–13[312]iii.247.
– 1814a. *Minutes of Evidence Taken before the Select Committee of the House of Commons on Petitions Related to East-India-Built Shipping.* Parliamentary Papers 1813–14[115]viii.1.
– 1814b. *Report from the Committee on the State of Gaols of the City of London &c.* Parliamentary Papers 1813–14[157]iv.249.
– 1815a. *The Parliamentary Debates from the Year 1803 to the Present Time: Forming a Continuation of the Work Entitled "The Parliamentary History of England from the Earliest Period to the Year 1803."* Vol. 30. London: T.C. Hansard.
– 1815b. *The Parliamentary History of England from the Earliest Period to the Year 1803.* Vol. 24. London: T.C. Hansard.
– 1822a. *Returns of the Names of the Commissioners and Officers Employed in the Court for the Relief of Insolvent Debtors in England, Rules and Orders, and Other Matters Relating to the Said Court.* Parliamentary Papers 1822[276] xxii.203.
– 1822b. *The Statutes of the Realm.* 9:256–8. London: Dawsons of Pall Mall.
Griffin, D. 2014. *Authorship in the Long Eighteenth Century.* Newark: University of Delaware Press.
Grossman, H. 1948. "W. Playfair, the Earliest Theorist of Capitalist Development." *Economic History Review* 18 (1–2): 65–83. https://doi.org/10.1111/j.1468-0289.1948.tb00741.x.
Guibert, F.T. 1790. *Mémoire pour le Sieur Troussier Guibert et ses différentes requêtes présentées depuis le 27 mai à Leurs Hautes Puissances, les Etats de Brabant.*
Halloran, B.M. 1996. "The Scots College, Paris, 1653–1792." PhD diss., University of St. Andrews.
Harris, J. 1844. *Diaries and Correspondence of James Harris, First Earl of Malmesbury.* Vol. 3. London: Richard Bentley.
Hart, H.G. 1840. *The New Annual Army List, with an Index.* London: John Murray.
Henderson, H.O. 1957. "The Anglo-French Commercial Treaty of 1786." *Economic History Review* n.s. 10 (1): 104–12. https://doi.org/10.2307/2600065.

Higgins, D. 2006. "The New Monthly Magazine." *The Literary Encyclopedia*. https://www.litencyc.com/php/stopics.php?rec=true&UID=1682.
Hildebrandt, K. 2003. "Le concours de l'Académie de Turin sur la statistique (1803–1805)." In *Arithmétique politique dans la France du XVIIIe siècle*, edited by Thierry Martin, 453–90. Paris: Institut national d'études démographiques.
Hills, R.L. 1993. *James Watt*. Vol. 2, *The Years of Toil, 1775–1785*. Ashbourne, UK: Landmark Publishing.
Historical Manuscripts Commission. 1927. *Report on the Manuscripts of J.B. Fortescue. Esq. Preserved at Dropmore*. Vol. 10. London: Her Majesty's Stationery Office.
Hopkins, E. 1989. *Birmingham: The First Manufacturing Town in the World 1760–1840*. London: Weidenfeld & Nicolson.
Hutton, C. 1795. *A Mathematical and Philosophical Dictionary*. London: Johnson and Robinson.
Ireland, Parliament, House of Commons. 1790. *Parliamentary Register: Or History of the Proceedings and Debates of the House of Commons of Ireland, the Sixth Session of the Fourth Parliament in the Reign of His Present Majesty*. Vol. 9. Dublin: Byrne and Porter.
Ireland, Public Record Office, Oriel, J.F., and J.H. Sheffield. 1976. *An Anglo-Irish Dialogue: A Calendar of the Correspondence between John Foster and Lord Sheffield, 1774–1821*. Belfast: Record Office.
Jevons, W.S. 1886. *Letters and Journal*. London: Macmillan.
Jones, C. 2013. "French Crossings IV: Vagaries of Passion and Power in Enlightenment Paris." *Transactions of the Royal Historical Society*. Ser. 6. 23 (December): 3–35. https://doi.org/10.1017/S0080440113000029.
Jones, P.M. 1999. "Living the Enlightenment and the French Revolution: James Watt, Matthew Boulton, and Their Sons." *Historical Journal* 42 (1): 157–82. https://doi.org/10.1017/S0018246X98008139.
Jones, R.F. 1975. "William Duer and the Business of Government in the Era of the American Revolution." *William and Mary Quarterly* 32 (3): 393–416. https://doi.org/10.2307/1922131.
Junius. 1770. *The Letters of Junius*. London: J. Wheele.
Kuhn, R. 1995. *The Media in France*. New York: Routledge.
Laidler, D. 2000. "Highlights of the Bullionist Controversy. Department of Economics Research Reports 2000–2." London, ON: Department of Economics, University of Western Ontario.
– 2008. "Thornton, Henry (1760–1815)." *The New Palgrave Dictionary of Economics*. London: Palgrave Macmillan.
Lakanal, J. 1793. *Rapport sur le télégraphe du citoyen Chappe, fait par Lakanal, au nom du comité d'instruction publique et de la commission nommée par le décret du*

27 avril dernier. Imprimé par ordre de la Convention nationale. (25 juillet 1793.). Paris: Imprimerie nationale.

– 1794. *Rapport sur le télégraphe fait au nom du Comité d'instruction publique, réuni à la commission nommée par le décret du 27 avril dernier (vieux style).* Paris: Imprimerie nationale.

Leach, P. 2021. "Playfair, James (1755–1794), Architect." *Oxford Dictionary of National Biography.* https://doi-org.proxy1.lib.uwo.ca/10.1093/ref:odnb/37857.

LeCocq, G. 1881. *La prise de la Bastille et ses anniversaires d'après des documents inédits.* Paris: Charavay Frères.

Le martyrologe Belgique: l'an de fer 1790. 1791.

Levasseur, E. 1894. "The Assignats: A Study in the Finances of the French Revolution." *Journal of Political Economy* 2 (2): 179–202. https://doi.org/10.1086/250201.

Liévyns, A., J.M. Verdot, and P. Bégat. 1844. *Fastes de la Légion-d'honneur: biographie de tous les décorés accompagnée de l'histoire législative et réglementaire de l'ordre.* Deuxième éd. Tome 4. Paris: Bureau de l'administration.

Life, P. 2021. "Wright, John (1770/71–1844)." *Oxford Dictionary of National Biography.* https://doi-org.proxy1.lib.uwo.ca/10.1093/ref:odnb/30038.

Lord, J. 1966. *Capital and Steam-Power 1750–1800.* 2nd ed. London: Frank Cass.

Louw, H.J. 1987. "The Rise of the Metal Window during the Early Industrial Period in Britain, c. 1750–1830." *Construction History* 3:31–54.

Lowe, W.C. 2004. "Douglas [formerly Stewart], Archibald James Edward, First Baron Douglas (1748–1827)." *Oxford Dictionary of National Biography.* https://doi-org.proxy1.lib.uwo.ca/10.1093/ref:odnb/7874.

Macpherson, D. 1805. *Annals of Commerce, Manufactures, Fisheries and Navigation with Brief Notices of the Arts and Sciences Connected with Them.* Vol. 3. London: Nichols and Son.

Maguire, W.A. 2002. *Living like a Lord: The Second Marquis of Donegall, 1769–1844.* Belfast: Ulster Historical Foundation.

Maidment, J. 1841. *Liber conventus S. Katherine Senensis prope Edinburgum.* Edinburgh: Abbotsford Club.

Malcomson, A.P.W. 2000. "The Irish Peerage and the Act of Union, 1800–1971." *Transactions of the Royal Historical Society* 10:289–327. https://doi.org/10.1017/S0080440100000141.

– 2009. "Foster, John Baron Oriel." *Dictionary of Irish Biography.* http://dib.cambridge.org/viewReadPage.do?articleId=a3339.

Martin, T., and C. Behar. 2003. *Arithmétique politique dans la France du XVIIIe siècle.* Paris: INED.

Mavidal, M.J., M.E. Laurent, L. Claveau, and C. Pionnier. 1893. *Archives Parlementaires de 1787 à 1860 Première série (1787 à 1799) Tomes VIII, XVII, XVIII, XXXVIII, XL*. Paris: Paul Dupont. Online at the French Revolution Digital Archives. https://frda.stanford.edu.

Maxsted, I. n.d. *Exeter Working Papers in Book History: Biographical and Bibliographical Information on the Book Trades*. https://bookhistory.blogspot.com/.

May, T.E. 1844. *A Treatise upon the Law, Privileges, Proceedings and Usage of Parliament*. London: George Knight.

McClellan, A. 2000. "The Life and Death of a Royal Monument: Bouchardon's *Louis XV*." *Oxford Art Journal* 23 (2): 3–27. https://doi.org/10.1093/oxartj/23.2.1.

McNeil, I. 1968. *Joseph Bramah: A Century of Invention, 1749–1851*. New York: Augustus M. Kelley.

Mendelsohn, C.J. 1940. "Blaise de Vigenère and the 'chiffre carré.'" *Proceedings of the American Philosophical Society* 82:103–29.

Middleton, J. 1798. *View of the Agriculture of Middlesex: With Observations on the Means of Its Improvement, and Several Essays on Agriculture in General*. London: Macmillan.

Miles, W.A. 1890. *The Correspondence of William Augustus Miles on the French Revolution 1789–1817*, Vol. 1, edited by Charles Popham Miles. London: Longmans, Green.

Mitchell, W.C. 1927. *Business Cycles: The Problem and Its Setting*. New York: National Bureau of Economic Research.

Mitchison, R. 2015. "Sinclair, Sir John, First Baronet (1754–1835)." *Oxford Dictionary of National Biography*. Oxford: Oxford University Press. https://doi.org/10.1093/ref:odnb/25627.

Montefiore, J. 1803. *A Commercial Dictionary: Containing the Present State of Mercantile Law, Practice, and Custom Intended for the Use of the Cabinet, the Counting-House, and the Library*. London: Self-published.

Monthly Review. 1788a. "Joseph and Benjamin: A Conversation. Translated from a French Manuscript." Art. 47. 78:257.

– 1788b. "The Commercial and Political Atlas." Art. 12. 78:505–9.

– 1789. "An Essay on the National Debt with Copper-plate Charts, for Comparing Annuities with Perpetual Loans." Art. 12. 78:414–17.

– 1795. "Letter to the Right Hon. the Earl Fitzwilliam Occasioned By His Two Letters to the Earl of Carlisle." Art. 28. n.s. 17:99.

– 1800. "Strictures on the Asiatic Establishments of Great Britain; With a View to an Enquiry into the True Interest of the East-India Company." Ser. 2. 33:403–5.

– 1807. "An Enquiry into the Permanent Causes of the Decline and Fall of Powerful and Wealthy Nations." Art. 1. n.s. 53:225–38.

– 1822. "A Letter on Our Agricultural Distresses, Their Causes and Remedies. n.s. 99:79–88.
Moreau-Zanelli, J. 2000. *Gallipolis: histoire d'un mirage américain au XVIII siècle.* Paris: l'Harmattan.
Morellet, A., and E.G. Fitzmaurice. 1898. *Lettres de l'abbé Morellet à Lord Shelburne, depuis Marquis de Lansdowne, 1772–1803*. Paris: Plon, Nourrit.
Mui, H.-C., and L.H. Mui. 1963. "The Commutation Act and the Tea Trade in Britain 1784–1793." *Economic History Review* 16 (2): 234–53. https://doi.org/10.1111/j.1468-0289.1963.tb01728.x.
Muirhead, J.P. 1854. *The Origin and Progress of the Mechanical Inventions of James Watt Illustrated by His Correspondence with His Friends and the Specifications of His Patents*. Vol. 3. London: John Murray.
Munch-Petersen, T. 2008. "Count d'Antraigues and the British Political Elite, 1806–1812." *Napoleonica. La Revue* 2 (2): 121–35. https://doi.org/10.3917/napo.082.0007.
– 2013. "The Secret Intelligence from Tilsit in 1807." *Napoleonica. La Revue* 18 (3): 22–67. https://doi.org/10.3917/napo.133.0022.
Namier, L., and J. Brooke. 1964. *The History of Parliament: The House of Commons 1754–1790*. Vol. 2, *Members A–J*. London: Her Majesty's Stationery Office.
Nangee, B.C. 1934. *The Monthly Review First Series 1749–1789: Indexes of Contributors and Articles*. Oxford: Clarendon.
Newmarch, W. 1855. "On the Loans Raised by Mr. Pitt during the First French War; with Some Statements in Defence of the Methods of Funding Employed." *Journal of the Statistical Society of London* 18 (3): 242–84. https://doi.org/10.2307/2338317.
Nuki, G., and P.A. Simkin. 2006. "A Concise History of Gout and Hyperuricemia and Their Treatment." Supplement, *Arthritis Research & Therapy* 8:S1. https://doi.org/10.1186/ar1906.
Oddy, J.J. 1805. *European Commerce: Shewing New and Secure Channels of Trade with the Continent of Europe*. London: W.J. and J. Richardson.
– 1810. *A Sketch for the Improvement of the Political, Commercial, and Local Interests of Britain: As Exemplified by the Inland Navigations of Europe in General, and of England in Particular*. London: Stockdale.
Original Security Bank. 1796. *Plan of the Original Security Bank, Established in Norfolk Street, Strand: London*. London: Bateson.
Ormal-Grenon, J.-B., and N. Pomier. 2004. *Concise Oxford-Hachette French Dictionary*. Oxford: Oxford University Press.
Oxford University Press and University of Waterloo. Centre for the New Oxford English Dictionary. 2001. *Oxford English Dictionary*. Oxford: Oxford University Press. https://www.oed.com/.
Paine, T. 1792a. *Rights of Man: Being an Answer to Mr. Burke's Attack on the French Revolution*. London: Jordan.

– 1792b. *Rights of Man. Part the Second. Combining Principle and Practice*. London: Jordan.

Palsky, G. 1996. *Des chiffres et des cartes: Naissance et développement de la cartographie quantitative française au XIXe siècle*. Paris: Comité des travaux historiques et scientifiques.

Patterson, A.T. 1948. "Luddism, Hampden Clubs, and Trade Unions in Leicestershire, 1816–17." *English Historical Review* 63 (247): 170–88. https://doi.org/10.1093/ehr/LXIII.CCXLVII.170.

Peltier, J. 1793. *Dernier tableau de Paris: ou récit historique de la révolution du 10 août 1792, des causes qui l'ont produite, les évenémens qui l'ont précédé, et les crimes qui l'ont suivi*. London: Peltier.

Philipson, J., and P. Isaac. 1990. "A Case of Economic Warfare in the Late 18th Century." *Archaeolgia Aeliana* Ser. 5. 18 (5): 151–60.

Playfair, A.G. 1913. *Notes on the Scottish Family of Playfair*. Tunbridge Wells, UK: C. Baldwin.

Playfair, J. 1820. *Catalogue of the Library of the Late John Playfair, Esq*. Edinburgh: James Ballantyne.

Playfair, W. 1785. *The Increase of Manufactures, Commerce, and Finance, with the Extension of Civil Liberty: Proposed in Regulations for the Interest of Money*. London: G.G.J. and J. Robinson.

– 1786. *The Commercial and Political Atlas: Representing, by Means of Stained Copper-Plate Charts, the Exports, Imports, and General Trade of England, the National Debt, and Other Public Accounts*. London: Debrett and Robinson.

– 1787a. *The Commercial and Political Atlas: Which Represents at a Single View, by Means of Copper Plate Charts, the Most Important Public Accounts of Revenues, Expenditures, Debts, and Commerce of England*. London: Stockdale.

– 1787b. *The Commercial, Political, and Parliamentary Atlas, Which Represents at a Single View, by Means of Commercial Plate Charts, the Most Important Public Accounts of Revenues, Expenditures, Debts, and Commerce of England*. 2nd ed. London: Debrett, and G.G.J. and J. Robinson.

– 1787c. *An Essay on the National Debt: With Copper Plate Charts, for Comparing Annuities with Perpetual Loans*. London: Debrett, and G.G.J. and J. Robinson.

– 1787d. *Joseph and Benjamin, A Conversation*. London: Logographic, for J. Murray.

– 1789. *Tableaux d'arithmétique lineéaire, du commerce, des finances, et de la dette nationale de l'Angleterre*. Paris: Chez Barrois l'aîné.

– 1790a. *Lettre II. d'un Anglais à un Français sur les assignats. Par l'auteur de Qu'est-ce que le papier-monnoie*.

– 1790b. *Qu'est-ce que le papier-monnoie?: Lettre d'un Anglois a un François*. Paris: L.M. Cellot.

– 1791. "Playfair to Walker," 18 March. Ohio History Connection Archives VFM 2110.

– 1792a. *Inevitable Consequences of a Reform in Parliament*. London: Stockdale.
– 1792b. *A Letter to the People of England: On the Revolution in France*. Paris.
– 1793a. *Better Prospects to the Merchants and Manufacturers of Great Britain*. London: Stockdale.
– 1793b. *A General View of the Actual Force and Resources of France, in January, M. DCC. XCIII: To Which Is Added, a Table, Shewing the Depreciation of Assignats, Arising from Their Increase in Quantity*. London: Stockdale.
Playfair, W. 1793c. *Thoughts on the Present State of French Politics and the Necessity and Policy of Diminishing France, for Her Internal Peace and to Secure the Tranquility of Europe*. London: Stockdale.
– 1794. *Peace with the Jacobins Impossible*. London: John Stockdale.
– 1795a. *The History of Jacobinism, Its Crimes, Cruelties and Perfidies: Comprising an Inquiry into the Manner of Disseminating under the Appearance of Philosophy and Virtue, Religion, Liberty and Happiness*. London: Stockdale.
– 1795b. *The History of the French Revolution. Proposals for Publishing by Subscription, the History of Jacobinism*. London.
– 1795c. *Letter to the Right Honourable the Earl Fitzwilliam, Occasioned by His Two Letters to the Earl of Carlisle*. London: Stockdale.
– 1796a. *For the Use of the Enemies of England: A Real Statement of the Finances and Resources of Great Britain: Illustrated by Two Copper-Plate Charts*. London: Whittingham.
– 1796b. *Playfair's Answer to Thomas Paine's Decline and Fall of the English System of Finance: In Three Letters. As They Appeared in the Oracle and Public Advertiser*. London.
– 1797a. *A Fair Statement of the Proceedings of the Bank of England, against the Original Security Bank on Tuesday the Seventh of March, 1797: Contained in Two Letters, to the Bank Directors, and to the Rt. Hon. the Lord Mayor*. London: Self-published.
– 1797b. *Letter to Sir W. Pulteney, Bart. M.P. &c. &c. &c. on the Establishment of Another Public Bank in London*. London: Self-published.
– 1798a. *The History of Jacobinism, Its Crimes, Cruelties and Perfidies from the Commencement of the French Revolution to the Death of Robespierre: Comprising an Inquiry into the Manner of Disseminating, under the Appearance of Philosophy and Virtue, Principles Which Are Equally Subversive of Order, Virtue, Religion, Liberty and Happiness*. London: Wright.
– 1798b. *Lineal Arithmetic: Applied to Shew the Progress of the Commerce and Revenue of England during the Present Century*. London: Self-published.
– 1799. *Strictures on the Asiatic Establishments of Great Britain: With a View to an Enquiry into the True Interests of the East India Company*. London: Self-published.
– 1801a. *The Commercial and Political Atlas: Representing, by means of Stained Copper-Plate Charts, the Progress of the Commerce, Revenues, Expenditure and*

Debts of England during the Whole of the Eighteenth Century. London: Burton and Wallis.

– 1801b. *The Statistical Breviary: Shewing, on a Principle Entirely New, the Resources of Every State and Kingdom in Europe … to Which Is Added, a Similar Exhibition of the Ruling Powers of Hindoostan*. London: Wallis.

– 1804. *Proofs Relative to the Falsification of the Intercepted Correspondence in the Admiral Aplin, Indiaman*. London.

– 1805. *An Inquiry into the Permanent Causes of the Decline and Fall of Powerful and Wealthy Nations: Illustrated by Four Engraved Charts*. London: Greenland and Norris.

– 1808. *A Letter to His Majesty's Ministers Relative to the Appointment of a Commander in Chief for the British Forces in Spain: Followed by a Letter from the Great Duke of Marlborough, in the Shades, to His Royal Highness the Duke of York, in the Sun-shine*. London.

– 1809b. *A Fair and Candid Address to the Nobility and Baronets of the United Kingdom: Accompanied with Illustrations and Proofs of the Advantage of Hereditary Rank and Title in a Free Country*. London: Published for the proprietors of Family Antiquity.

– 1809–11. *British Family Antiquity: Illustrative of the Origin and Progress of the Rank, Honours, and Personal Merit, of the Nobility of the United Kingdom. Accompanied with an Elegant Set of Chronological Charts*. London: T. Reynolds and H. Grace.

– 1810. *Second Address to the British Nobility: Accompanied with Illustrations and Proofs of the Advantage of Hereditary Rank and Title in a Free Country*. London: Published for the proprietors of Family Antiquity.

Playfair, W. 1813. *Outlines of a Plan for a New and Solid Balance of Power in Europe: With a Map and Statistical Table Shewing the Extent, Population, Forces*. London: Stockdale.

– 1813–14. *Political Portraits, in This New Æra*. London: Chapple.

– 1814. *A Letter to the Right Honourable and Honourable the Lords and Commons of Great Britain, on the Advantages of Apprenticeships*. London: Sherwood, Neely, and Jones.

– 1815. *A Statement, Which Was Made in October, to Earl Bathurst, One of His Majesty's Principal Secretaries of State: And in November, 1814, to the Comte de la Chatre, the French Ambassador, of Buonoparte's Plot to Re-usurp the Crown of France*. London: Stockdale.

– 1816a. *Supplementary Volume to Political Portraits, in This New Æra: With Explanatory Notes – Historical and Biographical*. London: Chapple.

– 1816b. "To the Editor of the Monthly Magazine." *Monthly Magazine* 42:28–9, 129–30, 292–3.

– 1817. "To the Editor of the Monthly Magazine." *Monthly Magazine* 43:6–8, 123–4.

– 1819. *France as It Is, not Lady Morgan's France*. London: Chapple.
– 1820. *The Advantages of Emigration to France: Clearly Shewn to Be Infinitely Superior to All Others*. London: Souter.
– 1821. *A Letter on Our Agricultural Distresses, Their Causes and Remedies: Accompanied with Tables and Copper-Plate Charts, Shewing and Comparing the Prices of Wheat, Bread, and Labour, from 1565 to 1821. Addressed to the Lords and Commons*. London: Sams.
– 1822a. *A Letter on Our Agricultural Distresses, Their Causes and Remedies: Accompanied with Tables and Copper-Plate Charts, Shewing and Comparing the Prices of Wheat, Bread, and Labour, from 1565 to 1821. Addressed to the Lords and Commons. Second Edition, with an Additional Chart*. London: Sams.
– 1822b. *A Letter on Our Agricultural Distresses, Their Causes and Remedies: Accompanied with Tables and Copper-Plate Charts, Shewing and Comparing the Prices of Wheat, Bread, and Labour, from 1565 to 1821. Addressed to the Lords and Commons. Third Edition, with an Additional Chart*. London: Sams.
– 1822 (?). *Can This Continue?: A Question, Addressed to Those Whom It May Concern: Accompanied by an Engraving*. London: Printed for W. Sams.
– 2005. *The Commercial and Political Atlas and Statistical Breviary*. Edited by H. Wainer and I. Spence. Cambridge: Cambridge University Press.
– n.d. "Memoirs." Unpublished manuscript in the possession of John Lawrence Playfair.
[Playfair, W.]. 1786. *Letters of Albanicus to the People of England on the Partiality and Injustice of the Charges Brought against Warren Hastings, Esq., Late Governor General of Bengal*. London: Debrett.
– 1792. *Short Account of the Revolt and Massacre Which Took Place in Paris, on the 10th of August 1792*. London: Stockdale.
– 1793. *Account of the Revolt and Massacre Which Took Place in Paris, on the 10th of August 1792*. 2nd ed. London: Stockdale.
– 1794. *Scylla More Dangerous Than Charybdis*. London: Stockdale.
– 1817. "Varieties." *Galignani's Messenger*, 8 August.
Playfair, W., and Donnant, D.F. 1802. *Élémens de statistique: Où l'on démontre, d'aprés entièrement neuf, les ressources de chaque royaume, état et république de l'Europe: suivis d'un état sommaire des principales puissances et colonies de l'Indostan*. Paris: Chez Batilliot.
Playfair, W., J. Stockdale, and H. Lloyd. 1793. *Thoughts on the Present State of French Politics: And the Necessity and Policy of Diminishing France, for Her Internal Peace, and to Secure the Tranquillity of Europe*. London: Printed for John Stockdale, Piccadilly.
Pollard, G. 1953. *The Earliest Directory of the Book Trade by John Pendred*. London: Bibliographical Society.
Price, R. 1771. *Observations on Reversionary Payments; On Schemes for Providing Annuities for Widows, and for Persons in Old Age; On the Method of Calculating*

the Values of Assurances on Lives; And on the National Debt. London: T. Cadell.
– 1789. *A Discourse on the Love of Our Country, Delivered on Nov. 4, 1789, at the Meeting-House in the Old Jewry, to the Society for Commemorating the Revolution in Great Britain*. London: Cadell.
– 1994. *The Correspondence of Richard Price*. Vol. 3. Edited by W.B. Peach and D.O. Thomas. Durham, NC: Duke University Press.
Probert, R. 2009. *Marriage Law and Practice in the Long Eighteenth Century: A Reassessment*. Cambridge: Cambridge University Press.
Prosser, R.B. (1881) 1970. *Birmingham Inventors and Inventions: Being a Contribution to the Industrial History of Birmingham*. Wakefield: S.R. Publications.
Puisaye, J. 1804. *Mémoires du Comte Joseph de Puisaye Lieutenant-Général, [etc. etc.,] qui pourront servir à l'histoire du parti Royaliste François durant la dernière révolution*: Tome 3. London: Cox, Fils, & Baylis.
Quarterly Review. 1817. "France. By Lady Morgan." Art. 11. 17: 260–86.
Quérard, J.M. 1870. *Les supercheries littéraires dévoilées*. Tome 2. Paris: Daffis.
Quotationize. n.d. "Oscar Wilde and Lack of Imagination in Many of His Works." https://quotationize.com/oscar-wilde-and-lack-of-imagination/.
Robinson, E. 1963. "Eighteenth-Century Commerce and Fashion: Matthew Boulton's Marketing Techniques." *Economic History Review* n.s. 16 (1): 39–60. https://doi.org/10.1111/j.1468-0289.1963.tb01716.x.
Rogers, C. 1872. *A Century of Scottish Life*. 2nd ed. London: Charles Griffin.
– 1887. *Four Perthshire Families: Roger, Playfair, Constable and Haldane of Barmony*. Edinburgh: Self-published.
Rogers, H.C.B. 1977. *The British Army of the Eighteenth Century*. London: Allen and Unwin.
Rosenberg, D., and A. Grafton. 2010. *Cartographies of Time*. New York: Princeton Architectural.
Ross, E.A. 1892. "Sinking Funds." *Publications of the American Economic Association* 7 (4/5): 9–106.
Rothschild, E. 2001. *Economic Sentiments: Adam Smith, Condorcet and the Enlightenment*. Cambridge, MA: Harvard University Press.
Rudd, A. 2011. *Sympathy and India in British Literature, 1770–1830*. New York: Palgrave Macmillan.
Rutt, J.T. 1831. *Life and Correspondence of Joseph Priestley, LL.D., F.R.S., &c.* Vol. 1. London: R. Hunter and M. Eaton.
Salvadori, N., and R. Signorino. 2014. "Adam Smith on Monopoly Theory: Making Good a Lacuna." *Scottish Journal of Political Economy* 61 (2): 178–95. https://doi.org/10.1111/sjpe.12040.
Schama, S. 1989. *Citizens: A Chronicle of the French Revolution*. New York: Vintage Books.

Schofield, R.E. 1963. *The Lunar Society of Birmingham: A Social History of Provincial Science and Industry in Eighteenth-Century England*. Oxford: Clarendon.

Schranz, K.M. 2014. "The Tipton Chemical Works of Mr James Keir: Networks of Conversants, Chemicals, Canals and Coal Mines." *International Journal for the History of Engineering & Technology* 84 (2): 248–73. https://doi.org/10.1179/1758120614Z.00000000049.

Seccombe, T., and M.C. Loughlin-Chow. 2008. "Phillips, Sir Richard (1767–1840)." *Oxford Dictionary of National Biography*. Oxford: Oxford University Press. https://doi.org/10.1093/ref:odnb/22167.

Servian, M.S. 1985. Eighteenth-Century Bankruptcy Law: From Crime to Process. PhD diss., University of Kent at Canterbury.

Seton, G. 1890. *The House of Moncrieff*. Edinburgh.

Sharpe, C.K. 1888. *Letters from and to Charles Kirkpatrick Sharpe*. Vol. 1. Edited by Alexander Allardyce. Edinburgh: William Blackwood and Sons.

Sheffield, John Holroyd, Earl of. 1784. *Observations on the Commerce of the American States*. Dublin: Luke White.

Shelton, T. 1814. "The State of Newgate." *Philanthropist, Or, Repository for Hints and Suggestions Calculated to Promote the Comfort and Happiness of Man* 4:88.

Sherer, M. 1836. *Military Memoirs of Field Marshall the Duke of Wellington*. Vol. 1. Philadelphia: Robert Desilver.

Sibley, W.G. 1901. *The French Five Hundred, and Other Papers*. Gallipolis, OH: Tribune.

Sinclair, J.G. 1796. *Colonel Sinclair's Letters to His Enemies*. Surrey, UK: Printed by the author.

– 1798. *Letter to His Royal Highness Monsieur: Brother to Louis XVIII. King of France &c &c. from Colonel John Gordon Sinclair Relative to His Trial in the Court of King's Bench*. London.

Smiles, S. 1895. *Josiah Wedgwood, F.R.S.: His Personal History*. New York: Harper and Brothers.

Smith, A. 1776. *An Inquiry into the Nature and Causes of the Wealth of Nations*. London: Strahan and Cadell.

Smith, A., R.H. Campbell, A.S. Skinner, and W.B. Todd. 1976. *An Inquiry into the Nature and Causes of the Wealth of Nations*. Oxford: Clarendon.

Smith, A., and W. Playfair. 1805. *An Inquiry into the Nature and Causes of the Wealth of Nations*. London: Cadell and Davies.

– 1995. *An Inquiry into the Nature and Causes of the Wealth of Nations*. London: W. Pickering.

Smith, F.B. 1979. *The People's Health 1830–1910*. London: Croom Helm.

Smith, J.C. 1825. *Chronological Epitome of the Wars in the Low Countries, from the Peace of the Pyrenees in 1659 to That of Paris in 1815*. London: T. Egerton.

Snape, M.F. 2008. *The Royal Army Chaplains' Department, 1796–1953: Clergy under Fire*. Woodbridge, UK: Boydell.

Sparrow, E. 1999. *Secret Service: British Agents in France, 1792–1815*. Rochester, NY: Boydell.
Spence, I. 2004. "Playfair, William (1759–1823)." *Oxford Dictionary of National Biography*. Oxford: Oxford University Press. https://doi.org/10.1093/ref:odnb/22370.
Spence, I. n.d. "William Playfair Remembers: Memoirs Edited and Annotated." Unpublished.
Spence, I., C.R. Fenn, and S. Klein. 2017. "Who Is Buried in Playfair's Grave?" *Significance* 14 (5): 20–3. https://doi.org/10.1111/j.1740-9713.2017.01071.x.
Spence, I., and H. Wainer. 1997a. "Who Was Playfair?" *Chance* 10 (1): 35–7. https://doi.org/10.1080/09332480.1997.10554796.
– 1997b. "William Playfair: A Daring Worthless Fellow." *Chance* 10 (1): 31–4. https://doi.org/10.1080/09332480.1997.10554795.
– 2000. "William Playfair (1759–1823): An Inventor and Ardent Advocate of Statistical Graphics." In *Statisticians of the Centuries*, edited by C.C. Heyde, 105–110. Voorburg, Netherlands: International Statistical Institute.
– 2005. "William Playfair and His Graphical Inventions: An Excerpt from the Introduction to the Republication of His *'Atlas'* and *'Statistical Breviary.'* *American Statistician* 59 (3): 224–9. https://doi.org/10.1198/000313005X54216.
Srinivasan, A. 2005. *Most Influential Businessman*. Chennai: Sura.
Stephens, H.M., and S.M. Fraser. 2006. "Campbell, Sir Neil (1776–1827)." *Oxford Dictionary of National Biography*. Oxford: Oxford University Press. https://doi.org/10.1093/ref:odnb/4529.
Stockley, A. 2001. *Britain and France at the Birth of America: The European Powers and the Peace Negotiations of 1782–1783*. Exeter, UK: University of Exeter Press.
Stuart, R. 1829. *Historical and Descriptive Anecdotes of Steam-Engines, and of Their Inventors and Improvers*. Vol. 2. London: Wightman and Cramp.
Suddaby, E., and P.J. Yarrow. 1971. *Lady Morgan in France*. Newcastle upon Tyne, UK: Oriel.
Sweet, R. 1998. "Freemen and Independence in English Borough Politics c.1770–1830." *Past & Present* 161:84–115.
Symanzik, J., W. Fischetti, and I. Spence. 2009. "Commemorating William Playfair's 250th Birthday." *Computational Statistics* 24:551–66. https://doi.org/10.1007/s00180-009-0170-z.
Tann, J., and A. Burton. 2013. *Matthew Boulton: Industry's Great Innovator*. Stroud, UK: History.
Tassin, S.H. 2007. *Pennsylvania Ghost Towns: Uncovering the Hidden Past*. Mechanicsburg, PA: Stackpole Books.
Thompson, J.M. 1949. "Napoleon's Journey to Elba in 1814. Part 1. By Land." *American Historical Review* 55 (1): 1–21. https://doi.org/10.2307/1841084.

Thorne, R.G. 1986. *The History of Parliament: The House of Commons 1790–1820*. Vols. 3–5. London: History of Parliament Trust.
Thornton, H., and F.A. Hayek. 1939. *An Enquiry into the Nature and Effects of the Paper Credit of Great Britain (1802)*. London: G. Allen & Unwin.
Tribe, K., and H. Mizuta. 2002. *A Critical Bibliography of Adam Smith*. London: Pickering & Chatto.
Tufte, E.R. 1983. *The Visual Display of Quantitative Information*. Cheshire, CT: Graphics.
Turnbull, G. 1795. *A Narrative of the Revolt and Insurrection of the French Inhabitants in the Island of Grenada*. Edinburgh: Archibald Constable.
Uglow, J. 2002. *The Lunar Men: The Men Who Made the Future 1730–1810*. London: Faber and Faber.
Vandermonde, A.T., C.L. Berthollet, and G. Monge. 1786. "Mémoire sur le fer considéré dans les différens états métalliques." *Histoire de l'Académie royale des sciences*. Année MDCCLXXXVI, 132–200.
Vauban, J.A., and J. Le Prestre Comte de. 1806. *Mémoires pour servir à l'histoire de la guerre de la Vendée*. Paris: Maison de commission en librarie.
Vesey, F. 1827. *Reports of Cases Argued and Determined in the High Court of Chancery from the Year 1789 to 1817. 39 to 57 Geo. III*. Vol. 4. London: Samuel Brooke, Butterworth and Son and J. & W.T. Clarke.
Viton de Saint-Allais, N. 1872. *Nobiliaire universel de France: recueil général des généalogies historiques des maisons nobles de ce royaume*. Tome 6. Paris: Bureau du Nobiliaire universel de France, Réimprimé à la Librairie Bachelin-Deflorenne.
Wainer, H. 1996. "Why Playfair?" *Chance* 9 (2): 43–52. https://doi.org/10.1080/09332480.1996.10555002.
– 2005. *Graphic Discovery: A Trout in the Milk and Other Visual Adventures*. Princeton, NJ: Princeton University Press.
Warren, A. 1896. *The Charles Whittinghams: Printers*. New York: Grolier Club.
Watkins, J. 1827. *A Biographical Memoir of His Late Royal Highness Frederick, Duke of York and Albany; Commander-in-Chief of the Forces of Great Britain*. London: Henry Fisher.
Wedgwood, J. 1782. "An Attempt to Make a Thermometer for Measuring the Higher Degrees of Heat, from a Red Heat up to the Strongest That Vessels Made of Clay Can Support." *Philosophical Transactions of the Royal Society of London* 72:305–26. https://doi.org/10.1098/rstl.1782.0021.
– 1784. "An Attempt to Compare and Connect the Thermometer for Strong Fire, Described in Vol. LXXII of the Philosophical Transactions, with the Common Mercurial Ones." *Philosophical Transactions of the Royal Society of London* 74:358–84. https://doi.org/10.1098/rstl.1784.0028.
Welbourne, E. 1932. "Bankruptcy before the Era of Victorian Reform." *Cambridge Historical Journal* 4 (1): 51–62. https://doi.org/10.1017/S1474691300003280.

Werkmeister, L.T. 1963. *The London Daily Press, 1772–1792*. Lincoln: University of Nebraska.
Whittaker, G.B. 1825. *A New Biographical Dictionary, of 3000 Contemporary Public, British and Foreign, of All Ranks and Professions*. 2nd ed. Vol. 1, Part 1. London: George Whittaker.
Whitworth, C. 1776. *State of the Trade of Great Britain in Its Imports and Exports, Progressively from the Year 1697*. London: G. Robinson.
Wilkinson, D. 2004. "Fitzwilliam, William Wentworth, Second Earl Fitzwilliam in the Peerage of Great Britain, and Fourth Earl Fitzwilliam in the Peerage of Ireland." *Oxford Dictionary of National Biography*. Oxford: Oxford University Press. https://doi-org.proxy1.lib.uwo.ca/10.1093/ref:odnb/9665.
– 2008. "Windham, William." *Oxford Dictionary of National Biography*. Oxford: Oxford University Press. https://doi.org/10.1093/ref:odnb/29725.
Windham, W. 1866. *The Diary of the Right Hon. William Windham, 1784 to 1810*. Edited by Cecilia Anne Baring. London: Longmans, Green.
Wolfe, P.J., and W.J. Wolfe. 1994. "Prospects for the Gallipolis Settlement: French Diplomatic Dispatches." *Ohio History Journal* 103:41–56.
Worsley, L. 2017. *Jane Austen at Home*. London: Hodder & Stoughton.
Young, A. 1792. *Travels in France during the Years 1787, 1788, 1789, Undertaken More Particularly with a View of Ascertaining the Cultivation, Wealth, Resources, and National Prosperity of the Kingdom of France*. Bury St. Edmond's, UK: Rackham and Richardson.
Zöllner, E. 2004. "Arnold, Samuel." *Oxford Dictionary of National Biography*. Oxford: Oxford University Press. https://doi-org.proxy1.lib.uwo.ca/10.1093/ref:odnb/682.

Index